S0-BLM-141

PHYSICS OF THE SUN

Volume III: Astrophysics and Solar-Terrestrial Relations

GEOPHYSICS AND ASTROPHYSICS MONOGRAPHS

PHYSICS OF THE SUN

Volume III:
Astrophysics and Solar-Terrestrial Relations

Edited by

PETER A. STURROCK
*Center for Space Science and Astrophysics,
Stanford University, Stanford, California, U.S.A.*

Associate Editors:

THOMAS E. HOLZER
*High Altitude Observatory,
National Center for Atmospheric Research, Boulder, Colorado, U.S.A.*

DIMITRI M. MIHALAS
*High Altitude Observatory,
National Center for Atmospheric Research, Boulder, Colorado, U.S.A.*

ROGER K. ULRICH
*Astronomy Department, University of California,
Los Angeles, California, U.S.A.*

D. REIDEL PUBLISHING COMPANY

A MEMBER OF THE KLUWER ACADEMIC PUBLISHERS GROUP

DORDRECHT / BOSTON / LANCASTER / TOKYO

Library of Congress Cataloging in Publication Data

CIP data will appear
on separate card

ISBN 90-277-1862-8 (Vol. III)
ISBN 90-277-1823-7 (Set)

Published by D. Reidel Publishing Company,
P.O. Box 17, 3300 AA Dordrecht. Holland.

Sold and distributed in the U.S.A. and Canada
by Kluwer Academic Publishers
190 Old Derby Street, Hingham, MA 02043, U.S.A.

In all other countries, sold and distributed
by Kluwer Academic Publishers Group.
P.O. Box 322, 3300 AH Dordrecht, Holland.

Printed in The Netherlands

TABLE OF CONTENTS

PREFACE

This volume, together with its two companion volumes, originated in a study commissioned by the United States National Academy of Sciences on behalf of the National Aeronautics and Space Administration. A committee composed of Tom Holzer, Dimitri Mihalas, Roger Ulrich and myself was asked to prepare a comprehensive review of current knowledge concerning the physics of the sun. We were fortunate in being able to persuade many distinguished scientists to gather their forces for the preparation of 21 separate chapters covering not only solar physics but also relevant areas of astrophysics and solar-terrestrial relations.

It proved necessary to divide the chapters into three separate volumes that cover three different aspects of solar physics. Volumes 1 and 2 are concerned with 'The Solar Interior' and with 'The Solar Atmosphere'. This volume, devoted to 'Astrophysics and Solar-Terrestrial Relations', focuses on problems of solar physics from these two different but complementary perspectives. The emphasis throughout these volumes is on identifying and analyzing the relevant physical processes, but each chapter also contains a great deal of descriptive material.

In preparing our material, the authors and editors benefited greatly from the efforts of a number of scientists who generously agreed to review individual chapters. I wish therefore to take this opportunity to thank the the following individuals for this valuable contribution to our work: S. K. Antiochos, E. H. Avrett, J. N. Bahcall, C. A. Barnes, G. Bicknell, D. Black, M. L. Blake, P. Bodenheimer, F. H. Busse, R. C. Canfield, T. R. Carson, J. I. Castor, J. Christensen-Dalsgaard, E. C. Chupp, A. N. Cox, L. E. Cram, P. R. Demarque, L. Fisk, M. A. Fowler, D. O. Gough, L. W. Hartmann, J. W. Harvey, R. F. Howard, P. Hoyng, H. S. Hudson, G. J. Hurford, C. F. Kennel, R. A. Kopp, A. Krueger, R. M. Kulsrud, R. B. Larson, H. Leinbach, R. E. Lingenfelter, J. L. Linsky, D. B. Melrose, M. J. Mitchell, A. G. Newkirk, F. W. Perkins, R. Roble, R. T. Rood, R. Rosner, B. F. Rozsynai, S. Schneider, E. C. Shoub, B. Sonnerup, H. Spruit, R. F. Stein, M. Stix, J. Tassoul, G. Van Hoven, G. S. Vaiana, A. H. Vaughan, S. P. Worden, R. A. Wolf, and J. B. Zirker.

On behalf of the editors of this monograph, I wish to thank Dr. Richard C. Hart of the National Academy of Sciences, Dr. David Larner of Reidel Publishing Company, and Mrs. Louise Meyers of Stanford University, for the efficient and good-natured support that we received from them at various stages of the preparation of this volume.

Stanford University,
July 1985

P. A. STURROCK

Peter A. Sturrock (ed.) Physics of the Sun, Vol. III, p. xi

CHAPTER 16

FORMATION OF THE SUN AND ITS PLANETS

WILLIAM M. KAULA

1. Introduction

The Sun formed 4.6 Gy ago, probably as a fragment of a collapsing gas and dust cloud in close association with other stars, from which cluster it was ejected rather early. Conditions in the Galaxy then were not remarkably different from what they are now. Material of the solar system was processed through at least two supernova events within ~200 My before its formation, the last of them recent enough (~2 My) that it plausibly influenced the collapse which formed then Sun.

The planets most likely came from the same cloud as the Sun, since subsequently the interstellar medium would have been much too sparse. Like other collapse phenomena, the Sun's formation was probably accompanied by formation of an accretion disk in which there is a net outward flow of angular momentum and inward flow of mass. This accretion disk, or nebula, is the plausible locus of planetary formation, by some combination of viscous and resonant effects, gravitational instabilities, and accretionary growth. The major problem is the formation of the hydrogen-rich planets, Jupiter and Saturn.

The early Sun was considerably more active, as evidenced by remanent magnetism and implanted inert gases in meteorites; however, isotope anomalies in meteorites also set a moderate upper limit on proton fluxes in the nebula. The relative roles of the solar wind and simple heating in removing gases from the nebula is unsure. Other problems are the high nebula temperatures, >1500 K, indicated by some chondritic meteorites, and hydromagnetic effects.

The Sun alone provides no clues to its formation. Almost the same can be said of the planets, down to bodies at least as small as the Moon. The evidence has been obliterated because all these bodies have evolved appreciably as consequences of energy sources which have been predominantly internal for at least 4.4 Gy. Hence, observations relevant to the origin of the solar system are of (1) small fragments of the system, or (2) comparable properties of parts of the system, or (3) the likely surrounding circumstances, or (4) similar phenomena, or (5) possibly analogous phenomena. Examples in category (1) are the asteroids, comets, and meteorites; in (2), the similarities and differences in composition of the solar atmosphere and meteorites; in (3), the age of the solar system relative to the Galaxy; in (4), the contemporary behavior of T-Tauri stars; and in (5), the spiral structure of galaxies. In some cases the constraints of observations on solar system origin are rather direct and inescapable: e.g. isotopic evidence of the separation

Peter A. Sturrock (ed.), Physics of the Sun, Vol. III, pp. 1–32.

of materials constituting the Earth and meteorites. In other cases the inferences are extremely indirect and entail appreciable assumptions and modeling: e.g. the variety of planetary systems which could be formed in association with a solar-sized star.

The variety of data and theoretical ideas pertinent to the origin of the solar system make the problem a motivator for research in several scientific disciplines, rather than the object of a single coherent discipline. The relevant research can, however, be divided into three general parts: astronomical, cosmochemical, and dynamical.

This review is organized in terms of these three parts, starting with the astronomical because it is most pertinent to formation of the Sun, as distinguished from the solar system. We conclude with a briefer and more conjectural section on the plausible inferences about the Sun and planets which can be drawn from these diverse research efforts.

2. Star Formation

Current astronomical research in the formation of stars is very active, both observationally and theoretically. Most of this research is motivated by considerations other than explaining the solar system, partly because of observational feasibility and partly because of a broader conceptual framework, the drives being more to understand galactic and stellar evolution in general. We report here on those observations and models which bear on solar system formation. We first describe the astronomical context: the structure of the Galaxy, characteristics of stars and the interstellar medium, before concentrating on the observations and models pertaining to star formation.

2.1. Galactic Structure

The solar system is located close to the main plane of the Galaxy about 8 kpc from the center (Oort, 1977). The observable Galaxy is a typical spiral galaxy of $\sim 10^{11}$ $M_\odot$ mass and ~15 kpc radius. About half of this mass is within 3 kpc of the center. The age of the Galaxy is estimated to be ~10 Gy, mainly from the distribution on the H–R diagram of stars in globular clusters. The Galaxy is rotating and appreciably flattened. At the Sun's distance from the center, the period of a single revolution is about 200 My. In the solar vicinity, the half thickness of the Galaxy is about 0.3 kpc. The greatest uncertainty is the amount of mass in burned-out stellar remnants in the galactic halo: perhaps as much as 2×10^{12} $M_\odot$ (Bok, 1981). The Galaxy appears to be evolving very slowly in the Sun's vicinity and probably had a general structure 4.6 Gy ago similar to what it has now.

In the Sun's part of the Galaxy there are irregularities in velocity of ±10% superimposed on the uniform motion around the center (~260 km s^{-1} for the Sun). In addition, there are probably variations in mass density as observed in other spiral galaxies: on the scale of the spiral arms, ~2 kpc wide, perhaps by a factor of 3 or so about the mean for the solar vicinity, ~75 $M_\odot$ pc^{-2}, or 0.125 $M_\odot$ pc^{-3} (neglecting possible invisible remnants in the halo). Of this mean density only ~5 $M_\odot$ pc^{-2} is in the form of interstellar gas. Variations in density of the gas are appreciably greater. Some molecular clouds have $\sim 10^6$ $M_\odot$ within ~50 pc of their centers, an enhancement of more than 200 in density. Some have *OB associations*: groups of massive stars $\lesssim$20 My in age.

The average intensity of the galactic magnetic field is estimated to be $0.2–1.0 \times 10^{-5}$ gauss. Variations of this intensity are correlated with variations in gas density.

Dynamical theory suggests that the kinematic irregularities plus gravitational attraction superimposed on general galactic revolution account for the spiral arms (Toomre, 1977; Lin and Lau, 1979). However, the greater local concentrations which are predominantly gaseous probably depend on some sort of hydromagnetic instability.

2.2. Stellar Properties

Main sequence stars range in mass from ~100 to ~0.08 $M_\odot$, and in main sequence lifetime from ~0.01 to ~1500 Gy.

In the vicinity of the Sun, the observed number density of stars, corrected for the death of larger stars, approximates the rule (Scalo, 1978):

$$N \approx 100 \exp[-1.1(\log M + 1.0)^2], \tag{2.1}$$

where N is in $\text{pc}^{-2}(\log M)^{-1}$ and M is in $M_\odot$. The ability to fit the power law implies that past stellar formation rates were not significantly greater than the current rate. However, the current nova ($M \lesssim 8\,M_\odot$) and supernova ($M \gtrsim 8\,M_\odot$) rates may be too low to account for the observed abundances of heavy elements in smaller stars, $M \lesssim 1.5\,M_\odot$. Models which reconcile these data generally conclude that in the first few 0.1 Gy of the existence of the Galaxy, when the density of interstellar gas would have been appreciably greater than at present, the rate of star formation would have been greater than now, and a larger portion of the mass would have gone into massive stars. Hence, by the time the solar system formed ~5 Gy later, most of the interstellar matter would have been former stellar material; the gas density, and hence stellar formation rate, would have been only moderately higher than at present; and the major part of the mass would have gone into smaller stars (Bierman and Tinsley, 1974). Hence, circumstances of star formation 4.6 Gy in the past were not very different from what they are now.

The rotation rates of main sequence stars, inferred from Doppler broadenings of their spectral lines, are functions of their mass, with a sharp drop-off in the rotation–mass relationship at ~2 $M_\odot$ corresponding, perhaps, to the change in the energy transfer mode from entirely radiative to convective in the outer layers. The spin-down of smaller main sequence stars is plausibly related to their stellar winds. The present solar wind carries away so little angular momentum that the decay time for solar rotation therefrom is $\sim 10^{10}$ yr. However, for moderately massive stars, 1.2 $M_\odot$, which are known to be rather young, there are correlations of spin rates and chromospheric emissions presumably dependent (like the wind) on the vigor of convection, indicative of decay times a few times 10^8 yr (Kraft, 1967).

Another important property of stars relevant to their formation is the frequency of their occurrence as members of multiple systems, bound in orbits around each other. Only ~20% of stars are single, like the Sun; ~50% are members of binary pairs, ~20% of triplets, ~5% of quadruple systems, etc. (Batten, 1973). The orbital angular momentum vectors of these multiple systems are random in direction with respect to the angular momentum vector of the whole Galaxy. In a study confined to the 135 stars which are (1) of spectral classes F3 through G2 (masses ~1.5–1.0 $M_\odot$), (2) of declinations $\delta > -20°$, (3) of apparent magnitudes <5.5 (hence within ~30 pc), and (4) primaries in multiple systems, two-thirds were found to have stellar companions. The secondary primary mass ratios divide into two populations according to the period of the binary. For periods greater than 100 yr, the frequency distribution of mass ratios is the same as predicted by a random selection from all stars smaller than the primary. For periods less than 100 yr, the frequency varies with the mass ratio, with a $\sim M^{1/3}$ proportionality. The smallest companion inferred (from the Doppler shifts of the primary's spectrum) was $\sim M_1/16$ in mass, close to the minimum mass star directly observable (Abt and Levy, 1976; Abt, 1978).

2.3. Planetary Indications

Extrapolation of the $M^{1/3}$ frequency to masses below the limit inferable as spectroscopic binaries suggests that all 1.0–1.5 $M_\odot$ stars have companions, the smallest 20% of which are planets, i.e. bodies too small to have the pressures necessary for hydrogen burning. Systematic efforts have been made to detect astronomically unseen companions or stars within ~10 pc: i.e. by periodic shifts of a star against its background. Evidences of shifts have been found associated with stars such as the closest of $\delta > 0°$, Barnard's (a 0.15 $M_\odot$ body 1.8 pc distant), by Van de Kamp (1975). However, the possibilities of systematic errors in this work are considerable (Gatewood, 1976).

2.4. Interstellar Clouds

The density of the interstellar medium (as inferred from absorption of starlight by the dust component or from the 21 cm radiation by the atomic hydrogen component) is about 10^{-24} g cm^{-3}, or 0.6 cm^{-3} for hydrogen atoms. The composition appears to be about the same as the Sun and other stars: by mass, 75% H and 1% dust. The temperature is typically 100 K, but highly variable.

Of more interest for star formation are *interstellar clouds*, concentrations of the interstellar medium by a factor of 100 or more in density, and of 100 $M_\odot$ or greater. Most common are *diffuse clouds*, which typically have radii of 3 pc, densities of 10^{-22} g cm^{-3}, and occur at intervals on the order of 30 pc. Such clouds are still sparse enough to be transparent, and hence are maintained at ~100 K by starlight. When a cloud is more dense, 10^{-21}–10^{-19} g cm^{-3}, it becomes a *dark cloud*, opaque to visible light, and the internal temperature can drop to ~10 K, inferred from radio observations of transitions in carbon monoxide, CO. Greater densities also lead to molecular composition becoming dominant. For sufficiently large combinations of mass ($\gtrsim 10^4$ $M_\odot$) and density, several other molecular transitions can be detected, and considerable structure can be mapped by the CO transitions: hence the name *molecular clouds*. In places they may have 10 K temperatures, but elsewhere their temperatures are appreciably elevated by the occurrence of O and B type stars. Two molecular complexes of ~10^5 $M_\odot$ each appear to extend for ~50 pc across the Orion Nebula (Thaddeus, 1977; Evans, 1978).

2.5. Observations of Forming Stars

The only plausible material from which to make the Sun and similar stars is, of course, the interstellar matter. The simplest considerations of a forming star suggest that: (1) it would be close to some concentrated source, such as the clouds described above; (2) its main compositional distinction would be a lithium abundance higher than main sequence stars, like chondritic meteorites; (3) it would have larger radius, because matter is still falling in; (4) it would be more luminous, at least in the later stages, because of the amount of gravitational energy to be radiated away; and (5) it might appear redder, because the visible radiating surfaces (e.g. dust) could still be an appreciable distance from the core, even after the core had achieved sufficient mass to induce H-burning. Hence, there have been searches for such objects associated with clouds appearing above the main sequence in the Hertzsprung–Russell diagram. Particularly valuable to these searches are infrared techniques, because the dust in a dense cloud will absorb visible light and reradiate it at infrared wavelengths (Werner *et al.*, 1977; Strom *et al.*, 1975; Cohen and Kuhi, 1979).

Three categories of objects have been found consistent with the above criteria for young, still-forming stars of approximately solar mass (Strom *et al.*, 1975).

2.5.1. *T-Tauri Variable Stars*

Approximately 630 of these sources have been identified within 1 kpc of the Sun. They are characterized by a variety of spectral phenomena suggesting youth and enhanced activity: more intense lithium lines; Doppler broadenings indicating rapid rotation; Doppler shifts and intensities indicating appreciable mass outflow in most cases (perhaps $\sim 10^{-7}\ M_\odot\ \mathrm{yr}^{-1}$), but inflow in others; various emissions, indicating either a significantly different physical regime than the photosphere of a main sequence star (like the Sun) or a surrounding absorbing medium; and irregular temporal variations, normally by one or two magnitudes (a factor of ~2.5–6 in luminosity L), but in three cases by as much as a six magnitude rise. Some T-Tauri stars are located in OB associations, but others are in the smaller dark clouds. The belief that most T-Tauri stars are 0.2–2.0 $M_\odot$ in size seems to be based partly on the computational models discussed below, and partly on statistics: wherever T-Tauri stars appear, most stars already on the main sequence are more massive, $\gtrsim 3\ M_\odot$, and there is a dearth of less massive stars. The estimated duration of the 10^{-7}–$10^{-8}\ M_\odot\ \mathrm{yr}^{-1}$ mass outflows is also based on a combination of statistics and modeling considerations: perhaps 10^6–10^7 yr (Strom *et al.*, 1975; Strom, 1977; Herbig, 1978).

2.5.2. *Nonemission, Nonvariable Pre-Main Sequence (PMS) Stars*

Almost 50% of the stars occupying the same region of the Hertzsprung–Russell diagram as the T-Tauri stars and associated with clouds do *not* show any exceptional emissions or variability in their spectra. The evidence for a high Li content in these stars is weak. The existence of these stars raises the possibility that the exceptional phenomena observed in T-Tauri spectra are not inherent in forming solar-mass stars, but depend on structural circumstances: most obviously, on being a member of a binary pair (Strom *et al.*, 1975; Strom, 1977).

2.5.3. *Herbig–Haro Objects*

These sources are more diffuse and redder than T-Tauri stars, and are strong in IR emission. Some also have been interpreted as having spectral Doppler shifts indicating mass outflow. They are at locations in cloud and stellar associations where the most recent star formation is expected to be occurring. Hence the evident interpretation is that they are dusty envelopes of matter infalling to a star. An alternative interpretation is that they are the reflections of starlight from surrounding clouds (Strom *et al.*, 1975; Strom, 1977; Herbig, 1978; Schwartz and Dapita, 1980).

2.6. Conditions for Cloud Collapse

We discuss here circumstances relevant to initiation of collapse to a solar-sized star: a fragment of a dark cloud, density $\sim 10^{-19}$ g cm^{-3}, temperature ~ 10 K. Significant cloud fragmentation at higher densities has not been observed to occur, and hence the properties at this stage have important effects on star formation processes, and constitute appropriate starting conditions for collapse models (Larson, 1977, 1978, 1981). The simplest model is a spherically symmetric cloud, without magnetic field, rotation, or random motions. The equations of motion then become

$$\rho \frac{\mathrm{d}^2 r}{\mathrm{d}t^2} = -\frac{\mathrm{d}p}{\mathrm{d}r} - \frac{\rho GM}{r^2}, \tag{2.2}$$

where ρ is density, p is pressure, and M is the mass contained within radius r. If the pressure gradient $\mathrm{d}p/\mathrm{d}r$ is zero, as would occur in an isothermal homogeneous cloud with a matching external pressure, then the first term on the right drops out and the equation is solvable for the time required for a cloud with initial radius r_0 to collapse completely:

$$t_f = \frac{\pi}{2}\left[\frac{r_0^3}{2GM}\right]^{1/2} = \left[\frac{3\pi}{32G\bar{\rho}_0}\right]^{1/2}, \tag{2.3}$$

know as the *free-fall time*.

A cloud in dynamic equilibrium will have the left of Equation (2.2) zero. If it is sparse enough to be isothermal ($\mathrm{d}T/\mathrm{d}r = 0$), then, from the perfect gas law

$$p = \rho kT/m = c^2 \rho/\gamma \tag{2.4}$$

(where k is the Boltzmann constant, m is mean molecular mass, c is the sound speed, and γ is the ratio of specific heats), there must be a gradient in density ρ in order to have a pressure gradient $\mathrm{d}p/\mathrm{d}r$ balancing the gravity term. Assume $\rho = Ar^{-n}$ and isothermality ($\gamma = 1$) and substitute in (2.2):

$$0 = c^2 nAr^{-n-1} - 4\pi G \frac{A^2}{3-n} r^{1-2n}, \tag{2.5}$$

whence $n = 2$ and $A = c^2/2\pi G = M(r)/4\pi r$. Thence, for an overall radius R

$$\bar{\rho} = 3c^2/2\pi GR^2, \tag{2.6}$$

and for collapse there is required

$$M \geq 2c^2 R/G$$

$$\geq 2 \left(\frac{kT}{mG}\right)^{3/2} (\xi \pi \bar{\rho})^{-1/2}$$

$$\gtrsim 10^{-11} \left(\frac{T^3}{\bar{\rho}}\right)^{1/2} M_\odot = M_J, \tag{2.7}$$

for $\bar{\rho}$ in g cm^{-3}. For dark cloud, values of 10 K for T and 10^{-19} g cm^{-3} for $\bar{\rho}$, $M \gtrsim 1\ M_\odot$. M_J is known as the *Jeans mass*.

In a real cloud there are the additional effects of magnetic field, rotation, other internal motions, and external pressure. All these effects, except pressure, act to increase the critical mass for collapse. The assumption of a magnetic field intensity based on freezing the galactic field to the gas and a rotation rate based on the present relative motions between stars would make these effects much more important than temperature in resisting collapse. However, the axial character of these effects emphasizes the unrealism of the spherical model: a magnetic field would not resist collapse along the field lines, and rotation would not resist collapse parallel to its axis. If a dark cloud is as cold as 10 K, then the ionization would plausibly drop sufficiently for the neutral matter to slip with respect to the magnetic field. Shocks would reduce the relative motions to something less than the sound of speed (Mestel, 1977; Mouschovias, 1977, 1978).

The common occurrence of star formation in clusters suggests that external pressure is a significant factor in initiating cloud collapse. Sources of such pressure might be: a strong stellar wind from a newly formed star; or a supernova explosion; or the expansion of an H II region. H II regions have been invoked to explain the progression in age of OB stars along a molecular cloud (Lada *et al.*, 1978).

The considerations of this subsection emphasize that cloud collapse should be highly nonhomologous and asymmetric. These effects will be enhanced by any magnetic field, external pressure, or other inhomogeneity in initial conditions (Mouschovias, 1978; Larson, 1981).

2.7. Models of Star Formation

Although similarity solutions (Shu, 1977; Cheng, 1978) yield significant insight, complications such as opacity and shocks require numerical integration to construct a plausible scenario of cloud collapse to form a star.

So far, only spherically symmetric models without rotation or magnetic fields have been explored in detail for the process complete to stellar densities. It is useful to examine these idealizations as reference models for more complicated computer experiments. For the 10 K, 10^{-19} g cm^{-3}, 1 $M_\odot$ ($R \sim 10^4$ AU) starting conditions, the principal stages of these models are as follows (Bodenheimer and Sweigart, 1968; Larson, 1969, 1977; Winkler and Newman, 1980; Boss, 1980a; Stahler *et al.*, 1980):

Isothermal collapse. For $\rho < 10^{-13}$ g cm^{-3}, the cloud is transparent. Although heat is generated by the compression consequent on collapse, it is absorbed and efficiently radiated away by the 1% dust component. For any reasonable starting conditions, the model develops a $\rho \propto r^{-2}$ density gradient, as inferred from Equation (2.5).

CORE DEVELOPMENT. When the density exceeds $\sim 10^{-13}$ g cm^{-3}, the central region becomes opaque to infrared radiation, and the temperature rises. The resulting pressures halt the collapse, causing the development of a hydrostatic core. When falling gas hits this core, a shock front develops resulting in the conversion of kinetic energy into heat, and thence into radiation. This energy transfer at the accretion shock is the principal topic on which various numerical integrations differ; a number of devices have been employed (Winkler and Newman, 1980). Within the core, convection should bring about an adiabatic temperature gradient:

$$\frac{dT}{dr} = (1 - 1/\gamma)\frac{T}{p}\frac{dP}{dr}, \tag{2.8}$$

where γ is the ratio of specific heats: 7/5 for diatomic molecules and 5/3 for atoms.

HYDROGEN DISSOCIATION. When the temperature reaches $\sim$2000 K, hydrogen dissociates, resulting in a reduction of the specific heat ratio γ and a lowering of the adiabatic gradient, so that further collapse occurs.

FREE-FALL REGION. Just outside the dense core, the motion becomes dominated by free-fall, i.e. at radius r the velocity is

$$v = -(2Gm_0/r)^{1/2}, \tag{2.9}$$

where m_0 is the core mass. This velocity, proportional to $r^{-1/2}$, together with the steady-state continuity equation

$$\frac{\partial}{\partial r}(r^2 \rho v) = 0, \tag{2.10}$$

leads to a density $\rho \propto r^{-3/2}$. Consistent with Equation (2.9), the attraction of the central core is dominant in determining the pressure gradient from Equation (2.2) with $d^2r/dt^2 = 0$, $\rho \propto r^{-5/2}$. These two proportionalities in the perfect gas law (Equation (2.4)), lead to $T \propto r^{-1}$. The same proportionality is obtained using the diatomic specific heat ratio, $\gamma = 7/5$, in the adiabatic law, Equation (2.8). These density, pressure, and temperature gradients are sometimes used in models of condensation from the nebula. However, they depend on the aforestated assumptions and neglect of radiative processes affecting the temperature gradient.

LATER DEVELOPMENT. The core continues to maintain a radius of several $R_\odot$ and attains a maximum luminosity of about 30 $L_\odot$ in its growth. The growth time to $L = L_\odot$, $R = 2.0\, R_\odot$, is stretched out to $\sim 4t_f \approx 8 \times 10^5$ yr, most of it close to the main sequence, the later part following the Hayashi (1966) track. See Figure 1.

To infer the effects of rotation, inhomogeneous mass distribution, and magnetic fields,

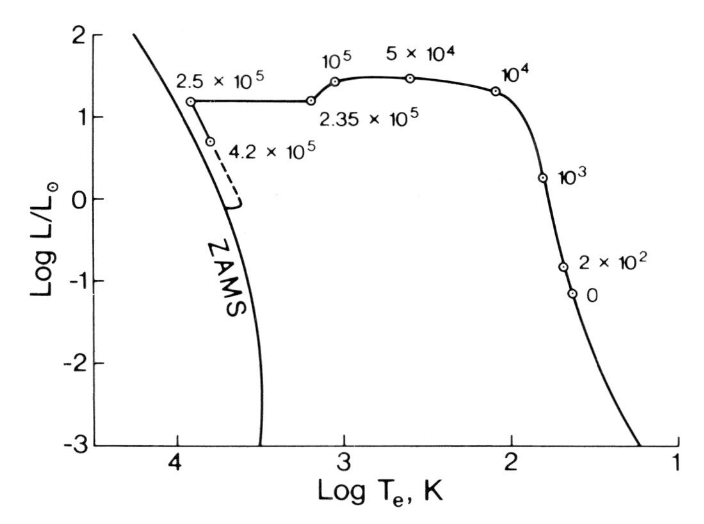

Fig. 1. Computed evolutionary path of a spherically symmetric model of a 1.0 $M_{\odot}$ star, by Winkler and Newman (1980). L is luminosity, T_e is brightness temperature, ZAMS is the zero-age main sequence, and the time markers are in years.

it is necessary to compute in two or three dimensions. All these phenomena would be expected to stretch the time-scale and to distort the shape of a collapsing protostellar cloud, and thus to interact with the thermal state. Hence, these models are commonly classified in terms of three initial energy ratios:

$$\left.\begin{aligned}\alpha &\equiv \text{(thermal energy)/(gravitational energy)} \approx 5kRT/2GMm,\\ \beta &\equiv \text{(rotational energy)/(gravitational energy)} \approx \Omega^2 R^3/2GM,\\ \delta &\equiv \text{(magnetic energy)/(gravitational energy)} \approx 5B^2/24GM^2R^2,\end{aligned}\right\} \tag{2.11}$$

where Ω is rotation rate, B is magnetic field strength, and the numerical coefficients correspond to a sphere homogeneous in ρ, T, Ω, and B. (Thus the Jeans condition (2.7) is equivalent to $\alpha = 1$, $\beta = \delta = 0$.) Relatively little computation of magnetic effects has been done, but there has been much recent work on the axisymmetric effects of rotation and on the interaction of rotation with perturbations to produce fragmentation. Pre-1978 work is reviewed by Bodenheimer and Black (1978) and Woodward (1978), so we summarize here the more recent results.

The principal debate over axisymmetric collapse of a rotating isothermal ($\gamma = 1$) cloud has concerned the conditions for formation of a toroid, or ring, rather than a spheroid. Calculations which allow angular momentum transfer (e.g. Larson, 1972; Boss, 1980b) obtain rings of radii and thicknesses similar to those predicted by equilibrium theory (Ostriker, 1964), while those which maintain local conservation of angular momentum (e.g. Norman *et al.*, 1980) obtain disks, a difference which, in hindsight, is not surprising. However, Boss (1980d) finds that for adiabatic ($\gamma > 1$) models the formation of a spheroid or toroid depends on the initial value of α. For $\gamma = 5/3$, $\alpha \lesssim 0.01$ is required to form a toroid; for $\gamma = 7/5$, $\alpha \lesssim 0.02$ is needed. In these calculations, the sum $\alpha + \beta$ evolves towards the value predicted by the virial theorem, 1/2, and the value of β alone towards consistency with fluid equilibrium theory, in which $\beta \approx 0.43$ is the dividing line between spheroid and toroid. Apparently the initial α must be appreciably less than 0.07 to obtain a ring and allow for the conversion of gravitational energy into heat.

In any case, a toroid seems rather unstable to nonaxisymmetric perturbation (NAP), so the model more relevant to the origin of the Sun is probably not a $\sim 1\ M_\odot$, low α and β cloud, but rather a many $M_\odot$ cloud which could have somewhat higher α and β. The questions then becomes the masses of the fragments produced in the collapse of such a cloud, and the partitioning of angular momentum between orbital motion and spin of these fragments. The answers have been sought in a variety of ways. Norman and Wilson (1978) and Cook and Harlow (1978) applied axisymmetric computations to rotating toroids, and find them to break up into two or three fragments in less than a rotation period, with about 10% of the angular momentum going into fragment spin. Lucy (1977), Gingold and Monaghan (1977), Larson (1978), and Wood (1981) applied Lagrangian techniques, following the motions of 'fluid particles'. This technique does not numerically damp smaller perturbations, but requires an artificial viscosity to cope with shocks. Narita and Nakazawa (1978), Rozyczka *et al.* (1980), Tohline (1980), Boss (1980c), and Bodenheimer *et al.* (1980b) have performed three-dimensional hydrodynamical integrations of the collapse of isothermal clouds with initial spherical shapes but specified NAPs. The practicable resolution limits the NAPs to long wavelengths, and most of the models have assumed only $m = 2$ perturbation modes. The main findings of the most recent and comprehensive publication (Bodenheimer *et al.*, 1980b) are:

1. For a duration of $\sim 1 t_f$, (2.3), the sphere collapses to a disk, and an NAP will not grow; in fact, for larger α it may appreciably damp.
2. The parameter α is most important: at higher α's corresponding to higher pressures and sound velocities, fragmentation is suppressed; for NAPs which are azimuthly sinusoidal density variations of ±50%, $\alpha \lesssim 0.4$ leads to fragmentation while $\alpha \geq 0.5$ leads to a toroid in computer runs of $\sim 2t_f$.
3. Diminishing the peak amplitude, or providing for a drop off in the magnitude with r and z, enhances the tendency to stability. Cases where the mass contained in a density enhancement was $\lesssim 0.2\ M_J$ (defined by (2.7)) and $\alpha \gtrsim 0.25$, led to spheroids.
4. The rotation parameter β had relatively little influence except near the marginal value of $\alpha \sim 0.4$, where enhancement of β to ~ 0.3 led to stabilization of a toroid.
5. The fragments themselves are unstable to collapse, typically with α and β less than 0.1, the latter corresponding to $\sim 20\%$ of the angular momentum going into spin of a fragment containing $\sim 15\%$ of the mass.

2.8. Modeling of Secondary Features

The only computation of magnetic effects so far has been for an axisymmetric, nonrotating, isothermal cloud (Scott and Black, 1980; Black and Scott, 1982). In all cases, there developed an appreciable flattening with equidensity contours at right angles to field lines, and field intensities in the core varying approximately as $\rho^{1/2}$. Similar results are obtained from equilibrium models at the verge of stability by Mouschovias (1976, 1978). The spectra of T-Tauri stars have not yet been adequately accounted for by formation models. The intensified emissions correspond to temperatures which are much too low to account for the outward motions indicated by Doppler shifts in the same spectra, *if* this motion is due to a hydrodynamically expanding stellar wind, similar to the present solar wind (Ulrich, 1978). In addition to the infalling envelope and the accretionary shock region, the effect on spectra of an *accretion disk* (the material expected to orbit the protostar in its equatorial plane, as discussed in Section 4.2 has been examined. The rate of energy dissipation, as well as the appreciable surface area, of an accretion disk should make it a major contributor to the luminosity of a forming star, with some characteristic differences in spectrum (Lynden-Bell and Pringle, 1974). There have been no inferences as to magnetic field intensity from T-Tauri or other PMS source spectra.

2.9. Dynamical Evolution

As discussed, it is very probable that stars form in clusters. Most small N-body configurations, $N \geq 3$, are unstable, and evolve by ejecting one member at a time. Usually the small bodies are ejected. Normally, ejection requires a close approach to another body in the system. For a cluster of radius R^*, total mass M, it is convenient to define a *crossing time* t_c

$$t_c = [8R^{*3}/GM]^{1/2}, \tag{2.12}$$

a number moderately larger than the free-fall time t_f (Equation (2.8)). The number of interest is the mean time to escape, t_e – a problem which so far has been attacked mainly by computing the dynamical histories of specific cases. The most systematic studies have been for two-dimensional $N = 3$ cases, with initial positions selected at random within a circle of radius R^*. Various assumptions can be made as to initial velocities; they make less difference than assumptions as to masses: as anticipated, equal mass systems are more stable than unequal, randomly selected masses. For unequal masses, the average t_e is $\sim 15t_c$, and the distribution of t_e's is Poisson-like. Thus, a system with $R^* = 4000$ AU and $M = 3\,M_\odot$ would have an average t_e of 10^6 yr, and a distribution of t_e's ranging from 10^5 to 3×10^6 yr (Standish, 1972).

3. Cosmochemistry

In cosmochemistry, *solar nebula* is a generic term for gas and dust in the vicinity of the Sun at the epoch of solar system formation. Compositions of solar system objects are generally ascribed to processes in the nebula, unless the evidence requires earlier processes: stellar or interstellar environments; or later: planetesimal or planetary environments. About 1972, the following nine items were substantially agreed:

1. The Sun and the nebula were identical in composition, based on the similarity in composition of the solar photosphere and carbonaceous chondrites, exceptions being the loss of lithium in the former and the loss of volatiles in the later.
2. The nebula was essentially homogeneous in composition, and hence had been well mixed, based on the similarity of nonradiogenic isotope ratios in all Earth, lunar, and meteoritic rocks.

3. Nucleosynthesis of solar system materials ceased $\sim 10^8$ yr prior to condensation in the solar nebula, based on products of ^{129}I and ^{244}Pu decay among the Xe isotopes.
4. The Earth, Moon, and stony meteorite parent bodies became separate entities at least 4.5 Gy ago, based on Rb/Sr and Pb/Pb isotope systematics.
5. The Earth's material lost its volatiles when in bodies no bigger than meteorite parent bodies, based on the similarities in depletion of the inert gases, He, Ne, Ar, Kr, and Xe (often called 'rare' or 'noble' gases).
6. The sizes of differentiated meteorite parent bodies were not more than ~ 300 km, based on the cooling rates inferred from nickel/iron ratios in the Widmanstätten crystallization patterns.
7. Variations in composition among chondritic meteorites arose mainly from equilibrium condensation at varying temperatures in the solar nebula.
8. Variations in composition among the terrestrial planets and smaller solar system bodies arose primarily from a decrease in nebula temperature with distance from the Sun proportional to $\sim 1/r$.
9. Exceptions in composition to the trend with solar distance arose from intraplanet differentiations and dynamical effects.

Most of items 1–8 were rather generally accepted, and cosmochemical debates were mainly about problems under item 9, such as the degree of refractory – volatile heterogeneity in the Earth's growth; the loss of iron by the Moon; the loss of silicates by Mercury; the mechanism for heating meteorite parent bodies; the loss of volatiles by the Moon, Io, and Europa; etc.

However, the turning to meteorites (most notably, the carbonaceous type 3, or C V, chondrite Allende which fell in Mexico in 1969) of the analytic capability developed for the Apollo project, plus mass spectrometer and gas chromatographic measurements of the Martian and Venerean atmospheres by the Viking and Pioneer probes, have since disrupted the consensuses on items 2, 3, 4, and 7, as well as enhancing the difficulties under item 9. This review will emphasize these developments of the last eight years, because they pertain more to the nebula as a whole – of greater interest relevant to the Sun – that to variations within the nebula. References for the other aspects include: for solar composition, Ross and Aller (1976); for elemental abundances, Cameron (1980b); for nucleosynthesis, Trimble (1975); for radiochronology, Kirsten (1978); for inert gases, Black and Pepin (1969), Mazor *et al.* (1970), Black (1972a, b), and Wasson (1974); for meteorite parent body size, Wood (1964) and Goldstein and Short (1967); for equilibrium condensation in the nebula, Grossman and Larimer (1974); and for inferences as to terrestrial planet composition, Lewis (1974), Morgan and Anders (1979, 1980), Ringwood (1979), Smith (1979), and Wood *et al.* (1981).

3.1. Chronology

Most measured crystallization ages of meteorites fall in the range 4.53–4.63 Gy, and could all be 4.57 Gy within observational uncertainty; a few older ages are suspect of systematic error (Kirsten, 1978). These ages employ all five of the significant long-lived decay systems in a variety of ways. 4.57 Gy is not much more than the separation age of the Earth, Moon and meteorites, and hence is customarily taken as the age of the solar system.

The time interval from the cessation of nucleosynthesis to the 'freezing' of a radioactive nuclide in a meteorite (or other rock) commonly called the *formation interval*, can be determined if (a) the ratio of a radioactive isotope to another isotope of the same chemical properties is known at the end of nucleosynthesis, and (b) there is a measurable daughter product of the radioactive decay in the meteorite. Requirement (a) entails not only the identification and quantification of nuclear reactions transforming nuclides, but also a model of the cycling of the material through stars and the interstellar medium throughout the lifetime of the Galaxy up to the cessation of nucleosynthesis $\gtrsim$4.57 Gy ago. Requirement (b) entails a decay time comparable to the formation interval and a mineral setting in which the daughter isotope can be identified.

The formation interval of $\sim 10^8$ yr stated in item 3 above depended on ratios of $\sim 2 \times 10^{-3}$ for $^{129}I/^{127}I$ and $\sim 3 \times 10^{-2}$ for $^{244}Pu/^{238}U$ at the end of nucleosynthesis. The events constituting the end of the interval are evidenced to be the solidification of the meteorites by the correlations of ^{129}Xe excesses with ^{127}I sites, and $^{131-136}Xe$ excesses with fission tracks. The weaknesses are the model assumptions behind the initial ratios and the assumption of identical chemistry for ^{224}Pu and ^{238}U.

However, extinct nuclides of much shorter half-life have now been measured, ^{26}Mg from ^{26}Al ($t_{1/2}$ = 0.76 My) in the C V chondrites Allende (Lee *et al.*, 1976, 1977: Hutcheon *et al.*, 1978) and Leoville (Lorin and Christophe Michel-Levy, 1978), and ^{107}Ag from ^{107}Pd ($t_{1/2}$ = 6.5 My) in the iron meteorite Santa Clara (Kelly and Wasserburg, 1978) and other ataxites (Kaiser and Wasserburg, 1981). These measurements depended on the identification of highly favorable sites: Al/Mg ratios of 100 to 700 in phases of the Allende Ca–Al rich inclusions, and a Pd/Ag ratio of $\sim 10^4$ in Santa Clara. In the Allende inclusions, radioactive decay is strongly indicated by $^{26}Mg/^{24}Mg$ correlating very highly with $^{27}Al/^{24}Mg$, (but not with $^{25}Mg/^{24}Mg$) while condensation in the nebula (rather than mixing of grains condensed earlier) is evidenced by $^{26}Mg/^{24}Mg$ versus $^{27}Al/^{24}Mg$ forming a systematic isochron from four different mineral phases throughout each $\sim$1 cm inclusion. In Santa Clara, radioactive decay is more weakly indicated by a higher than normal $^{107}Ag/^{109}Ag$ ratio, but association with the nebula is more strongly established by the phases being the consequence of solidification from a melt.

Creation of the ^{26}Al and ^{107}Pd by spallation in the solar nebula is implausible, because the required proton flux would have also produced other isotopes in abundances greatly in excess of observed. For example, the flux necessary to account for ^{26}Al in Allende inclusions would have produced $\sim 10^3$ times too much ^{22}Ne (Lee, 1978, 1979). Hence the simplest explanation for the ^{26}Al and ^{107}Pd is a second nucleosynthetic source (Cameron and Truran, 1977). The more obvious mechanisms (e.g. carbon burning) lead to $^{26}Al/^{27}Al$ of $\sim 1-2 \times 10^{-3}$ (Arnett and Wefel, 1978); hence the termination of this nucleosynthesis would have been $\sim$3.4 My before fixing of the typical 5×10^{-5} $^{26}Al/^{27}Al$ in the Allende inclusions. However, the amount of material which could be injected into the solar nebula from such a late supernova event is limited by other isotopic anomalies. For example, if $^{26}Al/^{27}Al$ of 2×10^{-3} were created by explosive carbon burning, there would also plausibly be *r*-process $^{129}I/^{127}I$ of $\sim$1 from the same event. The observed $^{129}I/^{127}I$ of $\sim 10^{-4}$ thus would limit the injected material to $\sim 10^{-4}$ of the nebula (Lee, 1979). But this estimate presumes homogeneity of the nebula, as well as correct nucleosynthetic models and absence of mass fractionation. Since all these effects also affect stable isotope ratios, we postpone their discussion to the end of the next subsection.

3.2. NUCLIDE VARIATIONS

The largest measured variations in isotope ratios occur among the inert gases. However, these data are difficult to interpret because of the extremely small quantities involved, uncertainty as to their associations with mineral phases, and obscurity of the relevant physics. More easily interpreted are the abundant active light gases C, N, O, of which oxygen is much more diagnostic because it has three stable isotopes, while carbon and nitrogen have only two each (Podosek, 1978).

The relevant data are given in Figure 2, which reflects work ongoing since Clayton

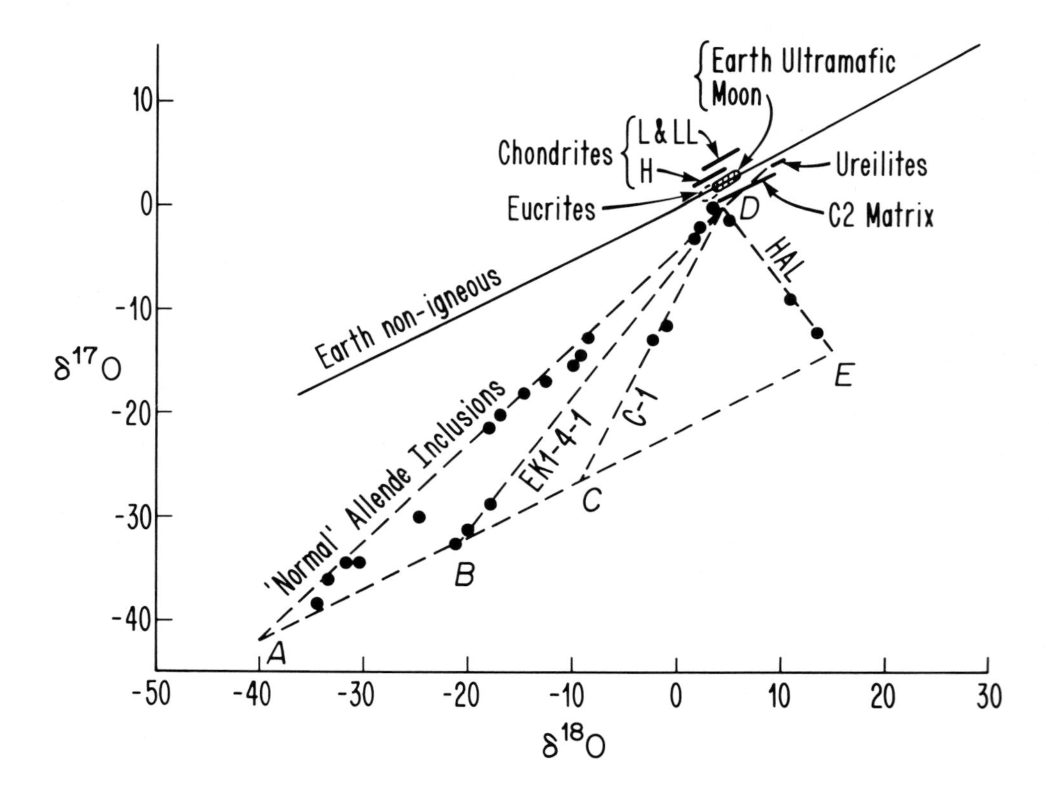

Fig. 2. Measured oxygen isotope ratios, from Clayton and Mayeda (1977) and Lee *et al.* (1980). $\delta^{I}O = 1000\,([{}^{I}O/{}^{16}O] - [{}^{I}O/{}^{16}O]_s)/[{}^{I}O/{}^{16}O]_s$, where the subscript s denotes SMOW: standard mean ocean water. Not shown are correlations with mineral phases in Allende inclusions: low ratios are in spinels; intermediate ratios, in pyroxenes; high ratios, in melilites.

et al. (1973). The 0.52 slope of the Earth line and other lines parallel to it is consistent with mass fractionation of isotopically homogeneous sources (the difference from 0.50 is a theoretically predicted nonlinear effect). Variations in other directions require varying proportions of material having differences arising from nuclear processes. Given the ^{22}Ne constraint, these processes cannot include significant spallation in the nebula.

The relatively small differences in location of the parallel Earth and chondrite lines are well above the measurement error. The Earth and Moon could come from the same isotopically homogeneous source, but to have the Eucrite parent body come from that source would require a modification unusually large for an igneous differentiation, and inconsistent with other indicators of temperature and oxygen fugacity in the Eucrite meteorites.

The ~1.0 slope of the line AD in Figure 2 is most simply explained as due to variations in pure ^{16}O, suggesting a minor source at or beyond A which was synthesized by $^{12}C(\alpha, \gamma)^{16}O$. All but three of more than 100 Allende inclusions examined scatter along the AD line. Small variations from the line are above measurement error, and explicable by physical effects – interphase diffusion, etc. – after the inclusions were formed. All inclusions with ^{26}Al excesses lie along the AD line.

The three inclusions with differing oxygen isotope ratios, EK1-4-1, C1, and HAL, have other isotope anomalies suggesting both fractionation and nuclear effects: hence the acronym 'FUN' for 'fractionation, unknown nuclear'. The clearly identified anomalies to date can be placed in three groups (Lee, 1979).

(1) *O, Mg, Si*. These elements all appear to exhibit small nuclear effects superposed on large mass fractionation effects. The oxygen isotopes are consistent with mixing of a source near D with mass fractionation products along a line DE. The $\Delta(^{26}Mg/^{24}Mg)$ are all slightly less than twice the $\Delta(^{25}Mg/^{24}Mg)$, while the $\Delta(^{30}Si/^{28}Si)$ are moderately less than twice the $\Delta(^{29}Si/^{28}Si)$ (where Δ denotes difference from standard ratio).

(2) *Ba, Nd, Sm*. These elements in the atomic mass range 130–154 are mixtures of *s*-process, *r*-process, and *p*-process products. If reference bases for each element are defined by their 'shielded' *s*-process isotopes, then in EK1-4-1 *s*-process isotopes show excesses, plus possibly *p*-process, but in C1 a *p*-process excess seems required. However, mass fractionation effects cannot be ruled out.

(3) *Xe, Sr, Ca, Ti*. These isotope anomalies appear more unsystematic and less explicable than the Ba, Sm, Nd. The xenon variations are discussed below in connection with those occurring in other meteorites. Calcium and titanium (11 isotopes, mass numbers 40 through 50) have anomalies which appear to be correlated, but after removal of simple mass fractionation, trends show such complex patterns as to lead to the suggestion of 'at least three nucleosynthetic components' (Niederer *et al.*, 1980). Strontium (4 isotopes, 84 through 88) is described as being light-isotope enriched in normal Ca–Al inclusions (Patchett, 1980), but either depleted in ^{84}Sr or enriched in ^{86}Sr in the FUN inclusions (Papanastassiou and Wasserburg, 1978).

An important difference of isotopic anomalies in the refractory elements from those in oxygen is sharp consistency among all phases within a given ~1 cm inclusion, which mitigates against the hypothesis that the inclusions are mixtures of grains which condensed in an earlier environment (Lee, 1979). Recent attempts to intrepret ^{26}Al and FUN anomalies still range widely: several supernovae sources (Reeves, 1979); preservation of ^{26}Mg excesses as interstellar grains (Clayton, 1977); various mechanisms of injection into the nebula (Margolis, 1979b; Herbst and Rajan, 1980); multiple mass and component fractionation in the nebula (Consolmagno and Cameron, 1980; Patchett, 1980); formation in an expanding supernova shell (Boynton and Cunningham, 1981); and disequilibrium condensation followed by crystallization and element redistribution (Kerridge, 1981; Stolper, 1981).

Because of the interpretive difficulties, investigations of the inert gases have concentrated on those with more than two isotopes: neon (20, 21, 22), xenon (9 isotopes from 124 to 136), and krypton (6 isotopes from 78 to 86). Observed variations in isotope ratios are explained as arising from different mixes of components which, themselves, usually have multiple isotopes. Progress is made by increasingly precise isolation of mineral phases, and matching of components therewith. With these refinements, more extreme isotope combinations are being found, and sometimes components are resolved into two. As might be expected, the more exotic components tend to be associated with the more refractory phases (Anders, 1981).

All neon in meteorites is, so far, explicable as a mix of four components normally characterized in terms of the ratios $^{20}Ne/^{22}Ne$ and $^{21}Ne/^{22}Ne$: B, or 'solar', 12.5 and 0.04; A, or 'planetary', 8.2 and 0.02; S, or 'spallation', 0.9 and 0.9; and E, or 'exotic', <0.2 and <0.003. Neon-E in the C I chondrite Orgueil has now been resolved into two components, the more extreme being Ne–E(L) with $^{20}Ne/^{22}Ne < 0.01$ and $^{21}Ne/^{22}Ne <$ 0.001 (Jungck and Eberhardt, 1979) strongly indicative of origin by radioactive decay of ^{22}Na ($t_{1/2}$ = 2.6 yr), since any plausible stellar nucleosynthesis would produce multiple isotopes of neon (Black, 1972b; Podosek, 1978).

At least seven kinds of meteoritic xenon have been suggested to date, some of them with their krypton equivalents. The most abundant component is 'primordial' (or 'planetary' or 'trapped') xenon; others are 'solar', a mass-fractionated modification of primordial; two radiogenic components; a spallogenic component; and two apparently exotic components, Xe–X and Xe–S. Xenon–X (or CCFXe) has excesses in both light and heavy isotopes, a V-shaped profile: $(^{I}Xe\text{–}X/^{I}Xe\text{–}O) - 1 \sim |I - 130|/6$, where Xe–O is solar wind xenon. It is found in acid-resistant residues of the matrix of Allende, comprising less than 1% of the meteorite. Hypotheses are (1) a combination of super-heavy nuclide ($Z > 110$) fission for the heavy isotopes with mass fractionation for the light, and (2) supernova processes: e.g. p for the light isotopes and r for the heavy. Xenon–S (or Q–Xe) is enriched in the isotopes 128, 130, 132 and was found in high-temperature fractions of the CM (or C2) meteorite Murchison (Srinivasan and Anders, 1978). Such an enrichment could be due to an s-process excess. Currently, the absence of other indicators of exotic causes (e.g. fission tracks, p- or r-type excesses of other nuclides) in the same phases as Xe–X and Xe–S tends to favor causes within the solar nebula (Anders, 1981).

Research in meteorite nuclides and their interpretation is now in a state of extreme flux, so that any review is obsolescent. A variety of hypotheses still abound, even including the single formation interval, necessarily combined with effects in expanding supernovae envelopes or the interstellar medium (Clayton, 1978, 1981). However, the investigators working most closely with nuclide data believe there are at least two sources, the strongest constraint being the combination of (1) the clear mineral separate Al–Ng isochron with (2) the firm evidence of ^{244}Pu, but not ^{248}Cm (Clayton, 1977), in the same specimens. Another common feature is the frequent absence of correlative stable isotope anomalies to confirm nucleosynthetic hypotheses for the origin of certain components, which suggests that much more needs to be understood about fractionation effects is both supernova envelopes and the solar nebula (Podosek, 1978; Lee, 1979).

3.3. Chemical Variations among Chondritic Meteorites

Chemical problems break down into two distinct categories: (1) the compositions of chondritic meteorites, and (2) the compositions of all other solar system objects (with the possible exceptions of comets and interplanetary dust). The reason for this discrimination is that chondrites are universally agreed to be polymict breccias whose constituents were condensed from gaseous media (and have since undergone variable degrees of metamorphism), while other objects accessible to examination are generally (but not universally) agreed to have been significantly affected by intrabody differentiations.

Given the consensus that the components of chondritic meteorites condensed from gaseous media, the debates are about (1) the locations of these condensations; (2) the degree to which the condensations were nonequilibrium; and (3) the history of the material between its condensation and collection into the meteorite parent body. The traditional view is that the condensations occurred in a solar nebula with $\partial T/\partial r < 0$ and $\partial T/\partial t < 0$ always, the range of temperatures affecting chondrites being ~1500 K to ~500 K (Anders, 1964; Wasson, 1978). Hence the characteristic structure consists of sizeable refractory inclusions in a finer grained matrix of more volatile composition. In particular, the apparently oldest refractory inclusions in Allende are ~1 cm in size (Grossman, 1980), $>10^4$ times as great as typical interstellar grains (Greenberg, 1979a). The properties expected of interstellar matter in chondrites – submicron sized grains in disequilibrium with each other – occur most markedly in the matrix of those chondrites having the greatest proportion of volatiles, carbonaceous type 1 or C I. But these grains are lacking in the spallogenic products expected to have been produced in an interstellar medium, quite different from those produced more recently by cosmic-ray effects on the meteorite (Wasson, 1978). More likely candidates for pre-nebula condensates are small grains of platinum group metals embedded within Ca–Al rich inclusions of Allende (Wark, 1979).

The condensation sequence is not sensitive to pressure over a range of pressures of $\sim 10^{12}$; other than that, iron appears before forsterite at $P \gtrsim 10^{-4}$ bar (Lattimer *et al.*, 1978). Hence the bulk composition is not strongly diagnostic of formation circumstances, and details of textures, configurations of the mineral phases, etc., must be examined. Such features as: alternating bands of differing refractoriness (Wark and Lovering, 1977); variations in content of rare earths (^{139}La through ^{175}Lu) correlated with relative volatility (Boynton, 1978); igneouslike textures (Blander and Fuchs, 1975) and fractured, as well as rounded, inclusions, indicate complex histories for parts of some meteorites. Heterogeneity of the nebula appears to be necessary for chemical, as well as nuclear, reasons. Also, parts of chondrites have undergone melting, as most strongly evidenced by chondrules. It is debatable, however, whether passage through a molten state is necessary to explain the chemistry of Allende inclusions (Blander and Fuchs, 1975; Grossman, 1980). Somewhat more speculative are suggestions that chemical variations depend significantly on plasma effects (Arrhenius, 1978) or processes in supernovae envelopes and the interstellar medium (Clayton, 1978). The certain identification of such effects will require more detailed analyses of ~1 μm grains, as well as better models of condensation in differing environments.

The greatest difficulty for models of condensation in the solar nebula is probably the very high temperatures, >1500 K required to account for refractory assemblages in some chondrites, such as the Allende inclusions: appreciably higher than indicated, further from the Sun than Mercury by most physical models in recent years, as discussed in Section 4. Consolmagno and Cameron (1980) suggest that the necessary temperatures, as well as mass fractionation effects, occurred in gaseous protoplanets – large gravitational instabilities – as discussed in Section 4. Morfill (1983) proposes that convection in a turbulent nebula disk circulated grains from hot to cold regions repeatedly.

3.4. Chemical Variations among Differentiated Objects

The evidence to be explained includes the compositions of Earth rocks, Moon rocks, and a wide variety of meteorites: achondrites, stony-irons and irons, spectroscopic observations of planetary atmospheres and surfaces; seismological observations of the Earth and Moon interiors; mean densities of all the planets and moments-of-inertia of five – Earth, Moon, Mars, Jupiter, and Saturn. To explain the compositional evidences pertaining to the major planets (Owen, 1978) and the Galilean satellites (Johnson, 1978) there is no need, as yet, to attribute other than solar composition minus varying degrees of volatiles. A considerable body of data developed in recent years is the reflectance spectra of asteroids, none of which can yet be said to be inconsistent with all categories of meteorites (Chapman, 1979). Although it is important to obtain data on the outer solar system, such as the H : He ratio in the Jovian atmosphere, at present the main problems are explaining the four terrestrial planets, the Moon, and the differentiated meteorites.

Historically, the attempt has been made repeatedly to explain these compositional differences among the predominantly silicate and iron bodies as arising from processes in the nebula: i.e. by interactions between gas (possibly ionized), dust, and planetesimals too small for internal energy sources to drive differentiation (Urey, 1952; Alfvén, 1954). Such models are still pursued by Anders and coworkers (Anders, 1977; Morgan and Anders, 1979, 1980). However, most workers now believe that differentiations internal to planetesimals, or associated with terrestrial planet formation, were important. The bases for this belief include: (1) imperfections in the correlation of composition with solar distance, particularly the Moon; (2) the inevitability, if planets formed, of an intermediate stage with a greater number of smaller bodies which must have collided; (3) the evidence of ~100 km parent bodies from iron meteorite cooling indicators (Wood, 1964; Goldstein and Short, 1967); (4) the agreement of Eucrite (calcium-rich basaltic achondrite) composition with what is petrologically predicted for partial melting of a small chondritic body (Stolper, 1977); (5) the plausibility of ^{26}Al as a heat source for small body differentiation (Lee *et al.*, 1977); and (6) the lack of an adequate physical hypothesis for separating iron from silicate material at the nebula stage (Banerjee, 1967).

The differences in solid body composition among the Earth, Venus, and Mars are readily attributable to differences in oxidation level (Ringwood, 1959, 1979), although this explanation leaves begging why the oxidation level does not vary systematically with solar distance. More difficult problems are the depletions of Mercury in silicates and of the Moon in iron. Mercury's composition may arise from rather high temperatures and pressures in its nebula zone (Grossman and Larimer, 1974) or from collisional processes (Weidenschilling, 1978); the Moon's composition may arise from its being formed by a major impact into the Earth after the Earth's core had formed (Ringwood, 1979; Kaula, 1977).

The greatest problem raised by the exploration of terrestrial planet atmospheres in recent years is the variation in inert gas abundance among the terrestrial planets. Assuming complete outgassing, the abundances proportional to planet mass of primordial argon, $^{36+38}Ar$ are 4×10^{-9} for Venus, 5×10^{-11} for Earth, and 2×10^{-13} for Mars. This range of 20 000–fold contrasts to that of radiogenic argon, ^{40}Ar, which is 20-fold. Some sort of solar wind implantation effect (Wetherill, 1981) seems more plausible than any sort of nebula gradient (Pollack and Black, 1979), but even so a remarkable inhibition of interzone mixing is required thereafter.

4. Planet Formation

Formation of the planetary system is inherently a more difficult problem than star formation, mainly because it is a secondary process involving a minor fringe. It is constrained by many residual evidences, but not by any observations of contemporary occurrence. The difficulties are enhanced by the necessity of considering at least six types of entities, some of which can be divided into subtypes. But the one outstanding difficulty is the formation of Jupiter; once it is hypothesized to exist, other features can be plausibly (although not yet rigorously) explained.

Many comprehensive hypotheses have been advanced to explain the origin of the planets. None, however, is complete enough in the necessary physics to be called a theory: a better term would be scenario. Scenarios are important to help decide research directions. However, understanding of planetary formation is still at such a primitive stage that the current efforts on physical modeling which will contribute most effectively in the long run are probably directed at either different problems (e.g. spiral galaxies, planetary rings) or much simpler idealizations. This review attempts to select such research investigations. For a summary of scenarios of solar system origin and several examples thereof, see the first nine papers in the volume edited by Dermott (1978).

The types of entities, and their subtypes, for purposes of organizing the discussion, are:

- C the Sun;
- D the solar nebula, or accretion disk, with parts: g gas; d dust; p planetesimals;
- P planets, with subtypes: f iron; s silicate; h hydrogen and helium, i ice;
- W the solar wind;
- X neighboring stars;
- B magnetic fields, with subtypes: c solar; x external.

In this review, the functional distinction of a planetesimal from dust is that the body is large enough to move separately from the gas and to exert perceptible gravitational attraction upon close approach; of a planet from a planetesimal, that the body exerts significant gravitational attraction at a distance.

The central body C, the Sun, is common to all dynamical models; different categories of models are distinguishable by different combinations of the types of entities. We first consider the mechanical and thermal interactions in the categories CDP, before taking up complications arising from the solar wind W, external bodies X, and magnetic fields B.

4.1. *gd* GAS–DUST INTERACTION

The nucleation, growth, and evaporation of dust grains and their interaction with the gas are important to opacity (Gaustad, 1963; Kellman and Gaustad, 1969), preservation of isotopic anomalies (Margolis, 1979a; Elmegreen, 1981), and the separate motion of solids from gas (Safronov, 1972; Goldreich and Ward, 1973; Slattery, 1978; Weidenschilling, 1980; Nakagawa *et al.*, 1981). As mentioned in Section 2, the submicron size of interstellar grains and their $\sim 1\%$ proportional abundance leads to opacity of an protostellar cloud at $\rho \sim 10^{-12}$ g cm^{-3}. Margolis (1979a) estimates that grains cannot grow to more than ~ 2 μm size before clumping, if isotopic heterogeneities such as observed in Allende inclusions are to be preserved. The transition from essentially comovement with gas to essentially separate motion covers an appreciable size range. However, at nebula densities ($\gtrsim 10^{-9}$ g cm^{-3}) the growth of grains (whether by accretion or nucleation) is estimated to take $\sim 10^3$ yr to reach ~ 1 cm size, which has an area : mass ratio low enough to allow separation on a $\lesssim 10^2$ yr time-scale.

Molecular absorption by H_2 also contributes significantly to opacity in the range 1500 K (silicate and iron vaporization) to 2000 K (hydrogen dissociation) (Alexander, 1975). While the dependence of dust properties and opacity on composition, pressure, and temperature is far from perfectly understood, at present this problem area is not the foremost limitation on understanding planetary formation as a whole.

4.2. *CD* DISK DYNAMICS

Angular momentum conservation considerations, as well as radiations associated with massive and energetic collapse phenomena such as black holes, make appealing the concept of an *accretion disk* (Shakura and Sunyaev, 1973; Lynden-Bell and Pringle, 1974). It is likely that the circumsolar material never was close to axial symmetry in mass distribution, due to both infall heterogeneities and its own instabilities. Nonetheless, the axisymmetric model is most convenient to examine some dynamical properties necessary to any solar nebula.

If the effect of any interaction between adjacent elements of matter is expressible by a kinematic viscosity ν, then there must exist a torque g exerted by a material ring at r on the ring at $r + \mathrm{d}r$

$$g = 2\pi r^3 \nu\sigma \,\mathrm{d}\Omega/\mathrm{d}r, \tag{4.1}$$

where σ is the surface density and $-r(\mathrm{d}\Omega/\mathrm{d}r)/2$ is the strain rate. If there is a differential torque, $\partial g/\partial r \neq 0$, then there must be a change in the specific angular momentum $h = \Omega r^2$ and a corresponding flux of matter F,

$$-\frac{\partial g}{\partial r} = 2\pi r\,\sigma\,\frac{\mathrm{d}h}{\mathrm{d}t} = F\,\frac{\mathrm{d}h}{\mathrm{d}r}, \tag{4.2}$$

assuming steady state, $\partial h/\partial t = 0$. Since $\mathrm{d}\Omega/\mathrm{d}r \to 0$ at the Sun and $\nu \to 0$ at the outer limit of the nebula, there must be a net inward flux of matter F and an outward flux of specific angular momentum, unless and until interactions are reduced to gravitational attraction without ejection from the system (Lynden-Bell and Pringle, 1974).

To maintain the viscosity ν there must be turbulence: $\nu \propto v_R l$, where v_R is the turbulent velocity and the mixing length l is something less than the thickness (or scale height) of the disk. Such turbulence would be generated by *convection*, which would have occurred as a result of a temperature gradients, either radial, $\partial T/\partial r < 0$, or normal, $\partial T/\partial z < 0$. Full solutions for this convection have not yet been made, but steady-state models for $(\partial T/\partial z)$–dependent convection have been constructed by Cameron (1978), Lin and Papaloizou (1980), Lin (1981), Prentice (1978), and Cassen and Moosman (1981). In the Cameron model temperature is assumed to be a function of radius only and to be connected to the random velocity v_R through both the speed of sound c, assuming $v_R = \alpha c$, $\alpha \sim \frac{1}{3}$, and the energy dissipation rate D:

$$T^4 = T_0^4 + D/2s = T_0^4 + \left[\frac{1}{2\pi r} g \left(-\frac{\partial\Omega}{\partial r}\right)\right] \Big/ 2s, \tag{4.3}$$

where T_0 is the temperature of the surroundings and s is the Stefan–Boltzmann constant. An alternative hypothesis for estimating v_R (or ν) is to assume the Reynolds number, $Re = \Omega r^2/\nu$, to be at the critical value for instability, 10^2–10^3 (Lynden-Bell and Pringle, 1974). In the Lin and Papaloizou model, a power law opacity, $\propto T^\beta$, is assumed and a homologous contraction is constructed to prove the nebula may be intrinsically unstable against convection. They adopted a mixing-length model in which $\nu = v_c^2/\Omega$, where v_c is the convective velocity. Given T_0, the radiative + convective flux is integrated in the z-direction (normal to the disk) and the scale height H adjusted until zero flux is obtained at the mid-plane.

In both the Cameron and Lin models, either the external temperature T_0 or the mass infall rate $\dot{m}$ must be prescribed. Because of the viscous transfer of mass F, the surface density σ is proportional to $\dot{m}$. For instance, to obtain a surface density high enough for gravitational instability of the gas (see below), the effective time for accretion of the disk must be only a few percent of the free-fall time, Equation (2.3) (Cameron, 1978). An important unknown is the efficiency of conversion of infall energy; the steady-state solutions assume that the energy is lost by radiation at the shock front where infalling matter interacts with disk matter, so that the energy balance determining the temperature is that with viscous dissipation. Consequently, in all the models if it is assumed $T_0 <$ 100 K (as suggested by cloud properties), or if it is assumed the mid-plane $T < 200$ K at Jupiter, then even at Mercury the temperature is well under 1500 K, insufficient to account for Allende inclusion properties, etc.

For sufficient surface density σ, the nebular disk will be gravitationally unstable. If the dust grains are large enough that they can move separately from the gas, then this instability can occur in the dust layer alone (Safronov, 1972; Goldreich and Ward, 1973). Given a thin disk with prescribed density $\sigma(r)$ and rotation $\Omega(r)$, and assuming velocity departures u, v, density and potential perturbations $\delta\sigma$, δV all proportional to $\exp[i(\omega t + kr)]$ in the momentum, continuity, and Poisson equations, leads to a condition on wavenumber k for instability (Toomre, 1964):

$$\omega^2 = 2(2\Omega^2 + r\Omega\, \mathrm{d}\Omega/\mathrm{d}r) + \frac{c^2}{\gamma} k^2 - 2\pi G\sigma k < 0, \tag{4.4}$$

where the pressure appearing in the momentum equations has been replaced by $c^2 \rho/\gamma$ (Equation (2.4)) and $\delta\rho/\rho$ by $\delta\sigma/\sigma$. Thus the rotation stabilizes against low k, or long wavelength, instabilities, while random velocity, c or v_R, stabilizes against high k instabilities. Numerical experiments on the development of instabilities in disks have recently been carried out by Cassen *et al.* (1981).

If the present planets are smeared out into continuous layers, $v_R = 0$ assumed for solids but the sound velocity c assumed consistent with present equilibrium temperature, 278 K/$R^{1/2}$ (R in AU), then the densities are high enough and temperatures are low enough for formation of rocky planetesimals in the terrestrial planet zones and icy planetesimals in the major planet zones. However, the resulting planetesimals (after allowing for loss of relative momentum by gas drag (Goldreich and Ward (1973)) in the Jupiter and Saturn zones are much too small to capture appreciable hydrogen and helium, for which a $\sim$15 $M_\oplus$ body is required. Particles-in-box models (allowing for enhancement of relative velocities by mutual perturbations) for planet growth (Safronov, 1972) take much too long. Either (1) there was a much more massive nebula leading to gravitational instability of the gas, or (2) linear instability analysis is quite insufficient to infer the growth of Jupiter. Cameron (1978) chooses option (1), abetted by late formation of the Sun (reducing the Ω-dependent terms in Equation (4.4)), thereby entailing a rapid growth of the disk, followed by tidal disruption of the giant gaseous protoplanets and, after infall ends, appreciable loss of gas from the nebula. Explorations taking up option (2) are discussed below, in connection with planet–disk interaction.

There is a considerable literature dealing with the dynamics of nonuniform disks; recent reviews are by the Toomre (1977) and Lin and Lau (1979). However, this work is heavily influenced by the observed repeated occurrence of spiral structure in galaxies. The nebula could, on the one hand, have had a more primitive structure because of repeated disruption by infalls while, on the other, a more advanced structure, because in $\sim 10^3$ yr it underwent more revolutions in its denser parts than has the typical galaxy in 10^{10} yr. Hence, adaptation of ideas and methods from galactic dynamics to the nebula is risky.

4.3. *Cpp* PLANETESIMAL SWARMS

If gravitational instability of a dust layer led to the formation of many small bodies, then this population would have evolved in its distribution of sizes through collisions, and in its distribution of velocities through both close approaches and collisions. Analytically, these two evolutions have been treated only separately, adapting theory from colloidal chemistry for the mass distribution and from stellar dynamics for the velocity distribution (Safronov, 1972).

The principal modifications of colloidal chemical theory are to allow for breakup as well as growth from collisions, and for the enhancement of capture cross-sections by gravitational attraction. For a body of radius r and mass m, this cross-sectional area $a \approx \pi r^2 (1 + 2Gm/rv^2)$, where v^2 is the mean square approach velocity, provided by the separate velocity distribution theory. This enhancement leads eventually to the outstripping of the mass distribution $n(m) \propto m^{-q}$, $m \leq m_1$, within a zone by a single body of mass m', since a'/a_1 approaches r'^4/r^4 because v^2 is determined mainly by m' itself (Safronov, 1972).

Somewhat rigorous solutions of the velocity distribution problem have so far been made only for the case of a uniform population. Difficulties are: (1) properly accounting for angular momentum and energy transfers between random motions and reference motions which are coplanar circular orbits; and (2) estimating deviations from simple two-body gravitational close encounter due to the presence of the Sun. Analytic treatment has been by solution of the Boltzmann and Fokker–Planck equations using limited series expansions (Stewart and Kaula, 1980). So far, agreement within ~25% for the r.m.s. relative velocity is obtained with computer experiment (Wetherill, 1980a). Solutions for the velocity distribution given a mass distribution, $n(m) \propto m^{-q}$, have assumed equilibrium between velocity enhancement by gravitational perturbations and elastic collisions and velocity damping by inelastic collisions, and have neglected energy and momentum transfers between random and reference motion (Safronov, 1972; Kaula, 1979). Nonetheless, they also have obtained r.m.s. velocities on the order of the escape velocity of the mean sized body, and inverse correlation of random velocity with size.

Computer experiments simultaneously treating mass and velocity distribution evolution have started assuming uniform mass populations and terminated while both distributions were still evolving within a limited zone (Greenberg *et al.*, 1978).

The effects of one or more planets on a planetesimal are discussed below in the sections on planetary systems and terrestrial planet evolution.

4.4. *CPD* PLANET–DISK INTERACTIONS

Work directed at explaining planetary rings (Goldreich and Tremaine, 1979, 1980) or accretion disks around close binaries (Lin and Papaloizou, 1979a, b) or the distributions of a, e, I in the asteroid belt (Wisdom, 1980; Froeschlé and Scholl, 1981) may help elucidate the evolution of the circumsolar system after some planetary mass had accumulated, but before the disk had been dissipated. This work differs from most galactic structure research (e.g. Lynden-Bell and Kalnajs, 1972; Goldreich and Tremaine, 1978) in that there is a significant concentrated body (or external potential) in addition to the self-gravitation of the disk.

The common thread of planet–disk investigations is the occurrence of *resonances*,

$$\psi = m\Omega + \epsilon\kappa - [m\Omega_p + (l - m)\kappa_p] \approx 0, \tag{4.5}$$

where Ω, Ω_p are the angular rates of a test particle in the disk and the planet respectively, and κ, κ_p are epicyclic rates (galactic jargon) or anomalistic rates (celestial mechanical jargon). Since orbital eccentricities e probably are small and the disturbing function coefficient is proportional to $e^{|\epsilon|}$, particular attention is paid to the *Lindblad resonances*, $|\epsilon| = 1$. On the other hand, the longitudinal wavenumber m may be quite large; indeed, it must be if the planetary mass is not yet big compared to the disk mass, since $(R/R_p) \approx (\Omega_p/\Omega)^{2/3}$ by Kepler's law.

Applying a linear analysis – somewhat similar to that for gravitational instability, except that the wave is longitudinal: i.e. $(u, v, \delta\sigma, \delta V) \propto \exp[i(\omega t + k\lambda)]$, $(\dot{\lambda} = \Omega)$ – leads to torques which transfer angular mometum from an inner Lindblad resonance ($\epsilon = 1$) to the planet, and from the planet to an outer Lindblad resonance ($\epsilon = -1$). Corotation resonances ($\epsilon = 0$) lead to no net transfer of angular momentum.

When a planet mass is small compared to the disk mass, and the surface density varies appreciably with radius ($d\sigma/dR \ll 0$), the inner Lindblad resonances become a mechanism for moving the planet outward, and thus sweeping up additional matter. When a planet mass is comparable to the disk mass (as Jupiter must have been at one stage), then the inner resonances lead to a flux of disk matter towards the Sun (Goldreich and Tremaine, 1980). This flux has obvious implications for the removal of material from the zone of Mars and the asteroids, as well as volatiles from the entire inner solar system. But the outer Lindblad resonances will move matter outward, and may have contributed to the formation of Saturn and the icy planets.

The occurrence of a resonance in the disk could lead to sufficient local enhancement of surface density σ, for a gravitational instability, (4.4). This effect is damped by random motion, v_R or c, and rotational shear, $d\Omega/dr$. Goldreich and Tremaine (1980) apply a WKB approximation to bridge across the resonance width in cases which remain stable. Lin and Papaloizou (1979a, b) consider larger scale viscous effects on the removal of angular momentum from the disk between the Sun and planet, discussed in the preceding paragraph. They conclude that the relative importance of viscous disk effects (discussed in Section 4.2: Lynden-Bell and Pringle, 1974) and the inner Lindblad resonance effects depends on the magnitude of $q\,Re$, where q is the mass ratio $m_p/M_\odot$ and Re is the Reynolds number of the disk. If $q\,Re < 1$, disk matter will be pushed out past the planet; if $q\,Re > 1$, the planet can remove angular momentum rapidly enough to truncate the outer edge of the disk; if $q^2 Re > 1$, it can impose further truncations at the inner and outer $m = 1$ Lindblad resonances (2 : 1 commensurabilities), leading to density enhancements just beyond these limits (Lin and Papaloizou, 1979a, b; Franklin *et al.*, 1980).

4.5. *CPP, CPPp* PLANETARY SYSTEMS

A fairly obvious exercise is to assume a set of planets and orbits (plus possibly some dissipating effect to simulate close interactions with planetesimals), set them loose in a large computer, and see how the system evolves (e.g. Barricelli *et al.*, 1979). However, the computational expense of testing a significant range of possibilities in this manner provokes thought of how to remove short periodic oscillations in the orbit, which in turn leads (for noncrossing and nonresonant orbits) to *secular perturbation* algorithms, applied to the contemporary planetary system by Brouwer and Van Woerkom (1950) and earlier workers, and to the asteroid belt by Williams (1969). These algorithms take advantage of energy conservation and the separation of the in-plane and of-plane perturbations to order e^3 to obtain systems of ordinary differential equations which can be solved for sets of eigenfrequencies. At any time, the nodal and perihelion rates of a planet (or massless test particle) are linear combinations of these eigenfrequencies; if two eigenfrequencies become equal, a *secular resonance* occurs, leading to buildup of the eccentricity e until, in the mathematics, higher order terms must be considered, but in the solar system a collision may have occurred meanwhile.

The idea of applying secular perturbation theory to hypothetical planetary systems has been around for a decade or so, but the application has been to limited cases such as Mercury (Ward *et al.*, 1976) or Mars and the asteroid belt (Heppenheimer, 1980), mainly because of the difficulty of hypothesizing a realistic representation of the matter not in the planetary bodies, and correctly treating this representation and its evolution. The most extensive application has been by Ward (1981), who made the simplest possible assumption: the present planetary system plus an axisymmetric disk of surface density $\sigma \propto R^{-k}$, $k > 0$. The gradual dispersal of this disk leads to a sweeping of secular resonances across the inner solar system. The resulting buildups of eccentricities and inclinations are so fast that the dispersal time must be quite rapid, $\lesssim 10^{4.5}$ yr, in order for the observed values not to be exceeded.

The further application of secular perturbation theory should depend on the judicious application of inferences from models of planet–disk interaction discussed in Section 4.4.

4.6. *CP(f, s)p* TERRESTRIAL PLANET FORMATION

The volatile deficiencies of the terrestrial planets (especially the Moon, Mars, and Mercury) make attractive the idea that they formed by accretion of rocky planetesimals, the end products of the mass and velocity distribution evolutions discussed in Section 4.3. The hierarchy of outstripping of successive swarms by successively larger bodies with wider zones of accretion is not at all clear, and still the subject of appreciable debate (Weidenschilling, 1974; Wetherill, 1976, 1980b; Vityazev *et al.*, 1978; Safronov and Ruzmaikina, 1978; Horedt, 1979; Greenberg, 1979b). A considerable literature has accumulated, in part because of the abundance of evidential details and implications for terrestrial planet composition and evolution. However, it is rather clear that most of these complications have but slight implication for the formation and evolution of the Sun and the bulk of the planetary system. Hence, the reader is referred to Safronov (1972) and Öpik (1976) for dynamical principles, and to reviews by Wetherill (1980b) and Harris and Ward (1982) for recent dynamical developments. One aspect not treated in these reviews is the dynamics and energy conversions of impacts (Roddy *et al.*, 1977; O'Keefe and Ahrens, 1977; Orphal *et al.*, 1980), important to the early heating of the terrestrial planets (Safronov, 1978; Kaula, 1980).

Models of the formation of terrestrial planets which do have important implications for the evolution of the Sun and the gaseous nebula are those which hypothesize initial incorporation of considerable gases within the terrestrial planets, either by gravitational instability of the gaseous nebula (Cameron, 1978) or by growth from planetesimals within a gaseous medium (Hayashi *et al.*, 1977; Horedt, 1979; Ringwood, 1979). In the theory of Cameron (1978), the gaseous protoplanet is broken up by the tidal attraction arising from the subsequent growth of the Sun, but the details of gas loss from close to the remaining core are not discussed. Horedt (1980) favors loss by a solar wind, and finds a total wind of 0.1 $M_{\odot}$ sufficient; Sekiya *et al.* (1980) prefer heating by EUV radiation at a rate $\gtrsim 200$ times the contemporary rate. None of these authors addresses the similarity of Earth and meteorite inert gas depletions mentioned in Section 3.

4.7. *P, Pg* GASEOUS PROTOPLANET CONTRACTION

Appreciable work has been done in recent years on models of isolated Jovian-sized gaseous bodies in spherically symmetric collapse. These models include homogeneous bodies in computations essentially the same as those for stellar collapse described in Section 2, but starting at densities of 10^{-10} to 10^{-11} g cm^{-3}. A Jovian body is large enough to attain temperatures sufficient for hydrogen dissociation and thence dynamic collapse, regaining hydrostatic equilibrium with a density ~1 g cm^{-2} (DeCampli and Cameron, 1979; Bodenheimer *et al.*, 1980a). Auxiliary problems of these models are the rain-out of an iron or mineral core (Slattery, 1978; Slattery *et al.*, 1980) are the effect of the surrounding nebula (Cameron, 1980a; Cameron *et al.*, 1981). Spherically symmetric models hypothesizing a pre-existing iron–silicate core have been developed by Mizuno *et al.* (1978) Mizuno (1980), and Grossman *et al.* (1980) with similar end results.

4.8. *WD, XD, BD* SOLAR AND EXTERNAL EFFECTS ON THE NEBULA

Research of a decade or two ago placed greater emphasis on the effects of the Sun on the nebula, through the T-Tauri wind and the magnetic field (e.g. Hoyle, 1960; Cameron, 1962). This emphasis was to some extent stimulated by the similarity in magnitude of the angular momentum excess of the nebula with the angular momentum deficiency of the Sun, the need to expel the excess volatiles from the nebula, the inferences of mass outflow from T-Tauri spectra, and the need for a means to heat meteorite parent bodies, for which electromagnetic induction seemed feasible (Sonett and Colburn, 1968). In more recent years, only Alfvén and Arrhenius (1975) have maintained this emphasis. Most workers have concentrated on thermomechanical models of the nebula, with the solar influences being mainly those which exist now: gravitation and visible radiation. The reasons include: the realization of alternative mechanisms for angular momentum transport, gas loss from the nebula, and small body heating; the lack of an adequate ionization model; the paucity of observational constraints on early solar behavior; the alternative interpretations of T-Tauri spectra; and the 7° tilt of the solar rotation axis to the angular momentum vector of the solar system, as well as Occamish idealism.

Nonetheless, solar effects remain possibilities to be used by those who need to explain the loss of primordial atmospheres, great variations among terrestrial planets in inert gas retention, meteorite magnetizations, and other observations or ideas. The most recent searching examination of the interaction between the solar wind and the nebula does not obtain the naive expectation of blasting nebula gas out of the solar system, but rather the inference that, because of the instability of the wind–nebula shock interface, it will mainly enhance the turbulence of the gas and thus enhance the Lynden-Bell and Pringle (1974) effect, leading to spiraling of matter into the Sun (Elmegreen, 1978). It is not clear how much this result depends on the assumption of no intrinsic angular momentum in the wind. In any case, more interesting is the solar wind which goes beyond the orbit of a growing Jupiter, thereby increasing $q\ Re$ and inhibiting that body from pushing gas away from itself.

Similar to solar effects, effects from outside the solar system can plausibly be evoked to explain certain phenomena. It has long been the consensus that passing stars were necessary to raise the perihelia of comets beyond planetary reach (Oort, 1950). More recently, the discovery of ^{26}Al products in Allende has led to the hypothesis of injections from a supernova event not long before solar system formation (Cameron and Truran, 1977). Various hypotheses of how this injection occurred are discussed by Margolis (1979b) and Herbst and Rajan (1980). Not so much demanded by observations, but of comparable physical plausibility are two hypotheses of angular momentum redistribution in the nebula: magnetic braking by the interstellar medium (Mouschovias and Paleologou, 1980), and tidal effects of a passing star (Kobrick and Kaula, 1979).

5. Implications for the Formation of the Sun and Planets

The formation of the Sun is properly regarded as a subproblem of star formation. As yet, there is no indication of how unusual is the Sun in having a planetary system, other than that it is among the ~20% of stars which are single. It is not yet known which type of observed PMS star is closest to the Sun in character. It is possible that the variable emissions associated with T-Tauri stars depend on their being members of binary pairs, in which the collapse of an accretion disk onto the star is inhibited by tidal effects. Continued observations of PMS stars at higher resolution should clarify this matter in the next few years. The Sun probably formed in a cluster. Further isotopic measurements of chondritic meteorites, together with better models of nucleosynthesis and mass fractionation, should elucidate how many stellar sources are required and their respective formation intervals. Understanding the fragmentation and subsequent evolution of a collapsing cloud will come more slowly, since it requires not only more detailed algorithms in bigger computers, but also more heuristic analyses and incorporation of hydromagnetic and external pressure effects.

Solar formation differs from star formation in that there is a veritable zoo of data in the planetary system. Direct evidence of early solar properties therein is relatively slight: the remanent magnetization (Brecher and Arrhenius, 1974; Lanoix *et al.*, 1977), and probably the remarkable variations in inert gas distribution (Black, 1972a, b), of meteorites. But the indirect evidence is considerable, ranging from the leading properties of the planetary system itself down to nuclide and chemical variations of chondrites. Most of this evidence pertains to the solar nebula. But the distinction between a forming star and its accretion disk is somewhat semantic, and probably most of the variations of star formation from spherically symmetric are associated with the nebula (which, in turn, probably always had significant longitudinal variation).

The main problem of planetary formation is Jupiter. The phasing of its collapse relative to the Sun's is uncertain, and it is still not known whether the formation of Jupiter is better characterized as an exceptionally small fragmentation in a cloud collapse, or a gravitational instability in a gaseous nebula, or an accumulation from a combination of resonant and viscous effects. The distinctions, and overlaps, among these different approaches undoubtedly will be sorted out much better in a few years. Meanwhile, better convective models of the nebula are needed, as well as better solutions for a variety of idealized thermomechanical problems in disk–planet interaction, gravitational kinetic systems, gaseous protoplanets, planetary systems, and other areas. Some of these idealizations will eventually be shown to have little relevance to the actual solar system, but all will contribute to the requisite physical insight.

Acknowledgement

This work was supported by NASA Grant NGL 05–007–002. Helpful comments were received from G. S. Stewart, T. Lee, D. N. C. Lin, J. V. Smith, D. C. Black, A. G. W. Cameron, and two anonymous referees.

References

Abt, H. A.: 1978, in T. Gehrels (ed.), *Protostars and Planets*, Univ. of Arizona Press, Tucson, p. 323.
Abt, H. A. and Levy, S. G.: 1976, *Astrophys. J. Suppl.* **30**, 273.
Alexander, D. R.: 1975, *Astrophys. J. Suppl.* **29**, 363.
Alfvén, H.: 1954, *On the Origin of the Solar System*, Oxford Univ. Press, Oxford.
Alfvén, H. and Arrhenius, G.: 1975, *Structure and Evolutionary History of the Solar System*, D. Reidel, Dordrecht.
Anders, E.: 1964, *Space Sci. Rev.* **3**, 583.
Anders, E.: 1977, *Phil. Trans. Roy. Soc. London* **A285**, 23.
Anders, E.: 1981, *Proc. Roy. Soc. London* **A374**, 207.
Arnett, W. D. and Wefel, J. P.: 1978, *Astrophys. J.* **224**, L139.
Arrhenius, G.: 1978, in S. F. Dermott (ed.), *The Origin of the Solar System*, Wiley, New York, p. 521.
Banerjee, S. K.: 1967, *Nature* **216**, 781.
Barricelli, N. A., Clemetsen, T., Aashmar, K., and Bolviken, E.: 1979, *Moon & Planets* **21**, 419.
Batten, A. H.: 1973, *Binary and Multiple Systems of Stars*, Pergamon Press, New York.
Bierman, P. and Tinsley, B. M.: 1974, *Astron. Astrophys.* **30**, 1.
Black, D. C.: 1972a, *Geochim. Cosmochim. Acta* **36**, 347.
Black, D. C.: 1972b, *Geochim. Cosmochim. Acta* **36**, 377.
Black, D. C. and Pepin, R. O.: 1969, *Earth Planet. Sci. Lett.* **6**, 345.
Black, D. C. and Scott, E. H.: 1982, *Astrophys. J.* **263**, 696.
Blander, M. and Fuchs, L. H.: 1975, *Geochim. Cosmochim. Acta* **39**, 1605.
Bodenheimer, P. and Black, D. C.: 1978, in T. Gehrels (ed.), *Protostars and Planets*, Univ. of Arizona Press, p. 288.
Bodenheimer, P., Grossman, A. S., DeCampli, W. M., Marcy, G., and Pollack, J. B.: 1980a, *Icarus* **41**, 293.
Bodenheimer, P. and Sweigart, A.: 1968, *Astrophys. J.* **152**, 515.
Bodenheimer, P., Tohline, J. E., and Black, D. C.: 1980b, *Astrophys. J.* **242**, 209.
Bok, B. J.: 1981, *Sci. Am.* **244**(3), 92.
Boss, A. P.: 1980a, *Astrophys. J.* **236**, 619.
Boss, A. P.: 1980b, *Astrophys. J.* **237**, 563.
Boss. A. P.: 1980c, *Astrophys. J.* **237**, 866.
Boss. A. P.: 1980d, *Astrophys. J.* **242**, 699.
Boynton, W. V.: 1978, in T. Gehrels (ed.), *Protostars and Planets*, Univ. of Arizona Press, Tucson, p. 427.
Boynton, W. V. and Cunningham, C. C.: 1981, *Lun. Planet. Sci. Conf. XII*, p. 106.
Brecher, A. and Arrhenius, G.: 1974, *J. Geophys. Res.* **79**, 2081.
Brouwer, D. and Van Woerkom, A. J. A.: 1950, *Astron. Papers Amer. Ephem. Naut. Almanac* **13**(2), 79.
Cameron, A. G. W.: 1962, *Icarus* **1**, 13.
Cameron, A. G. W.: 1978, *Moon & Planets* **18**, 5.
Cameron, A. G. W.: 1980a, Harvard-Smithsonian Center for Astrophys. Prep. 1120.
Cameron, A. G. W.: 1980b, Harvard-Smithsonian Center for Astrophys. Prep. 1357.
Cameron, A. G. W., DeCampli, W. M., and Bodenheimer, P. H.: 1981, *Lun. Planet. Sci. Conf. XII*, p. 123.
Cameron, A. G. W. and Turan, J. W.: 1977, *Icarus* **30**, 447.
Cassen, P. M. and Moosman, A.: 1981, *Icarus* **48**, 353.
Cassen, P. M., Smith, B. E., Miller, R. H., and Reynolds, R. T.: 1981, *Icarus* **48**, 377.
Champman, C.: 1979, in T. Gehrels (ed.), *Asteroids*, Arizona, p. 25.
Cheng, A. F.: 1978, *Astrophys. J.* **221**, 320.
Clayton, D. D.: 1977, *Icarus* **32**, 255.
Clayton, D. D.: 1978, *Moon & Planets* **19**, 107.
Clayton, D. D.: 1981, *Lun Planet. Sci Conf. XII*, p. 151.
Clayton, R. N. Grossman, L., and Mayeda T. K.: 1973 *Science* **182**, 485.
Clayton, R. N. and Mayeda, T. K.: 1977, *Geophys. Res. Lett.* **4**, 295.
Cohen, M. and Kuhi, L. V.: 1979, *Astrophys. J. Suppl.* **41**, 743.

Consolmagno, G. J., and Cameron, A. G. W.: 1980, *Moon & Planets* **23**, 3.
Cook, T. L. and Harlow, F. H.: 1978, *Astrophys. J.* **225**, 1005.
DeCampli, W. M. and Cameron, A. G. W.: 1979, *Icarus* **39**, 367.
Dermott, S. F. (ed.): 1978, *The Origin of the Solar System*, Wiley, New York.
Elmegreen, B. G.: 1978, *Moon & Planets* **19**, 261.
Elmegreen, B. G.: 1981, *Astrophys. J.* **251**, 820.
Evans, N. J. II: 1978, in T. Gehrels (ed.), Protostars and Planets, Univ. of Arizona Press, Tucson, p. 153.
Franklin, F. A., Lecar, M., Lin, D. N. C., and Papaloizou, J.: 1980, *Icarus* **42**, 271.
Froeschlé, C. and Scholl, H.: 1981, *Astron. Astrophys.* **93**, 62.
Gatewood, G.: 1976, *Icarus* **27**, 1.
Gaustad, J. E.: 1963, *Astrophys. J.* **138**, 1050.
Gingold, R. A. and Monaghan, J. J.: 1977, *Monthly Notices Roy. Astron. Soc.* **18**, 375.
Goldreich, P. and Tremaine, S.: 1978, *Astrophys. J.* **222**, 850.
Goldreich, P. and Tremaine, S.: 1979, *Astrophys. J.* **233**, 857.
Goldreich, P. and Tremaine, S.: 1980, *Astrophys. J.* **241**, 425.
Goldreich, P. and Ward, W. R.: 1973, *Astrophys. J.* **183**, 1051.
Goldstein, J. and Short, J. M.: 1967, *Geochim. Cosmochim. Acta* **31**, 1733.
Greenberg, J. W.: 1979a, *Moon & Planets* **20**, 15.
Greenberg, R.: 1979b, *Icarus* **39**, 141.
Greenberg, R., Wacker, J. F., Hartmann, W. K., and Chapman, C. R.: 1978, *Icarus* **35**, 1.
Grossman, A. S., Pollack, J. B., Reynolds, R. T., Summers, A. L., and Graboske, H. C.: 1980, *Icarus* **42**, 358.
Grossman, L.: 1980, *Ann. Rev. Earth Planet. Sci.* 8, 559.
Grossman, L. and Larimer, J. W.: 1974, *Rev. Geophys. Space Phys.* **12**, 71.
Harris, A. W. and Ward W. R.: 1982, *Ann. Rev. Earth Planet. Sci.* **10**, 61.
Hayashi, C.: 1966, *Ann. Rev. Astron. Astrophys.* **4**, 171.
Hayashi, C., Nakazawa, K., and Adachi, I.: 1977, *Publ. Astron. Soc. Japan* **29**, 163.
Heppenheimer, T. A.: 1980, *Icarus* **41**, 76.
Herbig, G. H.: 1978, in S. F. Dermott (ed.), *The Origin of the Solar System*, Wiley, New York, p. 219.
Herbst, W. and Rajan, R. S.: 1980, *Icarus* **42**, 35.
Horedt, G. P.: 1979, *Moon & Planets* **21**, 63.
Horedt, G. P.: 1980, *Astron. Astrophys.* **92**, 267.
Hoyle, F.: 1960, *Quart. J. Roy. Astron. Soc.* **1**, 28.
Hutcheon, I. D., Steele, I. M., Smith, J. V., and Clayton, R. N.: 1978, *Proc. Lunar. Sci. Conf. 9th*, p. 1345.
Johnson, T.: 1978, *Ann. Rev. Earth Planet. Sci.* **6**, 93.
Jungck, M. H. H. and Eberhardt, P.: 1979, *Meteoritics* **14**, 439.
Kaiser, T. and Wasserburg, G. J.: 1981, *Lun. Planet. Sci. Conf. XII*, p. 525.
Kaula, W. M.: 1977, *Proc. Lunar Sci. Conf. 8th*, p. 321.
Kaula, W. M.: 1979, *Icarus* **40**, 262.
Kaula, W. M.: 1980, in D. W. Strangway (ed.), *The Continental Crust and its Mineral Deposits*, Geol. Assoc. Canada, Waterloo, p. 25.
Kellman, S. A. and Gaustad, J. E.: 1969, *Astrophys. J.* **157**, 1465.
Kelly, W. R. and Wasserburg, G. J.: 1978, *Geophys. Res. Lett.* **5**, 1079.
Kerridge, J. F.: 1981, *Lun. Planet. Sci. Conf. XII*, p. 534.
Kirsten, T.: 1978, in S. F. Dermott (ed.), *The Origin of the Solar System*, Wiley, New York, p. 267.
Kobrick, M. and Kaula, W. M.: 1979, *Moon & Planets* **20**, 61.
Kraft, R. F.: 1967, *Astrophys. J.* **150**, 551.
Lada, C. J., Blitz, L., and Elmegreen, B. G.: 1978, in T. Gehrels (ed.), *Protostars and Planets*, Univ. of Arizona Press, Tucson, p. 341.
Lanoix, M., Strangway, D. H., and Pearce, G. W.: 1977, *Proc. Lunar Sci. Conf. 8th*, p. 689.
Larson, R. B.: 1969, *Monthly Notices Roy. Astron. Soc.* **145**, 271.
Larson, R. B.: 1972, *Monthly Notices Roy. Astron. Soc.* **156**, 437.
Larson, R. B.: 1977, in T. DeJong and A. Maeder (eds), *Star Formation*, D. Reidel, Dordrecht, p. 249.
Larson, R. B.: 1978, *Monthly Notices Roy. Astron. Soc.* **184**, 69.

Larson, R. B.: 1981, *Monthly Notices Roy. Astron. Soc.* **194**, 809.
Lattimer, J. M., Schramm, D. N., and Grossman, L.: 1978, *Astrophys. J.* **219**, 230.
Lee, T.: 1978, *Astrophys. J.* **224**, 217.
Lee, T.: 1979, *Rev. Geophys. Space Phys.* **17**, 1591.
Lee, T., Mayeda, T. K., and Clayton, R. N.: 1980, *Geophys. Res. Lett.* **7**, 493.
Lee, T., Papanastassiou, D. A., and Wasserburg, G. J.: 1976, *Geophys. Res. Lett.* **3**, 109.
Lee, T., Papanastassiou, D. A., and Wasserburg, G. J.: 1977, *Astrophys. J.* **211**, L107.
Lewis, J. S.: 1974, *Science* **186**, 400.
Lin, C. C. and Lau, Y. Y.: 1979, *Stud. Appl. Math.* **60**, 97.
Lin, D. N. C.: 1981, *Astrophys. J.* **246**, 972.
Lin, D. N. C. and Papaloizou, J.: 1979a, *Monthly Notices Roy. Astron. Soc.* **186**, 799.
Lin, D. N. C. and Papaloizou, J.: 1979b, *Monthly Notices Roy. Astron. Soc.* **188**, 191.
Lin, D. N. C. and Papaloizou, J.: 1980, *Monthly Notices Roy. Astron. Soc.* **191**, 37.
Lorin, J. C. and Christophe Michel-Levy, M.: 1978, *4th Int. Conf. Geochron. Cosmochron. Iso. Geol.*, USGS Rept 78–701, Denver, p. 275.
Lucy, L. B.: 1977, *Astron. J.* **82**, 1013.
Lynden-Bell, D. and Pringle, J. E.: 1974, *Monthly Notices Roy. Astron. Soc.* **168**, 603.
Lynden-Bell, D. and Kalnajs, A. J.: 1972, *Monthly Notices Roy. Astron. Soc.* **157**, 1.
Margolis, S. H.: 1979a, *Moon & Planets* **20**, 49.
Margolis, S. H.: 1979b, *Astrophys. J.* **231**, 236.
Mazor, E., Heymann, D., and Anders, E.: 1970, *Geochim. Cosmochim. Acta* **34**, 781.
Mestel, L.: 1977, in T. DeJong and A. Maeder (eds), *Star Formation*, D. Reidel, Dordrecht, p. 213.
Mizuno, H.: 1980, *Prog. Theor. Phys. Kyoto* **64**, 544.
Mizuno, H., Nakazawa, K., and Hayashi, C.: 1978, *Prog. Theor. Phys. Kyoto* **60**, 699.
Morfill, G. E.: 1983, *Icarus* **53**, 41.
Morgan, J. W. and Anders, E.: 1979, *Geochim. Cosmochim. Acta* **43**, 1601.
Morgan, J. W. and Anders, E.: 1980, *Proc. Nat. Acad. Sci.* **77**, 6973.
Mouschovias, T. C.: 1976, *Astrophys. J.* **206**, 753.
Mouschovias, T. C.: 1977, *Astrophys. J.* **211**, 147.
Mouschovias, T. C.: 1978, in T. Gehrels (ed.), *Protostars and Planets*, Arizona, p. 209.
Mouschovias, T. C. and Paleologou, E. V.: 1980, *Moon & Planets* **22**, 31.
Nakagawa, Y., Nakazawa, K., and Hayashi, C.: 1981, *Icarus* **45**, 517.
Narita, S. and Nakazawa, K.: 1978, *Prog. Theor. Phys. Kyoto* **59**, 1018.
Niederer, F. R., Papanastassiou, D. A., and Wasserburg, G. J.: 1980, *Astrophys. J.* **240**, L73.
Norman, M. L. and Wilson, J. R.: 1978, *Astrophys. J.* **224**, 497.
Norman, M. L., Wilson, J. R., and Barton, R. T.: 1980, *Astrophys. J.* **239**, 968.
O'Keefe, J. D. and Ahrens, T. J.: 1977, *Proc. Lunar Sci. Conf. 8th*, p. 3357.
Oort, J. H.: 1950, *Bull. Astron. Inst. Neth.* **11**, 91.
Oort, J. H.: 1977, *Ann. Rev. Astron. Astrophys.* **15**, 295.
Öpik, E. J.: 1976, *Interplanetary Encounters*, Elsevier, Amsterdam.
Orphal, D. L., Borden, W. F., Larson, S. A., and Schultz, P. H.: 1980, *Proc. Lunar Sci. Conf. 11th*, p. 2309.
Ostriker, J. P.: 1964, *Astrophys. J.* **140**, 1067.
Owen, T.: 1978, *Moon & Planets* **19**, 297.
Papanastassiou, D. A. and Wasserburg, G. J.: 1978, *Geophys. Res. Lett.* **5**, 595.
Patchett, P. J.: 1980, *Earth Planet. Sci. Lett.* **50**, 181.
Podosek, F. A.: 1978, *Ann. Rev. Astron. Astrophys.* **16**, 293.
Pollack, J. B. and Black, D. C.: 1979, *Science* **205**, 56.
Prentice, A. J. R.: 1978, *Moon & Planets* **19**, 341.
Ringwood, A. E.: 1959, *Geochim. Cosmochim. Acta* **15**, 257.
Ringwood, A. E.: 1979, *Origin of the Earth and Moon*, Springer-Verlag, New York.
Reeves, H.: 1979, *Astrophys. J.* **235**, 229.
Roddy, D. J., Pepin, R. O., and Merrill, R. B. (eds): 1977, *Impact and Explosion Cratering*, Pergamon Press, New York.
Ross, J. E. and Aller, L. H.: 1976, *Science* **191**, 1223.

Rozyczka, M., Tscharnuter, W. M., Winkler, K. H., and Yorke, H. W.: 1980, *Astron. Astrophys.* **83**, 118.
Safronov, V. S.: 1972, *Evolution of the Protoplanetary Cloud and Formation of the Earth and Planets*, NASA Trans. TTF-677, Washington.
Safronov, V. S.: 1978, *Icarus* **33**, 3.
Safronov, V. S. and Ruzmaikina, T. V.: 1978, *Astron. Zh.* **55**, 107 (transl. *Soviet Astron. AJ* **22**, 60).
Scalo, J. M.: 1978, in T. Gehrels (ed.), *Protostars and Planets*, Univ. of Arizona Press, Tucson, p. 265.
Schwartz, R. D. and Dapita, M. A.: 1980, *Astrophys. J.* **236**, 543.
Scott, E. H. and Black, D. C.: 1980, *Astrophys. J.* **239**, 166.
Sekiya, M., Nakazawa, K., and Hayashi, C.: 1980, *Prog. Theor. Phys. Kyoto* **64**, 1968.
Shakura, N. I. and Sunyaev, R. A.: 1973, *Astron. Astrophys.* **24**, 337.
Shu, F. H.: 1977, *Astrophys. J.* **214**, 488.
Slattery, W. L.: 1978, *Moon & Planets* **19**, 443.
Slattery, W. L., DeCampli, W. M., and Cameron, A. G. W.: 1980, *Moon & Planets,* **23**, 281.
Smith, J. V.: 1979, *Miner. Mag.* **43**. 1.
Sonett, C. P. and Colburn, D. S.: 1968, *Phys. Earth Planet. Interiors* **1**, 326.
Srinivasan, B. and Anders, E.: 1978, *Science* **201**, 51.
Stahler, S. W., Shu, F. H., and Taam, R. E.: 1980, *Astrophys. J.* **241**, 637.
Standish, E. M.: 1972, *Astron. Astrophys.* **21**, 185.
Stewart, G. R. and Kaula, W. M.: 1980, *Icarus* **44**, 154.
Stolper, E.: 1977, *Geochim. Cosmochim. Acta* **41**, 587.
Stolper, E.: 1981, *Lun. Planet. Sci. Conf. XII*, p. 1049.
Strom, S. E.: 1977, in T. DeJong and A. Maeder (eds), *Star Formation*, D. Reidel, Dordrecht, p. 179.
Strom, S. E., Strom, K. M., and Grasdalen, G. L.: 1975, *Ann. Rev. Astron. Astrophys.* **13**, 187.
Thaddeus, P.: 1977, in T. DeJong and A. Maeder (eds), *Star Formation*, D. Reidel, Dordrecht, p. 37.
Tohline, J. E.: 1980, *Astrophys. J.* **235**, 866.
Toomre, A.: 1964, *Astrophys. J.* **139**, 1217.
Toomre, A.: 1977, *Ann. Rev. Astron. Astrophys.* **15**, 437.
Trimble, V.: 1975, *Rev. Mod. Phys.* **47**, 877.
Ulrich, R. K.: 1978, in T. Gehrels (ed.), *Protostars and Planets*, Univ. of Arizona Press, Tucson, p. 718.
Urey, H. C.: 1952, *The Planets: Their Origin and Development*, Yale Univ. Press, New Haven.
Van de Kamp, P.: 1975, *Ann. Rev. Astron. Astrophys.* **13**, 295.
Vityazev, A. V., Perchnikova, G. V., and Safronov, V. S.: 1978, *Astron. Zh.* **55**, 107 (transl. *Soviet Astron. AJ* **22**, 60).
Ward, W. R.: 1981, *Icarus* **47**, 234.
Ward, W. R., Colombo, G., and Franklin, F. A.: 1976, *Icarus* **28**, 441.
Wark, D. A.: 1979, *Astrophys. Space Sci.* **65**, 275.
Wark, D. A. and Lovering, J. F.: 1977, *Proc. Lunar Sci. Conf. 8th*, p. 95.
Wasson, J. T.: 1974, *Meteorites*, Springer-Verlag, New York.
Wasson, J. T.: 1978, in T. Gehrels (ed.), *Protostars and Planets*, Univ. of Arizona Press, Tucson, p. 488.
Weidenschilling, S. J.: 1974, *Icarus* **22**, 426.
Weidenschilling, S. J.: 1978, *Icarus* **35**, 99.
Weidenschilling, S. J.: 1980, *Icarus* **44**, 172.
Werner, M. W., Becklin, E. E., and Neugebauer, G.: 1977, *Science* **197**, 723.
Wetherill, G. W.: 1976, *Proc. Lunar Sci. Conf. 7th*, p. 3245.
Wetherill, G. W.: 1980a, in D. W. Strangway (ed.), *The Continental Crust and its Mineral Deposits*, Geol. Assoc. Canada, Waterloo, p. 3.
Wetherill, G. W.: 1980b, *Ann. Rev. Astron. Astrophys.* **18**, 77.
Wetherill, G. W.: 1981, *Icarus* **46**, 70.
Williams, J. G.: 1969, 'Secular Perturbations in the Solar System', Univ. of California, Los Angeles (Ph.D. Thesis).
Winkler, K.-H. A. and Newman, M. J.: 1980, *Astrophys. J.* **236**, 201.
Wisdom, J.: 1980, *Astron. J.* **85**, 1122.
Wood, D.: 1981, *Monthly Notices Roy. Astron. Soc.* **194**, 201.

Wood, J. A.: 1964, *Icarus* **3**, 429.
Wood, J. A., Anderson, D. L., Buck, W. R., Kaula, W. M., Anders, E., Consolmagno, G. J., Morgan, J. W., Ringwood, A. E., Stolper, E., and Wänke, H.: 1981, *Basaltic Volcanism on the Terrestrial Planets*, Pergamon Press, New York, p. 634.
Woodward, P. R.: 1978, *Ann. Rev. Astron. Astrophys.* **16**, 555.

Dept of Earth and Space Sciences,
Univ. of California,
Los Angeles, CA 90024,
U.S.A.

CHAPTER 17

THE SOLAR NEUTRINO PROBLEM: GADFLY FOR SOLAR EVOLUTION THEORY

MICHAEL J. NEWMAN

1. Introduction

Prior to the pioneering work by the team from Brookhaven National Laboratory led by Raymond Davis, Jr, solar evolution theory was regarded as a solved problem. The world of astrophysics confidently awaited Davis's announcement that the Sun was indeed a thermonuclear reactor functioning as everyone knew it did. Initial results showed instead an upper limit for the flux of high-energy neutrinos far below the standard predictions. Then as advances in the experimental precision served for a number of years only to reduce that upper limit, consternation reigned, and the problem of the evolution of the Sun was re-examined in the greatest detail. Every measurement and every assumption that enters the calculation of present solar interior conditions was questioned in an effort to find the root cause of the discrepancy between theory and observations, without apparent success. Parker in his review of thermonuclear reactions in the solar interior for this study reports the current discrepancy as a factor of 3 to 4 overprediction of the counting rate for the ^{37}Cl experiment by the best modern models of the Sun. A current possibility, in view of the recent excitement following reports by Reines concerning a possible nonzero rest mass for the neutrino, is that the discrepancy may well be due to neutrino oscillations, or some other aspect of neutrino physics, and may not really be symptomatic of a problem with our understanding of the physics and evolution of the solar interior. Even if this should turn out to be the case, those of us who have struggled with the problem over the years will not feel that our efforts have been wasted, for our understanding of the solar interior and our confidence in our knowledge of the consequences of every input datum have grown considerably as the quest progressed, and have reached levels to which we would not have been driven without the stimulus of the enigmatic low reported neutrino counting rate. The degree of concordance in standard models of the Sun produced by different workers is striking. Rood, in his review (Rood, 1976), has put it very well: "For those suspicious of the results of large computer programs, I am aware of eight completely independent programs that give the same result to 20%, and much closer if they are trying to be the same." Few other numerical problems of similar complexity have had as much effort and thought put into them as has that of the evolution of the Sun, and it is to the solar neutrino experiment that this happy circumstance is due. Due to its temperature sensitivity, the ^{37}Cl experiment has forced theorists to examine ways to reduce the temperature in the central regions of the Sun as well as to question a number of fundamental assumptions about solar structure. There

Peter A. Sturrock (ed.), Physics of the Sun, Vol. III, pp. 33–45.

are a number of *ad hoc* ways of resolving the dilemma, as we shall see, but none of them is very convincing. Further progress seems to require additional experimental measurements of the low-energy (less-model-dependent) portion of the solar neutrino spectrum.

The status of the theory of solar evolution and the search for the missing neutrinos has been reviewed often and well (see, e.g. Iben, 1969; Reines and Trimble, 1972; Kuchowicz, 1973, 1976; Ulrich, 1974, 1975; Rood, 1976, 1978; Bahcall and Davis, 1976; Bahcall, 1978, 1979; Newman, 1977). Consequently, this reviewer feels no need to be exhaustive, and the present chapter will explore some representative contributions to solar evolution theory without attempting a complete bibliography of what has been a large and active field. A more recent summary of the status of solar neutrino research can be found in Cherry *et al.* (1985).

2. Standard Theory of Solar Evolution

The problem of the evolution of the Sun, which is usually considered to be a fairly common sort of star in the middle of its main sequence lifetime, is a rather simple application of basic stellar evolution theory as described in the standard textbooks (see, e.g., Chandrasekhar, 1939; Schwarzschild, 1958; Cox and Giulli, 1968; Clayton, 1968). The assumptions entering the calculation have been critically discussed in the reviews referenced in Section 1, and are particularly well stated in Ulrich (1974). The technique for constructing a solar model and a detailed exposition of the characteristics of a 'standard' solar model (the term has come to mean a model in which no neutrino-quenching tricks have been invoked) are discussed in Newman (1975).

The physical state of a stellar interior is conventionally described by four partial differential equations (cf. Clayton, 1968) specifying conservation and continuity of mass, hydrostatic pressure-balance equilibrium, conservation of energy, and a description of energy transport, usually by specification of the run of the temperature gradient. The time enters through the transmutation of thermonuclear fuel (hydrogen) into waste products (helium). Detailed considerations of microscopic physics (cf. Cox and Giulli, 1968) are required to furnish the equation of state, coefficient of opacity for radiative transport, and thermonuclear energy generation rates. The mixing-length theory furnishes a poor alternative to a satisfactory description of convective energy transport, but Newman (1975) has shown that the treatment of convection is not important for conditions deep in the solar interior. Similarly, the boundary conditions applied at the surface (e.g. at $r = R_\odot$, $L(r) = L_\odot$, $M(r) = M_\odot$, $p(r) = 0$ or a value determined from conditions in the photosphere) are not crucial for conditions in the interior. At the center $M(r) = 0 = L(r)$ for $r = 0$. A calculation for a standard solar model usually begins with a spherical configuration in hydrostatic equilibrium of uniform composition with ratio $Z/X \approx 0.019$ of mass fraction of elements heavier than helium to that of hydrogen fixed by observations, and evolves it for 4.7×10^9 y. The process is repeated, the initial helium mass fraction $Y = 1 - X - Z$ and the parameter α = ratio of mixing length to pressure scale height of the mixing-length treatment of convection being varied until $L = L_\odot$ and $R = R_\odot$ at the present solar age. Newman (1976) has shown how to facilitate this process, perturbing the end model of an evolutionary sequence to change the input parameters without the expense of repeating the full sequence of models. Prior to the advent of the solar neutrino dilemma, the effects of rotation and magnetic fields, to say nothing of the more exotic complexities we shall encounter in the following sections, were conventionally neglected.

The detailed numerical models confirm the results of simple dimensional analysis of the stellar structure equations (Cox and Giulli, 1968): the present solar interior should have central temperature near 15×10^6 K, density about 150 g cm^{-3}, pressure of order 10^{17} dyne cm^{-2}.

3. The Missing Solar Neutrinos

The prediction for the counting rate which should be observed in the ^{37}Cl neutrino experiment is then unambiguous: $\Sigma\, \sigma\varphi$ is expected to be significantly higher than is observed. ($\Sigma\, \sigma\varphi$ is the usual notation for the cumulative counting rate for a specific neutrino detector, in this case the Pontecorvo reaction $^{37}Cl + \nu \rightarrow {}^{37}Ar + e^-$, which is given as the sum over neutrino-producing reactions of flux φ weighted by detection cross-section σ.) The calculation of the total neutrino flux expected from the Sun is painfully straightforward (Bahcall, 1979): in the conversion of four protons to form one ^{4}He nucleus, resulting in the release of about ϵ = 25 MeV of energy, two protons must be converted into neutrons, with the production of one neutrino each. If the solar luminosity L (expressed in MeV s^{-1}) is due to this energy release, then there is no escaping the conclusion that at a distance r = 1 AU from the Sun we should receive a flux $\varphi = 2L_\odot/(4\pi r^2 \epsilon) \approx 6 \times 10^{10}$ neutrino $cm^{-2} s^{-1}$.

But the great strength (as it was once thought) of the ^{37}Cl experiment, which is also its great weakness (as is now more commonly perceived), is that the current Brookhaven experiment is not sensitive to the great majority of these neutrinos. The neutrinos from the basic p–p reaction $^1H + {}^1H \rightarrow {}^2H + e^+ + \nu$, which cannot be avoided if the Sun's power is to come from hydrogen fusion, are below threshold for the ^{37}Cl reaction, and the basic p–p chain PP I is completed without the emission of detectable neutrinos ($^2H + {}^1H \rightarrow {}^3He + \gamma$, $^3He + {}^3He \rightarrow {}^4He + 2\,{}^1H$). The neutrinos from the weak alternative to deuterium production $^1H + {}^1H + e^- \rightarrow {}^2H + \nu$ (the pep reaction) are above threshold, as are the ^{7}Be neutrinos from the PP II branch ($^3He + {}^4He \rightarrow {}^7Be + \gamma$, $^7Be + e^- \rightarrow {}^7Li + \nu$, $^7Li + {}^1H \rightarrow 2\,{}^4He$), but these reactions are rare in comparison to the PP I completions, and the counting rate expected from these sources in standard solar models is small (in fact, they sum to less than the level $\Sigma\, \sigma\varphi$ = 2.2 SNU Davis currently reports observing). The ^{13}N and ^{15}O neutrinos from the CN cycle, which is expected to play a minor role in the Sun, are also above threshold, but these neutrinos produce no conflict with observations. The entire problem with the predicted yield is due to the extremely rare ^{8}B neutrinos from the PP III completion ($^7Be + {}^1H \rightarrow {}^8B + \gamma$, $^8B \rightarrow {}^8Be^* + e^+ + \nu$, $^8Be^* \rightarrow 2\,{}^4He$), which excite a superallowed transition in the ^{37}Cl–^{37}Ar system (Bahcall, 1978), and dominate the predicted counting rate although their expected flux is low. This very weak branch is extremely temperature-sensitive, and it was hoped that this feature would allow the ^{37}Cl experiment to function as a thermometer, measuring rather precisely the temperature in the central regions of the Sun. Instead, as we shall see, it has allowed the theorists to escape the dilemma by depressing the central solar temperature by a variety of mechanisms; none of them, unfortunately, is very convincing.

The status and future of solar neutrino research, as it appeared in 1978, is reviewed in the excellent conference proceedings (Friedlander, 1978), and a more recent treatment is found in Cherry *et al.* (1985).

4. Have We Left Something Out?

The low counting rate observed for the ^{37}Cl neutrino experiment has caused a careful re-examination of the physics which enters a calculation of solar evolution. These efforts will be briefly reviewed before we pass on to some of the more desperate suggestions which have appeared in the literature.

4.1. Microscopic Physics

The great benefit derived from the excitement over the solar neutrino problem has been a most thorough checking and refinement of the experimental and theoretical microscopic physics database on which the details of the calculation rest. An entire chapter of the present study has been devoted to the state of understanding of the radiative processes in the deep solar interior, and another to thermonuclear processes; see the reviews by Heubner and Parker in Volume I of this work. The problem does not seem to lie in these areas, but our confidence in the quality of the microscopic physics input has been markedly increased by the tireless efforts of many dedicated individuals.

4.2. Rotation

Standard solar models were traditionally constructed in spherical symmetry, neglecting such unpleasant facts of life as rotation. This luxury, of course, could not be afforded in the face of the neutrino discrepancy – unless and until a given complication could be shown not to contain the solution. Ezer and Cameron (1965) showed that including the effects of fast differential rotation could lower the ^{8}B flux, and so $\Sigma\, \sigma\varphi$, by about a factor of 4, if it induced total mixing of the solar core throughout the whole main sequence lifetime, bringing additional thermonuclear fuel from the cooler outer regions. The usual role of rotation was to lessen the burden on thermal pressure, producing the same internal pressure at lower temperature. Rotating models were investigated by Bahcall *et al.*, (1968), Shaviv and Beaudet (1968), and Roxburgh (1974). Demarque *et al.* (1973a, b) proposed rotating models with extremely rapidly rotating cores, and ratios of centrifugal to gravitational forces as large as one-half. Roxburgh (1975) found that rapid differential rotation can lower the neutrino counting rate if the ratio of centrifugal force to gravity decreases outward. Monaghan (1974) raised objections concerning the approximation method of Demarque *et al.* (1973a, b) and Roxburgh (1974). Rood and Ulrich (1974) pointed out the difficulty of obtaining low counting rates in rapidly rotating models consistent with modern solar oblateness measurements (Dicke and Goldenberg, 1967; Hill and Stebbins, 1975), and this is the stumbling block for most of the explanations involving rotation.

4.3. MAGNETIC FIELDS

Magnetic fields can play a role similar to rotation, providing a nonthermal source of internal pressure to allow pressure balance with lower central temperatures. However, Iben (1968) pointed out that introducting magnetic field pressure at fixed helium abundance Y requires a compensating reduction in Z, or at fixed Z a compensating increase in Y, which leads to higher, rather than lower, central temperatures. (The story of the search for the missing solar neutrinos is full of such anecdotes, as the solar interior is a coupled interdependent system, and feedback mechanisms often overcompensate for a given perturbation, producing precisely the opposite from the effect intended. But it is through such trial and error that we have improved our understanding of the behavior of the stellar interior under a range of perturbations that would never have been applied without the stimulus of the neutrino dilemma.) Bahcall and Ulrich (1971) found that with Z/X taken from observations, large magnetic fields do not diminish $\Sigma\, \sigma\varphi$. Chitre *et al.* (1973), however, found large reductions in the counting rate with magnetic fields strongly concentrated towards the center, although the origin of the very large fields required (order 10^9 gauss) is problematical. Bartenwerfer (1973) found very low counting rates by superimposing the effects of rotation and magnetic fields. The same can be said for many of the solutions which have been suggested – taking a little of this effect and a little of that in just the right proportions can work wonders. But such an approach is not very satisfying.

4.4. ACCRETION

Much attention has been given to the possibility that the presently observed solar surface composition is not representative of the original primordial solar composition, following the analysis by Bahcall *et al.* (1973) of the sensitivity of the neutrino fluxes to opacity effects and the realization that a solar model with low interior metallicity (the low-Z model) would not be in conflict with the ^{37}Cl experiment. How, then, can we explain the present metal-rich surface? Models with cores of exotic composition which have added a hydrogen-rich envelope subsequent to their formation will be discussed in Section 4.5. The present discussion will concern the possibility that the original composition has been masked by material added to the solar surface.

Joss (1974) accounted for a proposed systematic enhancement of heavy element abundances on the solar surface by appealing to the impact of solid bodies, comets, or asteroids into the Sun. There does not seem to be sufficient remnant debris in the solar system at the present time to allow this suggestion to work quantitatively, and current ideas of solar system evolution do not suggest that there has been sufficient material available in the recent past, but the idea remains intriguing.

Shapley (1921) and Hoyle and Lyttleton (1939) suggested that the passage of the solar system through dense interstellar clouds may serve as a trigger for terrestrial Ice Ages, through the luminosity enhancement resulting from gravitational energy release by accretion of cloud material onto the solar surface, as we have been reminded by McCrea (1975). Auman and McCrea (1976) and Newman and Talbot (1976) were quick to point out that this accretion might provide the required adulteration of primordial composition required to make the low-interior-Z model work, although the source of the low-metallicity material required for solar system formation remains an embarrassment in the context of current ideas of the chemical evolution of the Galaxy (e.g. Talbot and Arnett, 1971, 1973). Guthrie (1970, 1971a, b, 1972) and Kuchowicz (1974) addressed the systematic enhancement of heavy-element abundances in stellar surfaces in peculiar A stars, and Drobyshevskii (1974) discussed surface metal abundance enrichment in Am stars (as well as in the solar system). Kuchowicz (1973) suggested that composition gradients in the solar surface layers might be revealed by varying composition effects in weak and strong flare solar cosmic-rays, which presumably would be produced at different depths. Talbot and Newman (1976) discussed the statistics of star–cloud encounters, and concluded on the basis of available observations of the interstellar medium that it was unlikely that all stars suffer masking of their initial compositions by accretion from interstellar clouds during their main sequence lifetimes, but that some disk stars must do so, and the chances are only about 5% that the present solar surface, due to such effects, is not representative of the initial composition of the interior regions. However, the chances are good that the Sun has added about 10^{-4} of its present mass from the interstellar medium subsequent to its formation, approximately replacing its solar wind losses. Talbot *et al.* (1976) discussed possible climatic effects on the Earth during such accretion episodes, in addition to those pointed out by Shapley (1921) and Hoyle and Lyttleton (1939), and Butler *et al.* (1978) discussed the implications for other members of the solar system. Newman and Talbot (1977) discussed the obstacle represented by the solar wind, and the conditions required to overcome it, and Michaud (1977) and Gabriel and Noels (1977) have discussed the possibility that diffusion and gravitational settling effects might cause the loss of any temporary masking. Christensen-Dalsgaard *et al.* (1979) have considered 'dirty' solar models, and Vidal-Madjar *et al.* (1978) have presented observational evidence concerning the possibility that the Sun is currently entering an interstellar cloud of low density.

4.5. Star Formation

Implicit in the assumption of an initially uniform primordial composition for the construction of model evolutionry sequences representing the Sun is the assumption that the early Sun was well-mixed, in agreement with the work of Hayashi (Hayashi, 1966, 1970; Hayashi and Nakano, 1965; Narita *et al.*, 1970), who found an extended fully convective pre-main sequence phase. However, the plausibility of Hayashi's starting conditions has been questioned by Newman and Winkler (1980), who prefer the physicially reasonable initial conditions of Larson (1968, 1969a, b, 1972, 1973, 1978). A hydrodynamic calculation virtually free of additional simplifying assumptions (other than that of spherical symmetry) of the evolution from a diffuse isothermal interstellar cloud of uniform density to the stellar state (Winkler and Newman, 1980a, b, c) does not produce a fully convective stage, except perhaps briefly at the very end of the main accretion phase. The assumption of a well-mixed protosun is not strongly supported by the latter calculations. The nuclear energy release expected during the scouring out of light primordial thermonuclear fuels during star formation has been reviewed by Newman (1978b), and this process is currently being investigated in a full hydrodynamic context by the Los Alamos–Munich collaboration. The details of the final stages of mass assembly may well be significantly affected by these effects.

Other suggested solutions of the solar neutrino problem have centered on the process of star formation. Prentice (1973) appealed to small-scale early inhomogeneities in the protosun to produce a burned-out He core with energy generation by hydrogen burning occurring at lower temperature in the outer regions, yielding a much reduced ^{8}B neutrino flux. Hoyle (1975) invoked a primordial core of two different cosmological compositions which later added the outer envelope whose composition we observe. Thorne (1967) discussed anisotropic cosmological models with large amounts of primordial ^{2}H and ^{3}He, and Kocharov and Starbunov (1970) suggested that such fuels could still be an important energy source in the present Sun. But Abraham and Iben (1970) found that the ^{3}He abundance required exceeds that observed in the interstellar medium. Newman (1978a) surveyed exotic light element reactions involving primordial thermonuclear fuels in an attempt to find an alternative to the proton–proton chains. Rouse (1969a, b, 1974, 1975, 1977, 1979) has considered many aspects of the problem of solar evolution.

5. The Exotic Models

It is to a certain extent prejudice which dictates which investigations are regarded as legitimate questioning of features of the standard solar model (Section 4) and which are classed as 'exotic' desperation attempts to escape the dilemma (Section 5). It is perhaps significant, however, that more of the author's own work is represented in Section 5 than in Section 4.

5.1. Mixing

Ezer and Cameron (1965, 1968) discussed large-scale mixing induced by rapid differential rotation, as did Shaviv and Beaudet (1968). Bahcall *et al.* (1968) found that mixing is important only if it lasts an appreciable fraction of the Sun's main sequence lifetime, and extends over a large fraction of the solar mass. Schatzman (1969) suggested turbulent diffusion of ^{3}He from the center of the Sun to the surface, but Bahcall and Ulrich (1971) and Shaviv and Salpeter (1971) argued that such diffusion would more likely proceed in the opposite direction. Abraham and Iben (1971) found that changes in opacity cannot easily bring about thermal convection. Fowler (1972) suggested that the low observed counting rate may be due to a sudden change in solar structure occurring in the recent past, and pointed out that convective mixing of fresh hydrogen to the central region would allow the observed energy generation rate to be maintained at lower temperatures. The luminosity would then be expected to evolve slightly on a Kelvin–Helmholtz time-scale until the new configuration is fully relaxed. Dilke and Gough (1972a, b) found an overstable mode driven by the ^{3}He gradient, and suggested that the overstability causes the solar core to mix every few hundred million years, inducing terrestrial Ice Ages and temporarily reducing the flux of high-energy solar neutrinos. Kuchowicz (1976) is reminded of Eddington's discussion of the overstability of spherically symmetric oscillations of Cepheids. Ezer and Cameron (1972) and Rood (1972) calculated detailed models incorporating sudden mixing, and it was found that the mixing of large equilibrium amounts of ^{3}He from cool outer regions into the core produces an initial increase in the ^{37}Cl counting rate until the rapid energy release produces expansion of the core and the temperature is lowered. The expansion of the core is accompanied by a contraction of the envelope, producing a decrease in luminosity and radius followed by a slow recovery to a new steady state if sudden mixing does not recur. Ulrich and Rood (1973) disputed the suggestion that overstability driven by the ^{3}He gradient leads to mixing. Rosenbluth and Bahcall (1973) showed the stability of the Sun against nonspherical thermal instabilities, and Schwarzschild and Harm (1973) against spherical thermal instabilities. Christensen-Dalsgaard *et al.* (1974) confirmed that low-order gravity waves become unstable after some nuclear burning has occurred, driven by the sensitivity of the energy generation rate to the ^{3}He abundance, but this does not necessarily lead to mixing. The sudden mixing model does produce low neutrino-counting rates, but the sudden mixing has not been shown to occur, and the accompanying luminosity excursions are difficult to reconcile with the terrestrial paleotemperature data.

5.2. VARYING *G*

If one is willing to change the values of the fundamental physical constants, almost every unpleasant theoretical result can be made to go away, but there is some justification provided for a time-varying gravitational constant by the Brans–Dicke cosomology (Brans and Dicke, 1961). Solar models with *G* decreasing in time have been investigated by Pochoda and Schwarzschild (1964), Ezer and Cameron (1966), Roeder and Demarque (1966), and Shaviv and Bahcall (1969). It is found that $\Sigma\ \sigma\varphi$ actually increases relative to the standard models. Chin and Stothers (1976), on the other hand, found that if the gravitational constant increases with time, the Sun is effectively less old, and the predicted counting rate is reduced. Chin and Stothers (1975) considered solar models in the context of Dirac's hypothesis of additive creation, and found ^{37}Cl counting rates similar to standard models. Mikkelsen and Newman (1977) investigated solar models with a gravitational constant which depends on the separation between attracting bodies.

5.3. QUARK CATALYSIS

Libby and Thomas (1969) suggested that free quarks may act as a catalyst for fusion in the proton–proton reaction, but Salpeter (1970) pointed out that the CNO cycle would also be catalyzed, and the neutrino counting rate would not necessarily be reduced by such processes.

5.4. DEPLETED MAXWELL TAIL

Clayton (1974) pointed out that the high-energy neutrinos which provide the bulk of the predicted counting rate for the ^{37}Cl experiment arise from reactions far out on the tail of the relative velocity distribution, and even a slight deviation of the distribution function from a perfect exponential fall-off could be sufficient to quench the offending neutrinos. Clayton *et al.* (1975a) performed a parameter study with a general assumed form for the energy dependence of the cutoff, and Vasilev *et al.* (1975) found models with counting rates below 1 SNU with a more abrupt cutoff. It is clear that the collision rate in the dense solar interior is rapid enough to quickly restore equilibrium (Gould and Levy, 1976), but it has not been shown rigorously that the two-body Maxwell-Boltzmann velocity distribution can be projected out of the complex many-body system interacting via the long-ranged Coulomb interaction. Thus perhaps inelastic processes and plasma effects could conspire to produce a non-Maxwellian equilibrium distribution, but this has not been shown to occur.

5.5. Immiscible H–He

An effect similar to the cosmological cores of Section 4.5 could be produced if hydrogen and helium were at least partially immiscible at high pressures and moderate temperatures, as pointed out by Wheeler and Cameron (1975). Gravitational separation occurring in the protosun could result, after some hydrogen burning has occurred, in a nearly pure helium core with a hydrogen envelope. Very low counting rates for the ^{37}Cl experiment can result if helium and the heavier elements fractionate together. Williams and Crampin (1971) and Williams and Handbury (1974) have considered segregation of material in the solar nebula, for heavy elements in planetary system formation.

5.6. The Central Black Hole

It is inevitable in modern astrophysics that if the situation becomes desperate enough a black hole will eventually be invoked to save the day. Hawking (1971) entertained the possibility that a black hole might have collided with the Sun and become trapped, or perhaps that a microscopic black hole served as the nucleus for star formation. Stothers and Ezer (1973) found that the increased gravity in the solar core would lead to higher temperatures and an increased high-energy neutrino flux, but Clayton *et al.* (1975b) point out that the energy released as mass falls into the central discontinuity would supplement the energy output of nuclear reactions, and require lower central temperatures to maintain the observed solar luminosity. Thus the counting rate for the ^{37}Cl experiment could be considerably depressed (excluding the contribution of possible neutrino processes occurring in the vicinity of the black hole itself), depending on the relative role of the gravitational energy release. An amusing feature of this model is the tendency of the tiny black hole, whose mass and luminosity increase exponentially with time, to eventually swallow the Sun on a short time-scale ($\sim 10^8$ yr), and the need to carefully balance the initial mass of the black hole and the age of the Sun to arrange that the contribution of the black hole luminosity is just comparable to the nuclear component in our epoch.

5.7. Nonconventional Energy Transport

Newman and Fowler (1976b) pointed out that the mean temperature gradient in the solar interior, and so the central temperature, would be reduced if a parallel energy transport process to radiative transport existed. However, they were unable to produce a convincing alternative to the processes of convection, conduction, and radiation which transfer energy in the standard solar models.

6. Conclusions

Additional exotic solar models could be discussed, but they can be found in the standard reviews and would add little to the present discussion. The point is that all models consistent with the neutrino observations are *ad hoc* in some essential way, and that it is possible with the present solar neutrino detector to escape the dilemma by a wide variety of different mechanisms. The solar neutrino problem has been most beneficial in providing a springboard for innovative thinking, as we have seen, but to reap the full benefit of that effort and distinguish among the various solutions in the literature additional experimental information is required – we must have the flux and energy spectrum distribution of the solar neutrinos if further progress in our understanding of the deep solar interior is to be made. It is particularly important to measure the flux of neutrinos from the proton–proton reaction itself, for if these are not found in the expected numbers we are driven to far-out non-nuclear models like the central black hole, or models in which the present time is special, with nuclear reactions temporarily turned off. The proposed ^{71}Ga experiment, therefore, which is sensitive to the p–p neutrinos, should be an extremely high priority for solar evolution theorists.

Much has been made of the embarrassment of the solar neutrino problem, but solar evolution theory is not completely powerless awaiting its solution. Newman and Fowler (1976a) was able to use constraints of solar structure to demonstrate the implausibility of an experimental result reported from a nuclear physics laboratory, which was later found by independent experiment to have been due to misinterpretation of the data. Similarly, Chin and Stothers (1976) were able to use solar evolution considerations to place limits on the allowed variation of the gravitational constant with time, and Mikkelsen and Newman (1977) used models of the structure of the Earth and the Sun to constrain any possible variation of G with distance. Newman and Rood (1977) were able to place limits on the chemical evolution of the Earth's atmosphere required to compensate for the inexorable increase in solar luminosity as hydrogen is converted to helium in the Sun's interior. Other climatic effects of solar evolution have been discussed by Cameron (1973), Sagan and Young (1973), Talbot *et al.* (1976), Dearborn and Newman (1978), and Newman (1980).

Nor is the solar neutrino problem the only puzzle facing solar evolution theory, or neutrinos the only available probe of interior conditions. Li is depleted relative to its cosmic abundance in solar-type stars, and ^{6}Li is depleted relative to ^{7}Li in the Sun (Wallerstein and Conti, 1969). These are clearly temperature effects, ^{6}Li being more fragile than ^{7}Li, which itself is destroyed by thermonuclear reactions at relatively low temperatures, but the solar convective envelope (which contains the surface Li) has never been at high enough temperatures to burn Li in standard models. Various explanations have been proposed (e.g. Strauss *et al.*, 1976; Hoyle, 1976), but the Li problem remains an open question in solar evolution theory. The theory and observations of waves and oscillations in the sun (Hill *et al.*, 1975; Brookes *et al.*, 1976; Christensen-Dalsgaard and Gough, 1976; Iben, 1976; Iben and Mahaffy, 1976; Severny *et al.*, 1976) has a whole chapter of this study devoted to it, and promises to be an exciting diagnostic tool of solar structure.

The study of solar evolution, then, is alive and well, and one can hope that its future may be as exciting as its past.

Acknowledgements

My views of solar evolution have been shaped by many enjoyable discussions with D. D. Clayton, W. A. Fowler, R. Kippenhahn, R. T. Rood, R. J. Talbot, Jr, and R. K. Ulrich, and portions of this chapter were influenced by the thorough published reviews of B. Kuchowicz. I am grateful to P. D. Parker and R. T. Rood for their helpful comments on an early version of this manuscript.

This work was performed under the auspices of the U.S. Department of Energy.

References

Abraham, Z. and Iben, I., Jr: 1970, *Astrophys. J.* **162**, L125.
Abraham, Z. and Iben, I., Jr: 1971, *Astrophys. J.* **170**, 157.
Auman, J. R. and McCrea, W. H.: 1976, *Nature* **262**, 560.
Bahcall, J. N.: 1978, *Rev. Mod. Phys.* **50**, 881.
Bahcall, J. N.: 1979, *Sp. Sci. Rev.* **24**, 227.
Bahcall, J. N., Bahcall, N. A., and Shaviv, G.: 1968, *Phys. Rev. Lett.* **20**, 1209.
Bahcall, J. N., Bahcall, N. A., and Ulrich, R. K.: 1968, *Astrophys. Lett.* **2**, 91.
Bahcall, J. N. and Davis, R., Jr: 1976, *Science* **191**, 264.
Bahcall, J. N., Heubner, W. F., McGee, N. H., Merts, A. L., and Ulrich, R. K.: 1973, *Astrophys. J.* **184**, 1.
Bahcall, J. N. and Sears, R. L.: 1972, *Ann Rev. Astron. Astrophys.* **12**, 25.
Bahcall, J. N. and Ulrich, R. K.: 1971, *Astrophys. J.* **170**, 593.
Bartenwerfer, D.: 1973, *Astron. Astrophys.* **25**, 455.
Brans, C. and Dicke, R. H.: 1961, *Phys. Rev.* **124**, 925.
Brookes, J. R., Isaak, G. R., and Van der Raay, H. B.: 1976, *Nature* **259**, 92.
Butler, D. M., Newman, M. J., and Talbot, R. J. Jr: 1978, *Science* **201**, 522.
Cameron, A. G. W.: 1973, *Rev. Geophys. Space Sci.* **11**, 505.
Chandrasekhar, S.: 1939, *Introduction to the Study of Stellar Structure*, Univ. of Chicago Press.
Cherry, M. L., Lande, K., and Fowler, W. A. (eds.): *Solar Neutrinos and Neutrino Astronomy* (Homestake 1984), AIP Cont. Proc. No. 126 (1985).
Chin, C. W. and Stothers, R.: 1975, *Nature* **254**, 206.
Chin, C. W. and Stothers, R.: 1976, *Phys. Rev. Lett.* **36**, 833.
Chitre, S. M., Ezer, D., and Stothers, R.: 1973, *Astrophys. Lett.* **14**, 37.
Christensen-Dalsgaard, J., Dilke, F. W. W., and Gough, D. O.: 1974, *Monthly Notices Roy. Astron. Soc.* **169**, 429.
Christensen-Dalsgaard, J. and Gough, D. O.: 1976, *Nature* **259**, 89.
Christensen-Dalsgaard, J., Gough D. O., and Morgan, J. G.: 1979, *Astron. Astrophys.* **73**, 121.
Clayton, D. D.: 1968, *Principles of Stellar Evolution and Nucleosynthesis*, McGraw-Hill, New York.
Clayton, D. D.: 1974, *Nature* **249**, 131.
Clayton, D. D., Dwek, E., Newman, M. J., and Talbot, R. J., Jr: 1975a, *Astrophys. J.* **199**, 494.
Clayton, D. D., Newman, M. J., and Talbot, R. J., Jr: 1975b, *Astrophys. J.* **201**, 489.
Cox, J. P. and Giulli, R. T.: 1968, *Principles of Stellar Structure*, Gordon and Breach, New York.
Dearborn, D. and Newman, M. J.: 1978, *Science* **201**, 201.
Demarque, P., Mengel, J. G., and Sweigart, A. V.: 1973a, *Astrophys. J.* **183**, 997.
Demarque, P., Mengel, J. G., and Sweigart, A. V.: 1973b, *Nature Phys. Sci.* **246**, 33.
Dicke, R. H. and Goldenberg, M.: 1967, *Phys. Rev. Lett.* **18**, 313.
Dilke, F. W. W. and Gough, D. O.: 1972a, *Nature* **240**, 262.
Dilke, F. W. W. and Gough, D. O.: 1972b, *Nature* **240**, 293.
Drobyshevskii, E. M.: 1974, *Earth Planet. Sci. Lett.* **25**, 368.
Ezer, D. and Cameron, A. G. W.: 1965, *Can. J. Phys.* **43**, 1497.
Ezer, D. and Cameron, A. G. W.: 1966, *Can. J. Phys.* **44**, 593.
Ezer, D. and Cameron, A. G. W.: 1968, *Astrophys. Lett.* **1**, 177.

Ezer, D. and Cameron, A. G. W.: 1972, *Nature Phys. Sci.* **240**, 180.
Fowler, W. A.: 1972, *Nature* **238**, 24.
Friedlander, G.: 1978, *Proc. Brookhaven Solar Neutrino Conf.*, BNL Rept 50879.
Gabriel, M. and Noels, A.: 1977, *Astron. Astrophys.* **59**, 427.
Gould, R. G. and Levy, M.: 1976, *Astrophys. J.* **206**, 435.
Guthrie, B. N. G.: 1970, *Astrophys. Space Sci.* **8**, 172.
Guthrie, B. N. G.: 1971a, *Astrophys. Space Sci.* **10**, 156.
Guthrie, B. N. G.: 1971b, *Astrophys. Space Sci.* **13**, 168.
Guthrie, B. N. G.: 1972, *Astrophys. Space Sci.* **15**, 214.
Hawking, S.: 1971, *Monthly Notices Roy. Astron. Soc.* **152**, 75.
Hayashi, C.: 1966, *Ann. Rev. Astron. Astrophys.* **4**, 171.
Hayashi, C.: 1970, *Mem. Soc. Roy. Sci. Liege* **19**, 127.
Hayashi, C. and Nakano, T.: 1965, *Progr. Theor. Phys.* **34**, 754.
Hill, H. A. and Stebbins, R. T.: 1975, *Astrophys. J.* **200**, 471.
Hill, H. A., Stebbins, R. T., and Brown, T. M.: 1975, *Proc. 5th Int. Conf. At. Masses Fund. Const.*, p. 622.
Hoyle, F.: 1975, *Astrophys. J.* **197**, L127.
Hoyle, F.: 1976, Orange Aid Preprint 390, Caltech.
Hoyle, F. and Lyttleton, R. A.: 1939, *Proc. Camb. Phil. Soc.* **35**, 405.
Iben, I., Jr: 1968, *Phys. Rev. Lett.* **21**, 1208.
Iben, I., Jr: 1969, *Ann. Phys.* **54**, 164.
Iben, I., Jr: 1976, *Astrophys. J.* **204**, L147.
Iben, I., Jr and Mahaffy, J.: 1976, *Astrophys. J.* **209**, L39.
Joss, P. C.: 1974, *Astrophys. J.* **191**, 771.
Kocharov, G. E. and Starbunov, Y. N.: 1970, *Zh. Eksper. Teoret. Fiz. Pizma* **11**, 132.
Kuchowicz, B.: 1973, *Leningrad Conf.*, p. 147.
Kuchowicz, B.: 1973, *Astrophys. Lett.* **15**, 107.
Kuchowicz, B.: 1974, *Leningrad Conf.* p. 297.
Kuchowicz, B.: 1976, *Rep. Prog. Phys.* **39**, 291.
Larson, R. B.: 1968, Ph.D. Thesis, Caltech.
Larson, R. B.: 1969a, *Monthly Notices Roy. Astron. Soc.* **145**, 271.
Larson, R. B.: 1969b, *Monthly Notices Roy. Astron. Soc.* **145**, 297.
Larson, R. B.: 1972, *Monthly Notices Roy. Astron. Soc.* **157**, 121.
Larson, R. B.: 1973, *Fund. Cosmic Phys.* **1**, 1.
Larson, R. B.: 1978, in T. Gehrels (ed.), *Protostars and Planets*, Univ. of Arizona Press, Tucson, p. 43.
Libby, L. M. and Thomas, E. J.: 1969, *Nature* **222**, 1238.
McCrea, W. H.: 1975, *Nature* **255**, 607.
Michaud, G.: 1977, *Nature* **266**, 433.
Mikkelsen, D. R. and Newman, M. J.: 1977, *Phys. Rev.* **D16**, 919.
Monaghan, J. J.: 1974, *Monthly Notices Roy. Astron. Soc.* **169**, 13P.
Narita, S., Nakano, J., and Hayashi, C.: 1970, *Progr. Theor. Phys.* **43**, 942.
Newman, M. J.: 1975, Ph.D. Thesis, Rice Univ.
Newman, M. J.: 1976, *Astrophys. J.* **208**, 224.
Newman, M. J.: 1977, *Proc. IAU Gen. Assembly Grenoble*, p. 247.
Newman, M. J.: 1978a, *Proc. Brookhaven Solar Neutrino Conf.* Vol. 2, p. 145.
Newman, M. J.: 1978b, *Proc. 22nd Liege Int. Symp.*
Newman, M. J.: 1980, *Origins of Life* **10**, 105.
Newman, M. J. and Fowler, W. A.: 1976a, *Phys. Rev. Lett.* **36**, 895.
Newman, M. J. and Fowler, W. A.: 1976b, *Astrophys. J.* **207**, 601.
Newman, M. J. and Rood, R. T.: 1977, *Science* **198**, 1035.
Newman, M. J. and Talbot, R. J., Jr: 1976, *Nature* **262**, 559.
Newman, M. J. and Talbot, R. J., Jr: 1977, *Ann. N.Y. Acad. Sci.* **302**, 665.
Newman, M. J. and Winkler, K.-H. A.: 1980, *Proc. Conf. Ancient Sun*, p. 000.
Pochoda, P. and Schwarzschild, M.: 1964, *Astrophys. J.* **139**, 587.
Prentice, A. J. R.: 1973, *Monthly Notices Roy. Astron. Soc.* **163**, 331.
Reines, F. and Trimble, V.: 1972, *Rev. Mod. Phys.* **45**, 1.

Roeder, R. C. and Demarque, P.: 1966, *Astrophys. J.* **144**, 1016.
Rood, R. T.: 1972, *Nature Phys. Sci.* **240**, 178.
Rood, R. T.: 1976, *Lectures Summer School Erice.*
Rood, R. T.: 1978, *Proc. Brookhaven Solar Neutrino Conf.*, p. 175.
Rood, R. T. and Ulrich, R. K.: 1974, *Nature* **252**, 366.
Rosenbluth, M. N. and Bahcall, J. N.: 1973, *Astrophys. J.* **184**, 9.
Rouse, C. A.: 1969a, *Astron. Astrophys.* **3**, 122.
Rouse, C. A.: 1969b, *Nature* **242**, 1009.
Rouse, C. A.: 1974, *IAU Symp.* **66**, 249.
Rouse, C. A.: 1975, *Astron. Astrophys.* **44**, 237.
Rouse, C. A.: 1977, *Astron. Astrophys.* **55** 477.
Rouse, C. A.: 1979, *Astron. Astrophys.* **71**, 95.
Roxburgh, I. W.: 1974, *Nature* **248**, 209.
Roxburgh, I. W.: 1975, *Monthly Notices Roy. Astron. Soc.* **170**, 35P.
Salpeter, E.: 1970, *Nature* **255**, 165.
Sagan, C. and Young, A. T.: 1973, *Nature* **243**, 459.
Schatzman, E.: 1969, *Astrophys. Lett.* **3**, 139.
Schwarzschild, M.: 1958, *Structure and Evolution of the Stars*, Princeton Univ. Press.
Schwarzschild, M. and Harm, R.: 1973, *Astrophys. J.* **184**, 5.
Severny, A. B., Kotov, V. A., and Isap, T. T.: 1976, *Nature* **259**, 87.
Shapley, H.: 1921, *J. Geol.* **29**, 502.
Shaviv, G. and Bahcall, J. N.: 1969, *Astrophys. J.* **155**, 135.
Shaviv, G. and Beaudet, G.: 1968, *Astrophys. Lett.* **2**, 17.
Shaviv, G. and Salpeter, E. E.: 1971, *Astrophys. J.* **165**, 171.
Stothers, R. and Ezer, D.: 1973, *Astrophys. Lett.* **13**, 45.
Strauss, J. M., Blake, J. B., and Schramm, D. N.: 1976, *Astrophys. J.* **204**, 481.
Talbot, R. J., Jr and Arnett, W. D.: 1971, *Astrophys. J.* **170**, 409.
Talbot, R. J., Jr and Arnett, W. D.: 1973, *Astrophys. J.* **186**, 69.
Talbot, R. J., Jr and Newman, M. J.: 1976, *Astrophys. J. Suppl.* **34**, 295.
Talbot, R. J., Jr, Butler, D. M., and Newman, M. J.: 1976, *Nature* **262**, 561.
Thorne, K. S.: 1967, *Astrophys. J.* **148**, 51.
Ulrich, R. K.: 1974, *Astrophys. J.* **188**, 369.
Ulrich, R. K.: 1975, *Science* **190**, 619.
Ulrich, R. K. and Rood, R. T.: 1973, *Nature Phys. Sci.* **241**, 111.
Vasilev, S. S., Kocharov, G. E., and Levkovskii, A. A.: 1975, *Izvest. Akad. Nauk Ser. Fiz.* **39**, 310.
Vidal-Madjar, A., Larent, C., Bruston, P., and Audouze, J.: 1978, *Astrophys. J.* **233**, 589.
Wallerstein, G. and Conti, P. S.: 1969, *Ann. Rev. Astron. Astrophys.* **7**, 99.
Wheeler, J. C. and Cameron, A. G. W.: 1975, *Astrophys. J.* **196**, 601.
Williams, I. P. and Crampin, D. J.: 1971, *Monthly Notices Roy. Astron. Soc.* **152**, 361.
Williams, I. P. and Handbury, M. J.: 1974, *Astrophys. Space Sci.* **30**, 215.
Winkler, K.-H. A. and Newman, M. J.: 1980a, *Astrophys. J.* **236**, 201.
Winkler, K.-H. A. and Newman, M. J.: 1980b, *Astrophys. J.* **238**, 311.
Winkler, K.-H. A. and Newman, M. J.: 1980c, *Space Sci. Rev.* **27**, 261.

Univ. of California,
Los Alamos National Laboratory,
Los Alamos, New Mexico 87545
U.S.A.

CHAPTER 18

STELLAR CHROMOSPHERES, CORONAE, AND WINDS

J. P. CASSINELLI and K. B. MacGREGOR

1. Introduction

The three subjects listed in the title of this chapter were originally defined for phenomena that are observed in the Sun. Observations of other stars have now been made at a wide range of wavelengths and have revealed that one or more of these phenomena are present in stars of every class.

The atmospheric temperature rise that is characteristic of the solar chromosphere is detected in many other stars through the presence of emission cores in the strong optical and near ultraviolet lines of Ca II, Mg II, and H I. Very high temperature ($\sim 10^6$ K), gas like that in the solar corona has been discovered in most other classes of stars through X-ray emission detected by the Einstein satellite. Radial outflows or stellar winds which are much stronger than that from the Sun have been detected through ultraviolet observations made with the Copernicus and International Ultraviolet Explorer (IUE) satellites, and intensively studied using observations from ground-based optical, infrared, and radio observatories.

Figure 1.1 and 1.2 show H–R diagrams with summaries of some of the results of the various stellar surveys.

Figure 1.1 shows the X-ray luminosities of several stars, ranging from $\sim 10^{27}$ erg s^{-1} for A stars to $\sim 10^{33}$ for OB stars. X-ray emission has been detected from all classes of stars on the main sequence. For stars earlier than ~F5 V the X-ray luminosity L_{X} is correlated with the total luminosity, L_{bol}, while for stars later than ~F5 V the correlation is with the square of the rotation speed, v_φ. Unexpectedly high ionization stages of O VI through Si IV were discovered by the ultraviolet satellites in the winds of OB supergiants and these also provide information on the presence of coronae, as we shall see. The X-ray luminosity of late-type giants and supergiants remains undetected ($L_{\mathrm{X}} < 10^{-8.5} L_{\mathrm{bol}}$). There is a broad transition locus (indicated by TL) in the G + K giant region that separates stars of markedly different chromspheric line strengths and greatly different mass loss rates. Much of the work on cool stars has focused on this general region of the H–R diagram.

Many stars for which mass loss rates have been estimated are shown in Figure 1.2. The number associated with each star name is negative of logarithm of the stellar mass loss rate (i.e. $-\log \dot{M}$) in the units of solar masses per year. For example, for the Sun the mass loss rate is about $10^{-13.7}\, M_\odot\, \mathrm{yr}^{-1}$, this rate would yield a negligible total loss of mass over the life of the Sun. That is not the case for the very luminous O stars, Wolf-Rayet

Peter A. Sturrock (ed.), Physics of the Sun, Vol. III, pp. 47–123.

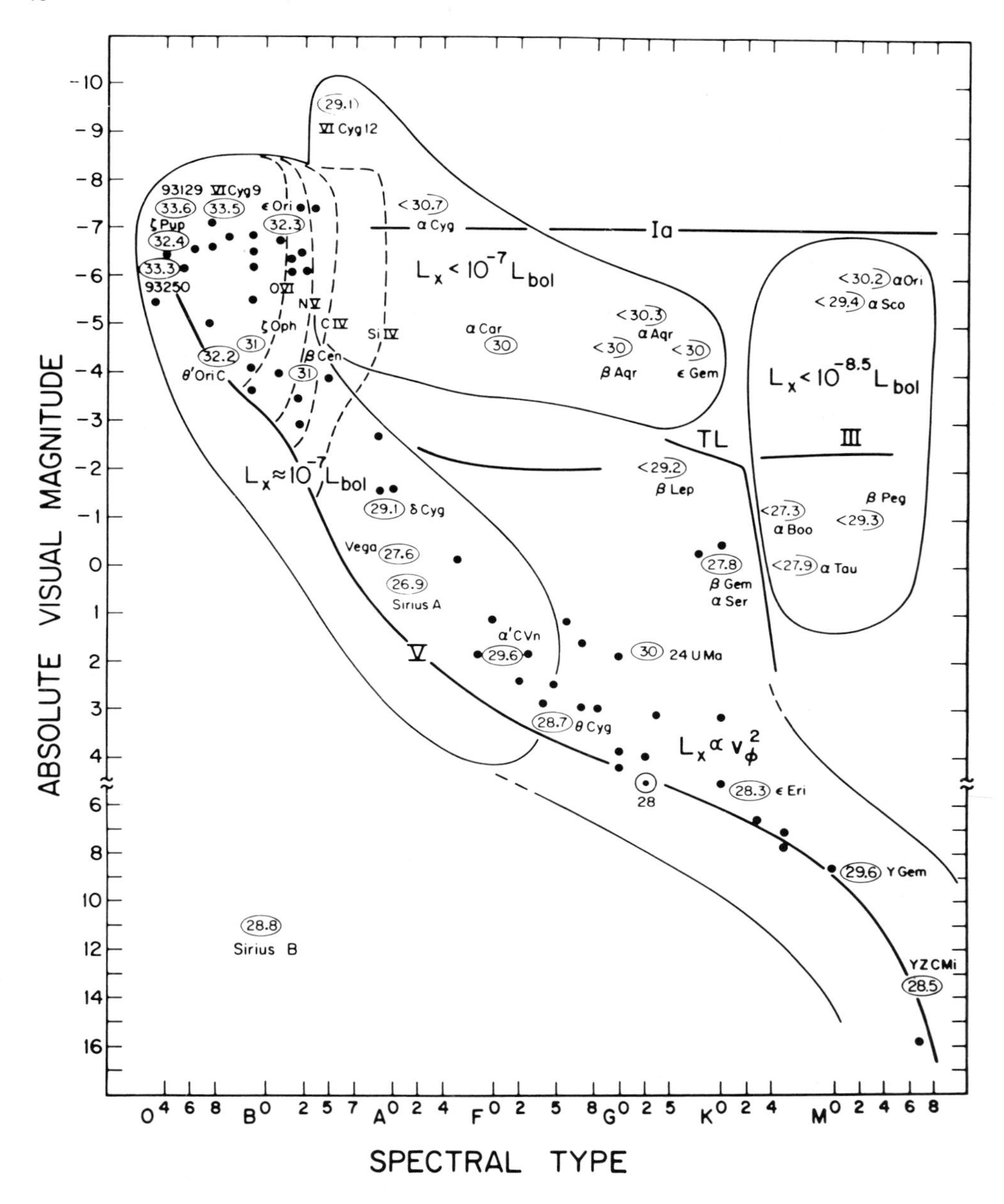

Fig. 1.1. Coronal emission in the H–R diagram. The solid dots indicate stars detected as X-ray sources by the Einstein satellite. The logarithm of the X-ray luminosity is given for several stars of each spectral class. Upper limits are also indicated for a few supergiants and very late giants. The extent of the presence superionization of O VI, N V, C IV, and Si IV are indicated by the dashed lines for the early-type stars. A rough indication of a transition locus is indicated by TL. For stars to the left and below TL, the UV spectra show high ionization stages such as C IV and usually some X-ray emission. For stars to the other side of TL these hot gas indicators are missing, but chromospheric emission occurs. For the late-type X-ray sources the X-ray luminosity is proportional to the rotation speed squared, as might be expected of the coronal emission depends on a dynamo mechanism. For other sources the X-ray emission is proportional to the bolometric luminosity. X-ray data are from the surveys of Long and White (1980), Vaiana *et al.* (1981), Cassinelli *et al.* (1981), and Pallavicini *et al.* (1981).

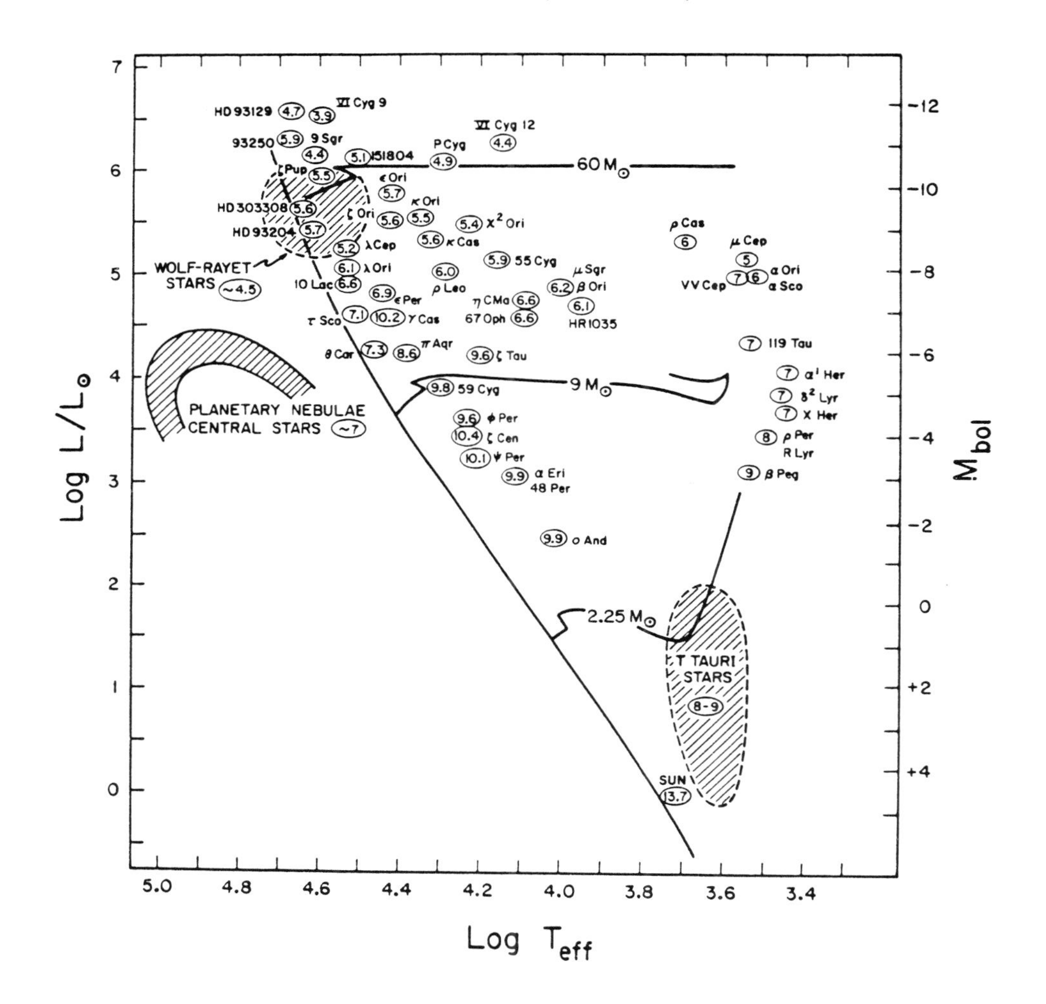

Fig. 1.2. Many stars in the H–R diagram for which mass loss rates have been estimated. The number in each oval is $(-\log \dot{M})$. The measured mass loss rates range from about 2×10^{-14} $M_{\odot}$ yr^{-1} for the Sun to $\sim 10^{-4}$ $M_{\odot}$ yr^{-1} for VI Cyg 9. For the early-type stars and Wolf-Rayet stars the data are from the radio observations of Abbott *et al.* (1980, 1981) and UV data of Lamers (1981). For the B stars, the estimates are those of Snow (1981). Only rough values are indicated for the late-type giants, pre-main sequence, T-Tauri stars, and central stars of planetary nebulae. Evolutionary tracks for stars of masses. 2.25, 9 (Iben, 1967), and 60 $M_{\odot}$ (without mass loss effects) by Stothers and Chin (1976).

stars and M supergiants. For these the mass loss rates may exceed 10^{-5} $M_{\odot}$ yr^{-1}. The consequences of the mass loss on stellar evolution is currently one of the most active areas in stellar interior theory, and is discussed in several papers in the symposia edited by Conti and de Loore (1979), and Stalio and Chiosi (1981). The recent book by de Jager (1980) focuses on the status of the research on the very luminous stars.

In the present chapter, we review the observational and theoretical results pertaining to the thermal and dynamical structure of early- and late-type stellar atmospheres. Our treatment is restricted to single stars either on the main sequence or in the post-main sequence stages of evolution. We are primarily interested in the physical properties of chromospheres, coronae, and winds and in the processes responsible for their formation and maintenance. In Section 2 the chromospheres, coronae and winds of late-type stars are discussed. Section 3 concerns the early-type stars and the interpretation of the observational data on their coronae and winds.

2. Late-Type Stars

2.1. INTRODUCTION

2.1.1. *Overview*

The discovery of chromospheres, coronae, and winds associated with stars other than the Sun affords a unique opportunity for the fruitful exchange of ideas between solar and stellar physicists. In particular, the observation and interpretation of solar-type phenomena in stars of different spectral types, luminosity classes, and ages should significantly increase our understanding of the Sun as a star by providing information concerning long-term evolution, variability, and the dependence of physical processes on stellar parameters. In the present section, we briefly review the observational evidence pertaining to the thermal and dynamical structure of the outer atmospheres of late-type stars (specifically, stars having spectral types F through M). These results are then compared with the predictions of theoretical models for the production of (1) atmospheric regions having gas temperatures in excess of the stellar effective temperature, (2) the continuous loss of mass in the form of a stellar wind, and (3) the spectral features characteristic of a nonradiatively heated, expanding atmosphere. In order to establish working definitions of the terms chromosphere, transition region, corona, and wind, we begin by describing a few of the physical properties of these regions in the specific case of the Sun.

2.1.2. *The Solar Case*

For conditions appropriate to the quiet Sun, Figure 2.1 depicts the observationally inferred temperature and total hydrogen number density as functions of height above the level at which the radial continuum optical depth at $\lambda = 5000$ Å is unity. Note that the occurrence of a temperature distribution which increases with height indicates that the classical assumption of radiative equilibrium is violated in the outer solar atmosphere. Since T becomes significantly greater than T_{eff}, non-LTE continuum effects such as the Cayrel mechanism cannot be responsible for the observed increase (see the discussion by Athay in Volume II). Hence, nonradiative heating mechanisms (presumably in the form of mechanical or magnetic energy dissipation, thermal conduction, or mass flows) must be operative throughout the solar chromosphere and corona.

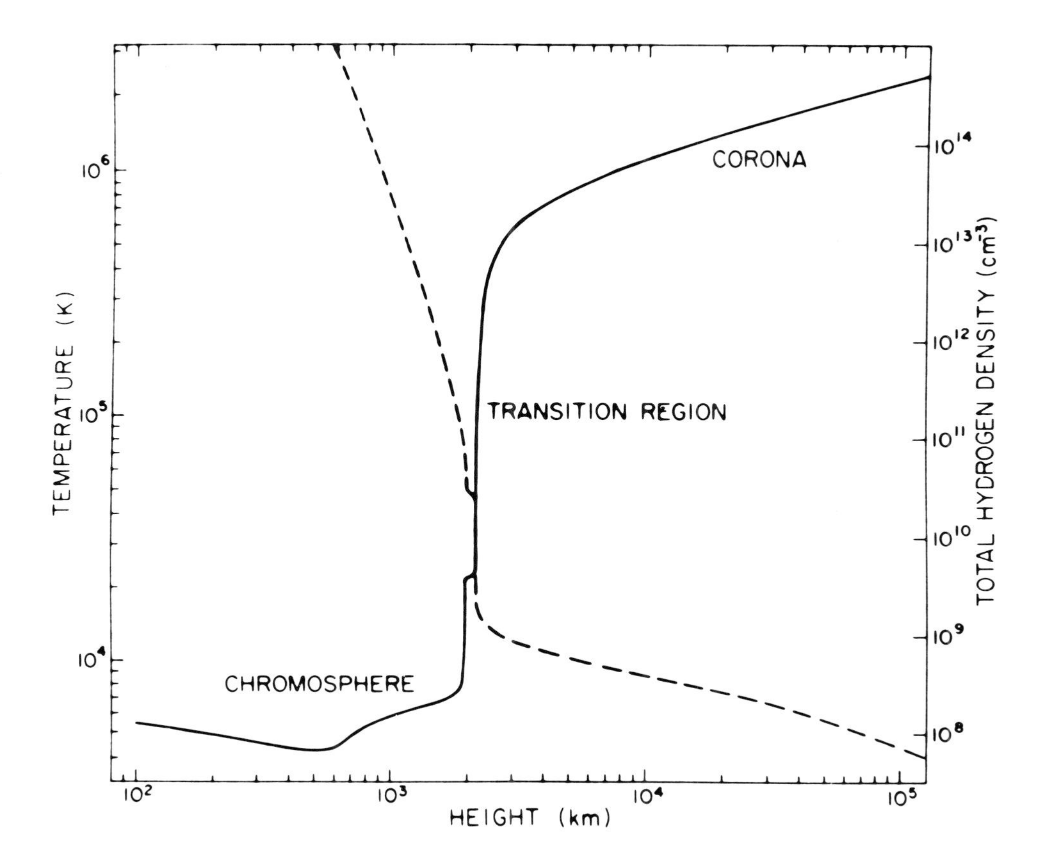

Fig. 2.1. Temperature and total hydrogen number density as functions of height in the quiet solar atmosphere. (From Withbroe and Noyes, 1977.)

The chromosphere consists of that region between the location of the temperature minimum ($h \simeq 600$ km) and $h \simeq 2300$ km over which the gas temperature varies between T_{min} and $T \simeq 3 \times 10^4$ K. It is conceptually useful to further divide the chromosphere into three regions which differ from one another on the basis of their temperature structure; namely, the low, middle, and upper chromosphere. The initial temperature rise takes place in the low chromosphere (extending upward from the base to a height $h \simeq$ 1000 km) in which hydrogen is predominantly neutral and free electrons are produced by the single ionization of metal atoms. Heating of the low chromosphere most likely occurs via the dissipation of mechanical energy with some radiative contribution due to, for example, the photo-ionization of H^- ions by the background photospheric radiation field. Optically thin H^- recombination radiation has usually been identified as the dominant source of cooling in the vicinity of T_{min}, although recent calculations suggest that H^- may act as a net radiative heating agent in this region (Avrett, 1981; Ayres, 1981; Vernazza *et al.*, 1981). At greater heights, energy loss due to spectral line emission becomes increasingly important, and Athay (1976) has estimated that the total radiative loss rate from the low chromosphere amounts to about 2×10^6 erg cm^{-2} s^{-1}. The middle chromosphere is characterized by a temperature scale height considerably in excess of the local density scale height, having $T \simeq 6 \times 10^3$ K near the lower boundary ($h \simeq 1000$ km) and $T \simeq 10^4$ K at the top ($h \simeq 2000$ km). Since the hydrogen Lyman continuum is very optically thick throughout much of this region, hydrogen ionization takes place via photo-ionization by the Balmer continuum radiation field from the collisionally populated $n = 2$ level. Nonradiative heating in the middle chromosphere is balanced by the cooling due to emission in effectively thin spectral lines, principally the collisionally excited resonance lines of Ca II and Mg II; other efficient coolants are the subordinate lines comprising the Ca II infrared triplet and the hydrogen Balmer lines and continua. The net emission due to these sources of radiative cooling is estimated by various authors to be approximately 2×10^6 erg cm^{-2} s^{-1} (Ulmschneider, 1974; Athay, 1976; Avrett, 1981). In the upper chromosphere (i.e. the region between $h \simeq$ 2000 km and $h \simeq 2300$ km) the gas temperature attains values in the range $10^4 \lesssim T \lesssim 3 \times 10^4$ K, the Lyman continuum optical depth becomes small, and hydrogen becomes fully ionized. Factors affecting the energy balance in the upper chromosphere are difficult to specify with certainty, primarily because of the existence of the temperature 'plateau' near $T \simeq 2 \times 10^4$ K (see Figure 2.1). Motivation for the introduction of this feature comes from the requirement that the hydrogen Lyman lines have optical thicknesses which are large enough to produce the central reversals observed in their profiles (see, e.g., Vernazza *et al.*, 1973). The largest single source of radiative cooling in this region is emission in the effectively thin Lyman-α line of neutral hydrogen. The loss rate due to this line is approximately 3×10^5 erg cm^{-2} s^{-1} (Athay, 1976; Avrett, 1981), an amount which may be balanced by local mechanical energy dissipation or by thermal conduction and mass flow from hotter overlying portions of the atmosphere. The reality of the Lyman plateau as a basic physical feature of the temperature profile in the solar atmosphere remains an open question, however, since it may result from the superposition of individual magnetically confined structures containing chromospheric gas, many of which are intercepted by the observer's line of sight (Avrett, 1981).

The interface between the chromosphere and the corona occurs in the geometrically thin layer termed the transition region (thickness several hundred km), throughout which the gas temperature increases from $T \simeq 3 \times 10^4$ K to $T \simeq 10^6$ K. From Figure 2.1, it is apparent that the characteristic scale length over which T varies in the transition region is significantly shorter than the local pressure scale height. The equilibrium temperature distribution in this layer has usually been attributed to the balance between the thermal energy supplied by downward conduction from the corona ($\sim 2 \times 10^5$ erg cm^{-2} s^{-1}; Withbroe and Noyes, 1977) and radiative losses due to emission in collisionally excited ultraviolet spectral lines of multiply ionized (by electron impact) C, N, O, Ne, Mg, and Si ($\sim 2 \times 10^5$ erg cm^{-2} s^{-1}; Athay, 1976). However, recent calculations (Pneuman and Kopp, 1977, 1978) indicate that the downward enthalpy flux associated with the return of spicular material to the chromosphere after being heated to coronal temperatures may be the dominant source of energy input to the transition region.

Although no distinct boundary exists between the top of the transition region and the base of the corona, for the purpose of illustration the corona can be identified as the region beyond $h \simeq 3000$ km in which the gas temperature becomes $\gtrsim 10^6$ K. Coronal energy losses are primarily the result of thermal conduction into lower layers of the atmosphere, the emission of radiation by highly ionized species at extreme ultraviolet and X-ray wavelengths, and the outward expansion of hot gas in the form of the solar wind (see below). For conditions appropriate to the average quiet Sun (cf. Figure 2.1) the conductive, radiative, and solar wind energy losses from the corona are about 2×10^5, 10^5, and $\lesssim 5 \times 10^4$ erg cm^{-2} s^{-1}, respectively (Withbroe and Noyes, 1977). It is important to note, however, that the structural inhomogeneities (determined largely by the solar magnetic field geometry) which have been ignored in the discussion thus far become increasingly pronounced at coronal levels. One consequence of this is that the magnetically open coronal holes (which at chromospheric levels are virtually identical to quiet regions) have loss rates due to conduction, radiation, and expansion of about 6×10^4, 10^4, and 7×10^5 erg cm^{-2} s^{-1}, respectively (Withbroe and Noyes, 1977). In each case, the inferred coronal energy losses must be balanced by a nonradiative heat input of between 3×10^5 and 8×10^5 erg cm^{-2} s^{-1}. While the dissipation of mechanical energy in the form of acoustic or MHD waves presumably plays a role in the heating and dynamics of gas in coronal holes, recent observational studies of propagating oscillations in the solar atmosphere suggest that the supply of wave energy in the quiet chromosphere is inadequate to heat the overlying corona (Athay and White, 1978b). In response to these and similar results, an alternative picture has emerged in which the corona outside the coronal holes is considered to be an assemblage of individual magnetically confined loops, each of which is heated in part by the dissipation of stored magnetic field energy (see, e.g., Vaiana and Rosner, 1978).

As noted above, the corona loses energy by virtue of the expansion of its outer layers into space to form the solar wind. The necessity of coronal expansion can be demonstrated by the following simple argument. Ignoring the effect of the solar magnetic field and assuming for simplicity that the corona is isothermal, the condition of hydrostatic equilibrium in a spherically symmetric atmosphere implies that the gas pressure at infinity has the finite value $P_\infty = P_0 \exp[-(GMm_p/2R_\odot kT)]$, where m_p is the proton mass the $P_0 = 2N_0kT$ is the gas pressure at a specified reference level r_0 ($\simeq R_\odot$) in the corona. Adopting $N_0 \simeq 4 \times 10^8$ cm^{-3}, $T \simeq 1.5 \times 10^6$ K, it follows that $P_\infty \simeq 7.5 \times 10^{-5}$ dyne cm^{-2}, a value far in excess of either the gas pressure ($\sim 10^{-12}$ dyne cm^{-2} for $N \sim 1$ cm^{-3}, $T \sim 10^4$ K) or the magnetic pressure ($\sim 4 \times 10^{-12}$ dyne cm^{-2} for $B \sim 10^{-5}$ G) in the local interstellar medium (a calculation using a more realistic temperature distribution yields essentially the same results). In view of the magnitude of the pressure mis-match, it must be concluded that the corona cannot be maintained in a state of hydrostatic equilibrium, but must expand continuously into space. Using a hydrodynamic description of the corona, Parker (1958, 1963) demonstrated that the solar wind solution which satisfies the boundary conditions of subsonic expansion in the low corona and vanishing gas pressure at infinity is accelerated smoothly (by the force due to the thermal pressure gradient) to a constant supersonic velocity far from the Sun – a prediction subsequently verified by direct observations from satellites in Earth orbit. Although detailed measurments of solar wind properties have revealed the presence of complex spatial structure and temporal fluctuations, the average quiet flow can be characterized by a streaming velocity $V_\infty \simeq 400$ km s^{-1} and proton flux $\simeq$ few $\times$ 10^8 cm^{-2} s^{-1} at the orbit of the earth, implying a rate of mass loss from the Sun of order $\dot{M}_\odot \sim 10^{12}$ g s^{-1} $\sim 10^{-14}$ $M_\odot$ yr^{-1}.

2.1.3. *Methodology*

In the following sections, we review the observational and theoretical results relating to the existence, formation, and maintenance of chromospheres, coronae, and winds in late-type stars. Adopting the Sun as a prototypical member of this class of objects, our discussion will be based in large part upon the use of the inferred solar atmospheric thermal structure, energy balance, and dynamical state to (1) suggest spectral diagnostics by means of which the emergent radiation from late-type stellar atmospheres can be analyzed to obtain evidence for the presence of hot ($T > T_{\rm eff}$) gas and mass motions, and (2) provide a basic conceptual framework within which observational results can be interpreted theoretically.

At this juncture it is worthwhile listing a number of references which contain material relevant to the study of the outer atmospheres of the Sun and late-type stars. The properties of the solar chromosphere and corona have recently been reviewed by Withbroe and Noyes (1977) and by Vaiana and Rosner (1978). The physical processes at work in the outer solar atmosphere are discussed in detail in books by Thomas and Athay (1961), Zirin (1966), and Athay (1976). The solar wind has been considered at length in books by Parker (1963), Brandt (1970), and Hundhausen (1972), and in a recent review by Holzer (1976). The subject of stellar chromospheres and coronae has been extensively reviewed by Wilson (1966), Linsky (1977, 1980a, b, 1981a, b) Mewe (1979), Ulmschneider (1979), Jordan (1980), Rosner (1980), and Vaiana (1980). Recent and detailed information concerning mass loss from late-type stars is available in review papers by Cassinelli (1979), Goldberg (1979), Dupree (1981), Dupree and Hartmann (1980), Hagan (1980), Holzer (1980), Castor (1981), Hartmann (1981a, b), and Linsky (1981c).

2.2. OBSERVATIONAL EVIDENCE FOR THE PRESENCE OF CHROMOSPHERES IN LATE-TYPE STELLAR ATMOSPHERES

2.2.1. *Spectral Diagnostics and Line Formation*

A considerable amount of observational effort has been expended in attempting to obtain evidence for the existence of chromospheres in the outer atmospheres of late-type stars. For the present discussion, the term 'chromosphere' is taken to refer to an atmospheric layer whose basic morphology is similar to that of the middle solar chromosphere – that is, a region in which the balance between nonradiative heating and radiative cooling gives rise to gas temperatures in excess of the stellar effective temperature, and in which T is a relatively slowly increasing function of height. For stars having effective temperatures in the range $3000 \lesssim T_{\rm eff} \lesssim 6500$ K, the presence of such regions is most frequently inferred from the detection of spectral lines whose formation requires temperatures and densities comparable to those found in the solar chromosphere. Among the most useful spectroscopic indicators of chromospheric gas are resonance lines such as the Ca II H (λ3968) and K (λ3934) lines, the Mg II h (λ2803) and k (λ2796) lines, and the Lyman-α (λ1216) line of H I. Important subordinate lines include the infrared triplet lines (λ8498, λ8542, λ8662) of Ca II, the He I λ5876 line and the Balmer series lines of H I (particularly in dwarf M stars). Chromospheric conditions may also be indicated by the presence of the λ10830 line of He I and the λ1640 line of He II in late-type stellar spectra, although the formation of these lines is thought to be influenced by radiation from overlying coronal layers in the atmosphere. Further evidence for the presence of chromospheres in late-type stars is obtained through the observation of ultraviolet lines due to species such as C I, C II, O I, Si II, S I, and Fe II. The reader is referred to the papers of Praderie (1973), Linsky (1977), Dupree (1978), and Ulmschneider (1979) for additional information concerning chromospheric spectral diagnostics.

Strong spectral lines (e.g. Ca II H and K, Mg II h and k) which originate in a chromospheric layer frequently have the characteristic double-peaked appearance shown schematically in Figure 2.2(a). Such lines (called 'self-reversed' or 'doubly reversed')

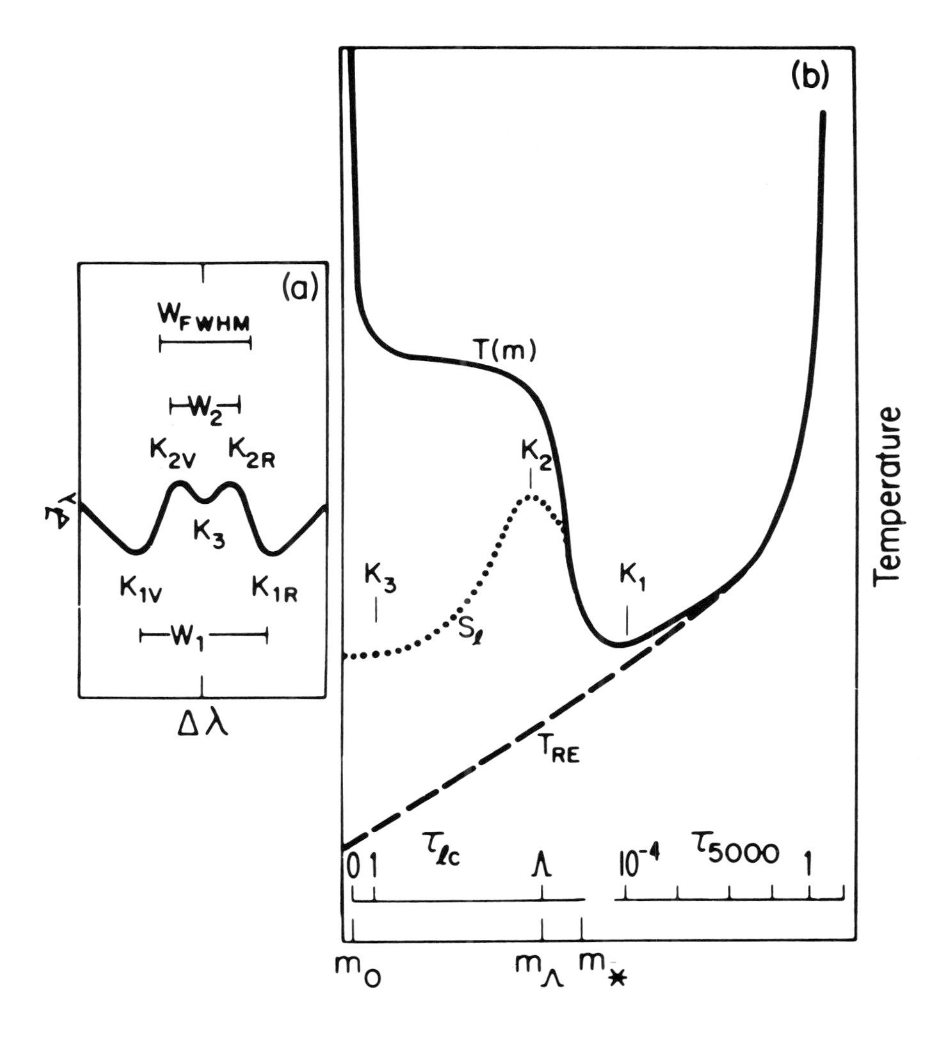

Fig. 2.2. (a) Schematic representation of a self-reversed chromospheric emission core. (b) Depth variation of the source function S_l for a collision-dominated line formed in an atmosphere containing a chromospheric temperature rise. (From Ayres, 1979.)

consist of a broad absorption feature with an emission core having a central absorption reversal. That the shapes of these lines can be explained as a consequence of the depth variation of the line source function in an atmosphere having a chromospheric temperature rise was first recognized by Jeffries and Thomas (1959). A qualitative description of the way in which emission cores are formed in the centers of some absorption lines is given below; detailed discussions of this problem are given in by Thomas and Athay (1961), Jeffries (1968), Athay (1972, 1976), and Mihalas (1978).

Consider an atom having two bound states plus a continuum. By assuming complete frequency redistribution and considering the various processes (both radiative and collisional) which contribute to the population of each level, the source function S_l for the line can be written in the form

$$S_l = \frac{\int \mathrm{d}\nu \varphi_\nu J_\nu + \epsilon B_\nu(T) + \eta B^*}{1 + \epsilon + \eta}, \tag{1}$$

in which φ_ν is the line profile function, J_ν is the mean intensity of the radiation field, and $B_\nu(T)$ is the Planck function evaluated at the local gas temperature T. Explicit expressions for the remaining terms in Equation (1) will not be given here (see the references listed above for details), but the physical interpretation of each quantity appearing in the result for S_l is as follows. The first term in the numerator of Equation (1) describes the effect of photon scattering – that is, photon absorption in the line followed by re-emission as the result of radiative de-excitation. The second term ($\epsilon B_\nu(T)$) gives the rate at which photons are created in the line by collisional excitation followed by radiative de-excitation. The term ηB^* is proportional to the product of the rate at which atoms in the lower state are ionized and the fraction of ions that recombine to the upper state. Because the upper state can subsequently decay radiatively, this term represents an additional source of line photons due to continuum processes. It is important to note that for the conditions of density and temperature which exist in most last-type stellar chromospheres, the photo-ionization rate exceeds the collisional ionization rate. Moreover, it is generally the case that the line-forming layers are transparent to the ionizing radiation field. Hence the source term ηB^* (unlike the term $\epsilon B_\nu(T)$) is essentially independent of the local gas temperature T, reflecting instead the conditions appropriate to a level deeper in the atmosphere (indeed, the quantity B^* can be interpreted as $B^* = B_\nu(T_\mathrm{r})$, where $T_r(\neq T)$ is a brightness temperature characterizing the continuum radiation field (see, e.g., Mihalas, 1978)). In the denominator of Equation (1), the quantities ϵ and η are sink terms, describing the ways in which line photons can be destroyed. The first of these terms represents the effects of collisional de-excitation of the upper bound level, while the second is proportional to the product of the ionization rate from the upper state and the recombination rate to the lower state. Since the upper state can be populated by the absorption of a line photon, this term accounts for the loss of such radiation due to continuum processes; like the source term ηB^*, it too is essentially independent of T.

A useful distinction between the possible types of chromospheric spectral lines has been suggested by Thomas (1957), based upon the relative magnitudes of the various source and sink terms appearing in Equation (1). Lines for which $\epsilon B_\nu(T) > \eta B^*$ and $\epsilon > \eta$ are termed 'collision dominated'; from the preceding discussion it is expected that the source function S_l for such lines will couple (to some extent) to the local gas temperature T. Lines for which $\eta B^* > \epsilon B_\nu(T)$ and $\eta > \epsilon$ are called 'photo-ionization dominated', and should not reflect the atmospheric temperature structure in the line formation region. Apart from the physical properties of the atmosphere, the classification to which a particular line belongs depends most strongly on atomic structure through the relevant ionization and excitation energies. Singly ionized metals (e.g. Ca II, Mg II) have ionization potentials in the energy range ($\gtrsim$10 eV) where the continuum radiation field in late-type stellar atmospheres is generally weak. Since these species have resonance lines with excitation energies on the order of several eV, they are therefore collisionally controlled (the Lyman-α line of H I also falls into this category). Alternatively, neutral metal atoms and H I in the $n = 2$ level have low ionization potentials, and are easily photo-ionized by the stellar continuum radiation field. Hence, the resonance lines of neutral metals and the H I Balmer series lines are usually photo-ionization dominated.

In view of the extent to which the character of the source function S_l can change depending upon whether a line is collision or photo-ionization dominated, it might be expected that the line profiles corresponding to the two categories would likewise be different. Numerous calculations have verified that this in fact is the case (for details, see the references listed above). Figure 2.2(b) depicts (schematically) S_l for a collision dominated line as a function of the optical depth τ_{lc} at the line center (or mass column density m) in an atmosphere containing a chromospheric temperature rise. At depths greater than the thermalization depth Λ (i.e. the depth at which the probability of photon escape from the atmosphere equals the probability of photon destruction by collisional or continuum processes) the source function is closely coupled by collisions to the outward temperature increase. For $\tau_{lc} \simeq \Lambda$, the influence of the atmospheric surface becomes important, and effects due to scattering in the line begin to dominate the behavior of S_l; for $\tau_{lc} < \Lambda$, S_l decreases outward. That the profile depicted in Figure 2.2(a) would result from the source function shown in Figure 2.2(b) can be seen qualitatively from the Eddington–Barbier relation. Specifically, assume that the normally emergent intensity at frequency ν in the line can be approximated as $I_\nu(\tau_\nu = 0) \simeq S_\nu(\tau_\nu = 1)$, where for a line profile function φ_ν which is independent of depth and is normalized to unity at the line center, $\tau_\nu \simeq \tau_{lc}\varphi_\nu$ (the reader is referred to the book by Athay (1972) for a complete discussion of the applicability of the Eddington–Barbier relation). Then a point corresponding to a particular frequency ν on the line profile I_ν reflects the value of S_l at a depth such that $\tau_\nu \simeq 1$, and the source function maps directly into I_ν. Thus, it can be seen from the figure that the central absorption reversal (K_3) at the line center originates near $\tau_{lc} \simeq 1$, the symmetric emission peaks (K_2) arise from the peak in S_l near $\tau_{lc} \simeq \Lambda$ and the K_1 absorption features reflect the value of S_l near the temperature minimum. For a photo-ionization dominated line, the insensitivity of the terms η and ηB^* (cf. Equation (3.1)) to variations in T gives rise to a source function which, in the atmospheric layers where the continuum optical depth is small, decreases monotonically outward. This behavior leads to the production of a pure absorption profile with no central emission core.

We note in closing that the preceding discussion has been carried out under the assumption of complete frequency redistribution, and has neglected the important effects of partial redistribution on chromospheric line profiles. Such processes are treated in detail in the book by Mihalas (1978). Furthermore, note that collision dominated lines with sufficient optical thickness that they thermalize at heights above the temperature minimum will appear with profiles similar to that shown in Figure 2.2(a). Hence, the construction of semi-empirical model chromospheres to match observed spectral features of this type can provide information about the temperature and density stratification in the stellar atmosphere (see, e.g., the review by Linsky, 1980a).

2.2.2. *Observational Summary and Location in the H–R Diagram*

Because of their accessibility to ground based optical observers, the H and K lines of Ca II have been among the most extensively used of the chromospheric indicators listed above. The presence of emission cores at the centers of the H and K lines was noted in the spectra of some bright stars as early as 1913 (Schwarzschild and Eberhard, 1913), although systematic observational searches for such features did not begin in earnest until the 1950s. A catalog of the stellar types in which Ca II emission is seen has been compiled by Bidelman (1954), and Wilson (1966) has reviewed the structural and evolutionary information which can be derived from analysis of the H and K lines in stellar spectra (see also Linsky, 1977). Recent observational work has focused primarily on attempts to obtain chromospheric line profiles and surface fluxes for a wide variety of late-type stars. Notable efforts in this regard include the work of Blanco *et al.* (1974, 1976), Baliunas and Dupree (1979), and Linsky *et al.* (1979a).

Among late-type stars on the main sequence, Ca II H and K emission is almost always present in the spectra of stars with spectral types in the range between late F and early M. As one proceeds towards earlier spectral types, the Ca II lines weaken and ultimately disappear, with the hottest dwarf star in which H and K line emission is seen having spectral type F0 (Linsky, 1980a). At the present time it is not clear whether the absence of H and K lines in stars of earlier spectral type results from the disappearance of solar-type chromospheres (due, e.g., to the lack of hydrogen convection zones in early-type stars and the consequently diminished supply of mechanical energy available for heating), or from a reduction in the visibility of chromospheric emission features due to the bright photospheric continuum at higher values of T_{eff} (see the discussion of Linsky, 1981b). At the extreme cool end of the main sequence, strong Ca II emission is present in the spectra of dMe stars (i.e. dwarf M stars showing Hα emission), but appears to be weak or absent in the spectra of some very late dM stars (Liebert *et al.*, 1979). Among late-type giants and supergiants, Ca II emission is exhibited by stars belonging to about the same range of spectral types as in the case of dwarfs, with the hottest supergiant in which the H and K lines are seen having spectral-type F0 (Linsky, 1980a). These features tend to vanish in the spectra of cool giants and supergiants for which measurements of intrinsic polarization and infrared excesses indicate the presence of circumstellar dust grains (Dyck and Johnson, 1969; Jennings and Dyck, 1972). Jennings (1973) has speculated that the disappearance of chromospheric emission lines in the spectra of the coolest low-gravity stars may result from a reduction in the atmospheric gas temperature due to the thermodynamic effects of grains.

Figure 2.3 depicts intensity tracings of the Ca II K line for two G dwarf stars and a G supergiant. The figure illustrates both the considerable range of central intensities which can occur for stars of the same spectral type and luminosity class and the difference in intrinsic line widths between supergiants and dwarfs (see 2.2.3 below).

Additional information regarding the structure of late-type stellar chromospheres can

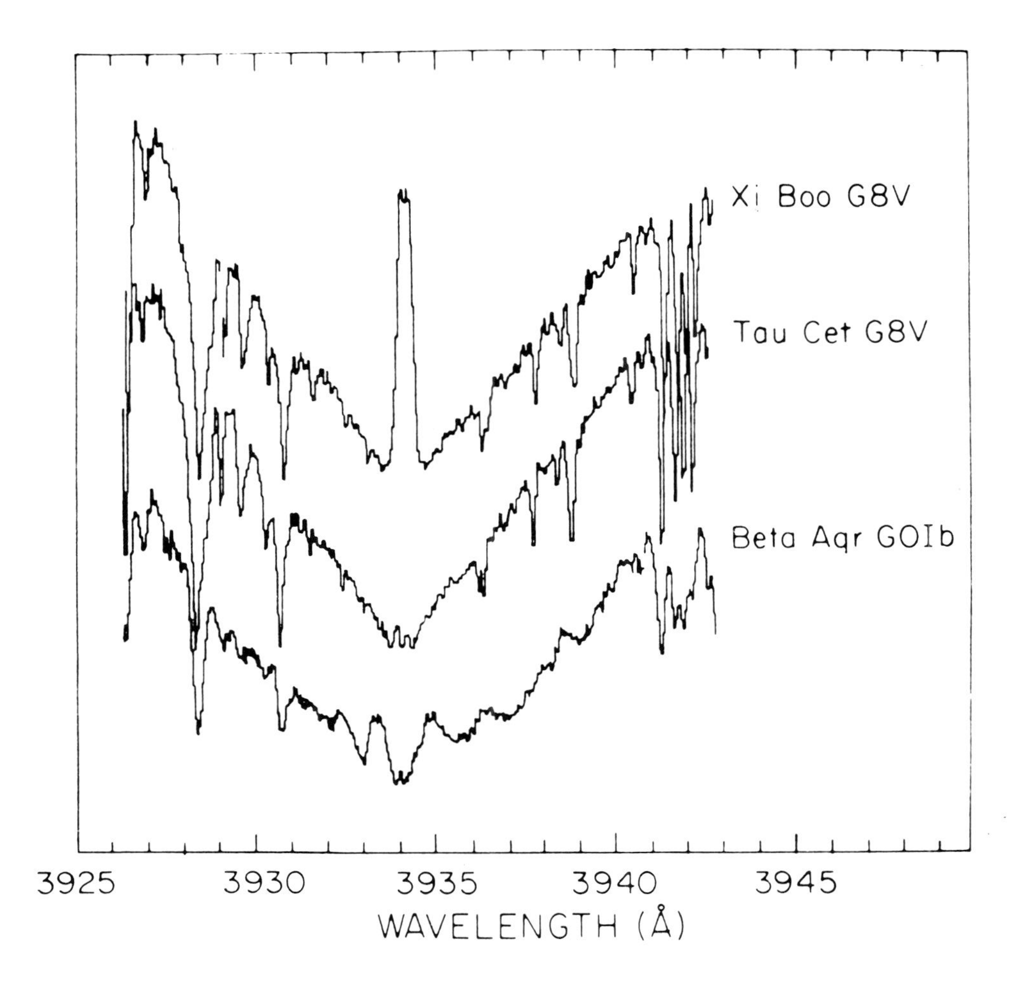

Fig. 2.3. Comparison of chromospheric Ca II K lines for two G dwarf stars and a G supergiant. (From Hartmann, 1980.)

be obtained by observations of the h and k lines of Mg II. Since the ionization potential and excitation energies for Mg II are somewhat greater than those for Ca II, the emission cores of these lines are formed at slightly higher, hotter levels of the chromosphere. Moreover, the larger abundance of Mg (approximately 15 times that of Ca), together with the fact that the background photospheric continuum in late-type stars is generally less intense at $\lambda = 2800$ Å than at $\lambda = 3950$ Å, enhances the visibility of the Mg II h and k lines and permits the detection of weaker emission. However, because the Earth's atmosphere is opaque to radiation with wavelengths near 2800 Å, observations of these features must be made with balloon or satellite-borne experiments.

Since 1970, ultraviolet spectra obtained by a variety of observers using the Balloon-borne Ultraviolet Stellar Spectrometer (BUSS), and the instrumentation on board the OAO-2, Copernicus (OAO-3), and International Ultraviolet Explorer (IUE) satellites have significantly increased our knowledge of both the nature and occurrence of chromospheres in late-type stars. Mg II emission in a number of stars having spectral types F through M has been studied by Doherty (1972, 1973) with the OAO-2 satellite, and by Kondo *et al.* (1972, 1975, 1976a, b, c), Van der Hucht *et al.* (1979), and de Jager *et al.* (1979) with the BUSS experiment. Further measurements of the h and k lines have been performed by Moos *et al.* (1974), McClintock *et al.* (1975a, b, 1978), Evans, Jordan, and Wilson (1975) (all of the preceding authors also studied the Lyman-α line of H I), Bernat and Lambert (1976a, b), and Weiler and Oegerle (1979) using the Copernicus satellite, and by Basri and Linsky (1979), Bohn-Vitense and Dettmann (1970), Stencel and Mullan (1980), and Stencel *et al.* (1980) using the IUE satellite.

Two general conclusions concerning the distribution and energetics of late-type stellar chromospheres can be drawn from analysis of the Mg II observations referenced above. First, the location of stars having chromospheres (as inferred from the detection of Mg II emission) in the H–R diagram is in good agreement with that derived from observations of Ca II H and K emission. In this regard, the results of an IUE survey of A, F, and G stars carried out by Bohm-Vitense and Dettmann (1980) indicate that among supergiants, Mg II emission occurs in the spectra of stars on the red side of the Cepheid instability strip (i.e. for stars with spectral types later than about F8), while for stars belonging to luminosity classes III–V, the h and k features are present for spectral types F2 and later. And second, the normalized chromospheric radiative loss rate due to the Mg II h and k lines, $F(\text{Mg II})/\sigma T_{\text{eff}}^4$, may decrease slowly with decreasing T_{eff}, but does not appear to be strongly dependent on stellar luminosity or surface gravity (Linsky and Ayres, 1978; Basri and Linsky, 1979; Stencel *et al.*, 1980). This can be seen from Figure 2.4, in which $F(\text{k}_1)/\sigma T_{\text{eff}}^4$ ($F(\text{k}_1)$ is the measured stellar surface flux between the Mg II k_1 features) is shown as a function of T_{eff} for a large number of late-type stars (including the Sun) observed with either IUE or Copernicus (see Basri and Linsky, 1979, for details). As is apparent from the figure, the values of $F(\text{k}_1)/\sigma T_{\text{eff}}^4$ corresponding to stars of similar effective temperature and luminosity class can differ by as much as an order of magnitude. Moreover, for a given value of T_{eff}, stars belonging to different luminosity classes can have Mg II fluxes which are nearly equal in magnitude. The implications of these results for the identification of the mechanism(s) responsible for chromospheric nonradiative heating will be discussed in a later section.

Although the observational results summarized in this section have been based largely upon the detection of resonance lines due to H I, Mg II, and Ca II in the spectra of late-type stars, many other spectroscopic indicators of chromospheric gas have been utilized. Surveys using individual lines belonging to the Ca II infrared triplet have been perofrmed by Anderson (1974) (λ8498) and by Linsky *et al.* (1979a, b) (λ8542). The λ10830 line of He I has been measured in a large number of both early and late-type stars by Zirin (1976), and has recently been detected in the spectra of α Boo (K2 III) and α^1 Her (M5 II) by O'Brien and Lambert (1979). While the physical processes involved in the formation of this line are not currently understood, its presence suggests the existence of chromospheric gas since most of the proposed mechanisms require hot plasma in the stellar atmosphere (see, e.g., the discussion of Linsky, 1977). Finally, the advent of ultraviolet spectroscopy from space has made it possible to study short wavelength spectral lines arising from ions of species such as C, N, O, Mg, Si, S, and Fe. Examples of the detection and interpretation of these features in ultraviolet emission

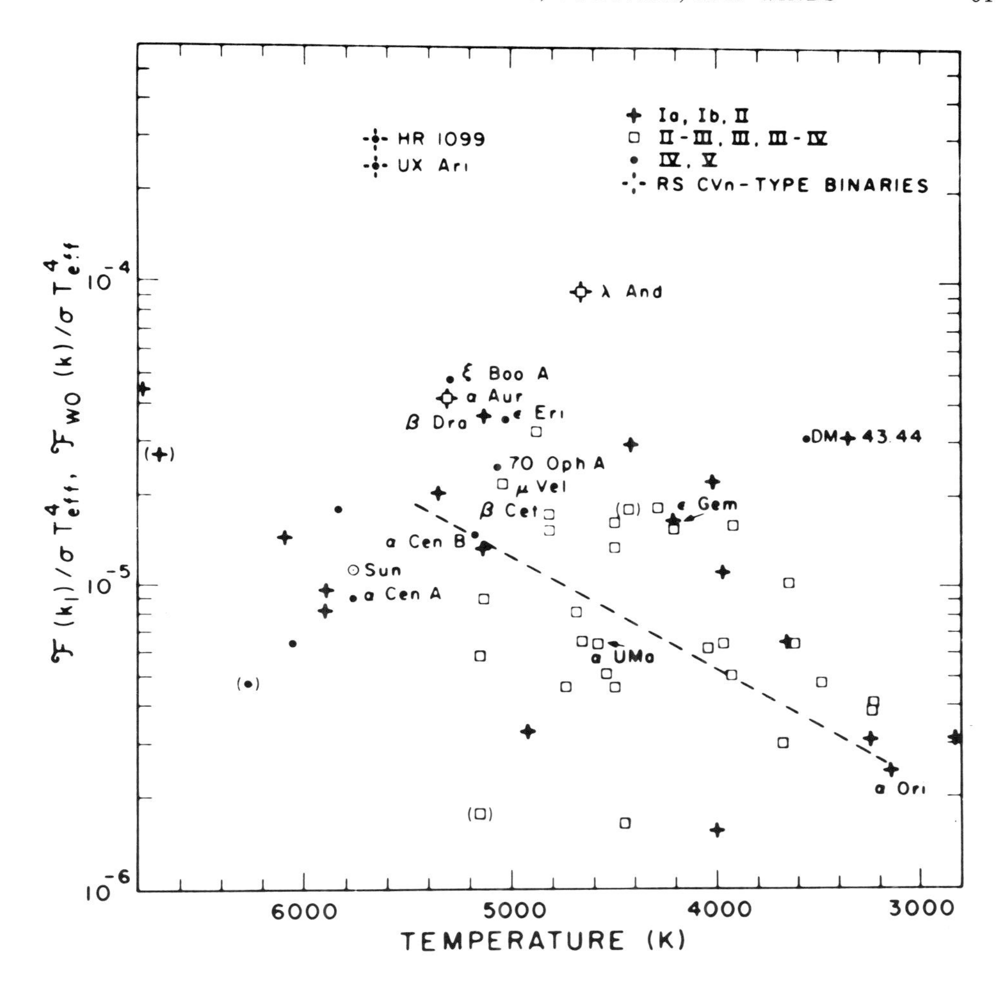

Fig. 2.4. Normalized Mg II k line surface fluxes plotted versus T_{eff}, as discussed in the text. (From Basri and Linsky, 1979.)

line spectra can be found in the work of Linsky *et al.* (1978), Linsky and Haisch (1979), Van der Hucht *et al.* (1979), de Jager *et al.* (1979), Carpenter and Wing (1979), and Hartmann *et al.* (1979).

2.2.3. *The Wilson–Bappu Effect*

In addition to providing information concerning chromospheric temperature and density distributions, quantitative analysis of the shapes of emission cores of collision dominated lines can also assist in the determination of other less readily observable stellar properties. Extensive studies of chromospheric emission features have led to the discovery of a number of correlations between observable measures of line structure and stellar parameters. Among the most interesting and remarkable of these empirically determined relations is that due to Wilson and Bappu (Wilson and Bappu, 1957; Wilson, 1959a, b; see also the recent detailed review by Linsky, 1980a). In their classic investigation of the Ca II H and K lines in the spectra of late-type stars, Wilson and Bappu found a correlation between the width of the line emission cores and the stellar luminosity. If W(Ca II K) (measured in units of km s^{-1}) is the full width at half maximum of the Ca II K emission component (see Figure 2.2(a)), The Wilson–Bappu relation can be expressed in the form

$$M_v = 27.59 - 14.94 \log W(\text{Ca II K}), \tag{2}$$

where M_v is the stellar absolute visual magnitude. The relation (2) is observed to apply over a range of approximately 15 magnitudes (corresponding to a range of about 10^6 in luminosity), and its validity appears to be independent of the strength of the emission. Apart from its obvious applications in the determination of stellar distances and luminosities, Equation (2) implies that the width of the Ca II K line emission core increases with increasing luminosity, as can be seen from inspection of the spectra in Figure 2.3. Subsequent studies have revealed that other chromospheric emission lines obey width– luminosity relations similar to that given by Equation (2) for Ca II. For example, from the results of their survey of Mg II emission in late-type stars, Weiler and Oegerle (1979) obtained an analogous relation between the full width of the k line emission core W(Mg II k) (units: km s^{-1}) and M_v, namely

$$M_v = 34.93 - 15.15 \log W(\text{Mg II k}) \tag{3}$$

(see also Kondo *et al.*, 1972, 1975, 1976b, c; McClintock *et al.*, 1975b; Dupree, 1976).

As stated, the observationally determined Wilson–Bappu relation (2) gives the width W(Ca II K) as a function of M_v. To interpret the correlation theoretically, it is convenient to re-express it in terms of fundamental stellar parameters such as the effective temperature T_{eff} and surface gravity g. This can be accomplished by using the methodology of Reimers (1973). First, M_v is converted to $M_{\text{bol}} \propto \log L$ $(L \propto R^2 T_{\text{eff}}^4)$ by applying the bolometric correction (which itself depends on T_{eff}). Then, using $g \propto M/R^2$ and an empirical relation between M and R, W(Ca II K) can be expressed as a function of g and T_{eff}; the result is a relation of the form $W(\text{Ca II K}) \propto g^{\alpha} T_{\text{eff}}^{\beta}$, where $-0.23 \lesssim \alpha \lesssim -0.20$ and $1.3 \lesssim \beta \lesssim 1.7$ (Ayres, 1979). W(Ca II K) may also be weakly dependent on stellar composition, an effect which has been ignored above (see Ayres, 1979, and Linsky, 1980a, for a more complete discussion).

Traditionally, most attempts to explain the width–luminosity correlation have assumed that the Ca II K line emission core is formed in the Doppler portion of the line profile function. The emission width is therefore a manifestation of the broadening due to turbulent motions in the stellar chromosphere (see, e.g., Goldberg, 1957; Athay and Skumanich, 1968, Fosburg, 1973; Scharmer, 1976). It is important to note that in order to account for the observed line widths in low-gravity giants and supergiants, such an interpretation requires the presence of highly supersonic turbulent velocities. Scharmer (1976) assumed that this is in fact the case, and that turbulent pressure contributes to the support of the atmosphere. By balancing the rate of turbulent energy dissipation against the radiative cooling rate, Scharmer (1976) obtained the relation $W(\text{Ca II K}) \propto g^{-0.25}\, T_{\text{eff}}$, in good agreement with the results described above. An alternative approach to the problem is motivated in part by observational results indicating that the Ca II K_1 absorption features (see Figure 2.2(a)) also obey a width–luminosity relation which is similar in form to that given by Equation (1) (Lutz *et al.*, 1973; Engvold and Rygh, 1978; Cram *et al.*, 1979). Several authors have proposed explanations for this effect, based on physical models which explicitly account for radiative transfer effects in the line (Ayres, *et al.*, 1975; Cram and Ulmschneider, 1978; Engvold and Rygh, 1978; Ayres, 1979). To illustrate this approach, we follow the discussion of Ayres *et al.* (1975) and assume that the K_1 dip originates at the local minimum of the line source function S, the location of which is coincident with the temperature minimum T_{min} (see Figure 2.2(b)). Moreover, assume that the K_1 feature is situated in the damping wings of the line profile function. Its position relative to the line center should therefore respond not to turbulent motions but to the mass column density above T_{min}. To demonstrate this, note that the dominant source of continuous opacity at T_{min} is bound-free absorption by the H^- ion, for which the mass absorption coefficient $K(H^-)$ (units: $cm^2\ g^{-1}$) is proportional to the electron pressure P_e. Since most of the free electrons in the vicinity of T_{min} are the result of metal ionization (see Section 2.1.2), $P_e \propto A_m gm$, where A_m is the metal abundance relative to hydrogen, m is the mass column density of the overlying material, and the proportionality follows from the assumption of hydrostatic equilibrium ($p = mg$). Integrating $K(H^-)$ to obtain the continuum optical depth τ_c^* at T_{min} (i.e. at $m = m^*$) yields $\tau_c^* \propto A_m g\ (m^*)^2$. If τ_c^* is taken to be independent of stellar surface gravity, $m^* \propto g^{-1/2}$, implying that chromospheric thickness increases with decreasing gravity. The optical depth at a wavelength λ in the damping wing of the Ca II K line is $\tau_\lambda \propto N(\text{Ca II})/(\Delta\lambda)^2 \propto A_{Ca} m/(\Delta\lambda)^2$, where N(Ca II) is the column density fo Ca II ions, A_{Ca} is the Ca abundance relative to H, and $\Delta\lambda$ is the wavelength displacement from the line center. By assumption, the K_1 feature is formed at T_{min}; from the Eddington–Barbier relation (see Section 3.2.1), the emergent intensity at $\Delta\lambda = \Delta\lambda_{K_1}$ reflects the value of S_l at $\tau_{K_1}^* \simeq 1$. Evaluating τ_λ at T_{min} it then follows that $\tau_{\Delta\lambda_{K_1}} = W(\text{Ca II K}_1) \propto (m^*)^{1/2} \propto g^{-0.25}$. Hence, the width–luminosity relation for the K absorption dip results from the variation of the atmospheric column density with gravity, assuming that this feature is formed in the damping wings of the line. The reader is referred to the papers by Ayres (1979), Linsky (1980a), and Basri (1980) for more detailed discussions of these effects.

We close this section by noting that in addition to the width–luminosity relation discussed above, the intensity of chromospheric Ca II emission is correlated with a number of other stellar parameters. Among these are stellar age (Wilson, 1963, 1966; Wilson and Skumanich, 1964), rotation rate (Skumanich, 1972; Soderblom, 1983, and references therein; Vaughn *et al.*, 1981), and, in the case of the Sun, magnetic field stregth (e.g. Skumanich *et al.*, 1975). Also, Wilson (1978) has measured Ca II H and K line fluxes for 91 late-type main sequence stars over time intervals of up to 11 yr. A significant fraction of the stars contained in this survey were found to exhibit cyclic variations in the level of chromospheric emission. These results constitute the first direct evidence for the occurrence of activity cycles in stars other than the Sun, and (together with the correlations listed above) provide valuable information about both the nature and time evolution of the physical processes responsible for the generation of mechanical energy and magnetic fields.

2.3. Observational Evidence for the Presence of Transition Regions and Coronae in Late-Type Stellar Atmospheres

2.3.1. *Transition Regions*

The observational evidence reviewed in the preceding section indicates that chromospheric regions are typically present in the atmospheres of stars having spectral types in the range between middle F and M. By analogy with the solar case, these results suggest that nonradiative heating of late-type stellar atmospheres leads to a temperature distribution in which T increases with height above the photosphere. In view of this qualitative correspondence, it seems reasonable to suppose that analogs of the solar transition region and corona should also exist in the outer atmospheres of some late-type stars.

As noted in Section 2.1.2, the solar transition region is a thin layer (i.e. narrower than the local pressure scale height) separating the chromosphere and corona, within which T increases from approximately $T \simeq 3 \times 10^4$ K to values of the order of 10^6 K. Studies of the transition region are performed primarily through the observation and analysis of ultraviolet emission lines due to multiply ionized elements (see, e.g., Noyes, 1971; Athay, 1976; Dupree, 1978). Among the most useful spectroscopic probes of the transition region gas are the strong lines of ions such as Si IV (λ1394, λ1403), C III (λ977, λ1175, λ1909), C IV (λ1548, λ1551), N V (λ1239, λ1243), O V (λ1371), and O VI (λ1032, λ1038). Since the flux of ionizing photons in stars having $T_{\text{eff}} \lesssim 6500$ K is generally too low to produce significant abundances of these species by photo-ionization, the detection of the lines listed above in the spectra of late-type stars should indicate the presence of collisionally ionized gas with temperatures in the range $3 \times 10^4 < T < 3 \times 10^5$ K.

Because the gas temperature is a rapidly increasing function of height throughout the intrinsically narrow transition region, a particular ionization stage attains maximum relative abundance in a layer whose spatial extent is quite small. As a result, the transition region is optically thin in most spectral lines, and measurements of the intensities of lines due to different ions can provide direct information about the temperature distribution. To see that the optical thickness of the transition region is small, consider a line with central frequency ν_0 and oscillator strength f formed in a layer of vertical size ΔZ. If the number density of absorbing ions is N_i, the line center optical thickness $\Delta\tau$ of the layer can be estimated from $\Delta\tau \simeq (\sqrt{\pi}\, e^2/m_e c) \cdot (f/\Delta\nu_D) N_i\, \Delta Z$, where $\Delta\nu_D = (\nu_0/c)(2kT/m_i)^{1/2}$ is the Doppler width. For the specific case of the C IV resonance doublet (average wavelength $\lambda_0 = 1549.1$ Å, $f = 0.286$), note that C IV is the dominant ionization stage of carbon (i.e. $N_{\text{C IV}} \simeq N_{\text{C}}$) for $T \simeq 10^5$ K (see, e.g., Cox and Tucker, 1969). Adopting $N_{\text{C}}/N_{\text{H}} = 3.72 \times 10^{-4}$ (Withbroe, 1971), it follows that $N_i = N_{\text{C IV}} \simeq 3.72 \times 10^{-4}\, N_{\text{H}}$ and that $\Delta\tau \simeq 2.1 \times 10^{-17}\, N_{\text{H}}\, \Delta Z$. Assuming $N_{\text{H}} \simeq 10^9$ cm^{-3} and $\Delta Z \simeq 500$ km, we find $\Delta\tau \simeq 1$, indicating that photons emitted at the line center in all likelihood escape directly from the layer. For a given line, the outward directed flux of such photons can be evaluated directly from the source function S_l (cf. Equation (1)). If Z_1 and Z_2 are, respectively, the lower and upper boundaries of the layer, the frequency-integrated flux F in the optically thin line is given by $F = 2\pi \int_{Z_2}^{Z_1} \mathrm{d}z\, k_l S_l$, where k_l is the line absorption coefficient divided by the profile function. For negligible line and continuum optical depths and in the absence of external radiation incident on the layer, Equation (1) for S_l can be approximated by $S_l \simeq \epsilon B$. In the case of an atom consisting of two bound levels, the quantities k_l and ϵ can be defined explicitly in terms of the rate coefficients for the radiative and collisional processes by means of which the levels are populated (see, e.g., Mihalas, 1978). Using these definitions together with the Einstein relations, it can be shoun that $F = \frac{1}{2} h\nu_{12} \int_{Z_1}^{Z_2} dZ\, C_{12} N_1$, where $E_{12} = h\nu_{12}$ is the energy of the transition, N_1 is the number density of ions in the lower level, and C_{12} (units: s^{-1}) is the collisional excitation rate. This result indicates that under the conditions prevailing in the transition region, the flux of escaping photons in a particular line is proportional to the rate at which the upper level of the transition is populated by collisions. Note in addition that unlike the emission cores of collision-dominated lines originating in the chromosphere, the line profile in this case does not reflect the way in which the emitting material is distributed in the atmosphere.

In recent years the development of instrumentation for performing ultraviolet observations from space has made it possible to study transition regions in a large number of late-type stars. In the paragraphs below, we briefly summarize the results of several such investigations, with emphasis placed upon observations which assist in determining the distribution of stars possessing transition regions in the H–R diagram. Although the interpretation of these observations by means of the construction of physical models will not be described in detail, extensive discussion of the techniques available for the analysis of transition region emission lines can be found in the reviews by Pottasch (1964), Athay (1966, 1976), Jordan and Wilson (1971), Dupree (1978), and Jordan (1980).

Evidence existence of transition regions in stars other than the Sun was first obtained by Dupree (1975) (see also Vitz *et al.*, 1976) who detected the λ1032 line of O VI in the star α Aur (G6 III + F9 III). This identification, together with an observationally determined upper bound on the O V (λ1218) line flux, enabled Dupree (1975) to infer a temperature $\gtrsim 3 \times 10^5$ K for the O VI emitting region. Detailed analysis of these observations by Haisch and Linsky (1976) indicate that the measured line fluxes can be explained by a model in which the lines are formed in a thin, dense, conductively heated transition region having a gas pressure $\simeq$ 10 times that of the quiet solar transition region (see also Doschek *et al.*, 1978). More recent observations of α Aur by Linsky *et al.* (1978) using the IUE satellite have revealed the presence of many additional high-temperature emission lines (e.g. lines of Si IV, C IV, N V), and Ayres and Linsky (1980b) have suggested that the F9 secondary of this system may be responsible for most of the observed transition region emission.

In addition to these results, one or more of the transition region lines listed previously have been detected in the spectra of α CMi (F5 IV) (Evans *et al.*, 1975), HR 1099 (G5 V + K1 IV), λ And (G8 III–IV), Eri (K2 V) (Linsky *et al.*, 1978), EQ Peg (dM3.5e + dM4.54), ξ Boo A (G8 V) (Hartmann *et al.*, 1979), α Cen A (G2 V), α Cen B (K1 V) (Ayres and Linsky, 1980a), and Proxima Cen (dM5e) (Haisch and Linsky, 1980). Of particular interest with regard to these observations is the broad range of line surface fluxes which can occur in stars showing evidence for the presence of transition region gas. For example, the measured surface fluxes of transition region lines for α Cen A and B (Ayres and Linsky, 1980a) are found to be nearly the same as those for the quiet Sun, while those for EQ Peg and ξ Boo A (Hartmann *et al.*, 1979) are enhanced by a factor $\gtrsim$ 10 over the corresponding quiet solar values.

The results of an IUE survey of A, F, and G stars conducted by Bohm-Vitense and

Dettmann (1980) indicate that the hottest dwarf star showing evidence (in the form of C IV emission) for the presence of gas at transition region temperatures is α Cae (F2 V). As was noted earlier in the case of chromospheric emission lines (cf. Section 2.2.2), the bright photospheric background makes detection of transition region lines difficult in stars with spectral types earlier than about F5. Taken together with the specific observations referenced above, this suggests that dwarf stars with spectral types from about middle F to middle M possess transition regions (as well as chromospheres). Among giant (luminosity class III) stars, some spectroscopic evidence for the presence of transition regions exists for stars as early as spectral type middle–late F, while for supergiant (luminosity class I) stars, transition region lines have been detected in the spectra of α Aqr (G2 Ib) and β Aqr (G0 Ib) (Hartmann *et al.*, 1980; Bohm-Vitense and Dettmann, 1980). The question of whether or not the atmospheres of the coolest giants and supergiants contain transition regions has been investigated in some detail by Linsky and Haisch (1979). In their IUE survey of 22 late-type stars, Linsky and Haisch (1979) noted that the spectra of giants cooler than about K2 III and supergiants cooler than about G8 Ib contained many spectral features characteristic of gas at chromospheric temperatures, but none indicative of transition region plasma (see also Dupree and Hartmann, 1980). On the basis of these observations, Linsky and Haisch suggested that a dividing line could be drawn in the H–R diagram (see Figure 2.5), separating stars with spectra in which transition region lines are present (indicating gas at $T \sim 10^5$ K) from those with spectra in which only chromospheric lines appear (indicating gas at $T \lesssim 10$ K). They further noted that the proposed transition region dividing line was located in the

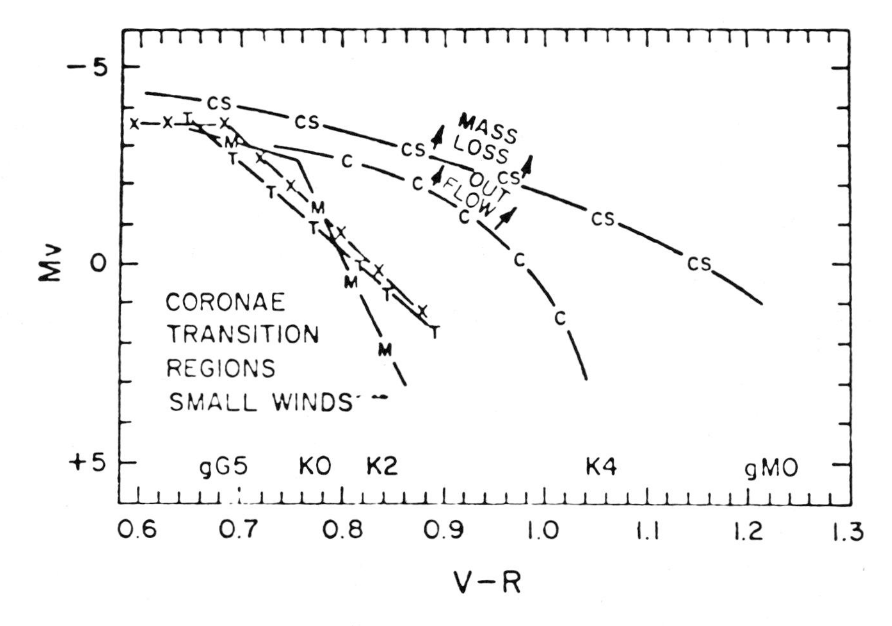

Fig. 2.5. Dividing lines in the H–R diagram separating stars with and without; transition region emission lines (T, cf. Section 2.3.1); coronal soft X-ray emission (X, cf. Section 2.3.2); $I_V/I_R < 1$ for Mg II k$_2$ features (M, cf. Section 2.5.3); $I_V/I_R < 1$ for Ca II K$_2$ features (C, cf. Section 2.5.3); circumstellar Ca II absorption lines (CS, cf. Section 2.5.2). (From Linsky, 1981b.)

same portion of the H–R diagram as the boundaries marking the occurrence of circumstellar absorption lines (Reimers, 1977a, b, c) and chromospheric emission line asymmetries (Stencel, 1978, Stencel and Mullan, 1980) in the spectra of late-type giants and supergiants. Since these features are generally interpreted as evidence for the existence of a massive stellar wind, Linsky and Haisch speculated that the apparent cessation of solar-type transition regions upon crossing the dividing line from left to right was related to the onset of mass loss at a rate significantly in excess of that due to the solar wind. Subsequent observations have led to the realization that the transition region dividing line is not sharp, but is characterized by some width. A number of stars located near the proposed boundary have been found to exhibit spectral features more representative of their counterparts on the opposite side of the dividing line (see, e.g., Dupree and Hartmann, 1980, Dupree, 1981; Hartmann, 1980; Stencel, 1980; Linsky, 1981a). For example, the spectra of the stars termed 'hybrid' by Hartmann *et al.* (1980) (see also Hartmann *et al.*, 1981) contain emission lines due to Si IV, C IV, and N V, and blue-shifted circumstellar absorption features due to Ca II and Mg II. These stars thus show evidence for the presence of both transition region gas and cool, massive ($\dot{M} \sim 10^{-8}\ M_\odot\ \mathrm{yr}^{-1}$) stellar winds. In the particular case of the hybrid star α TrA (K4 II), Hartmann *et al.* (1981) have suggested that the broad emission lines due to Si III, C III, and C IV are formed in the outflowing wind material, an interpretation which implies the existence of a 'transition region' whose structure and spatial extent differ considerably from those of the solar transition region.

2.3.2. *Coronae*

By analogy with the Sun, the detection of emission lines characteristic of gas at transition region temperatures provides indirect evidence for the presence of coronae in late-type stars. Direct observational evidence for the presence of coronal ($T \sim 10^6$ K) plasma is difficult to obtain, however, and definitive results regarding the existence and distribution of late-type stellar coronae have only recently become available. In the case of the Sun, the corona is studied through observations of line emission from highly ionized species at X-ray, ultraviolet, and visible wavelengths, continuum emission in the X-ray and radio portions of the spectrum, and photospheric radiation scattered by coronal free electrons. For late-type stars in general, the weakness of coronal emission at visible and radio wavelengths precludes the use of line and continuum measurements in these spectral regions to detect hot gas (see, however, Gary and Linsky, 1981). Moreover, while the solar coronal line spectrum is rich in detail at ultraviolet wavelengths, many of these features are located shortward (i.e. at wavelengths < 1000 Å) of the spectral range in which satellite-borne instruments have their greatest sensitivity. For these reasons, efforts to detect coronae in late-type stars have relied primarily upon the use of experiments designed to measure the emission from hot gas at X-ray wavelengths.

A crude estimate of the expected soft X-ray luminosity from a typical late-type stellar corona can be obtained in the following way. Assume for simplicity that the corona is hydrostatic and isothermal (with temperature T_C) and occupies a volume $4\pi R_*^2 h$, where R_* is the stellar radius and $h = kT_C/\mu g$ ($g = GM_*/R_*^2$) is the density scale height. Assuming further that the optical thickness of the corona in most lines and continua is small, the radiative cooling rate for the coronal gas can be expressed in the form $\Lambda = N_e N_H P_R(T_c)$ (units: erg cm^{-3} s^{-1}), where N_e and N_H are the electron and total hydrogen number densities, respectively, and $P_R(T)$ is the radiative loss function appropriate to an optically thin plasma in collisional ionization equilibrium (see, e.g., Cox and Tucker, 1969; Tucker and Koran, 1971; McWhirter *et al.*, 1975; Raymond *et al.*, 1976). The luminosity of the corona is then given by $L \simeq 4\pi R_*^2 h\, N_e N_H P_R(T_c)$ erg s^{-1}. Consider a star with $R_* = R_\odot$, $M_* = M_\odot$, and $T_c = 10^6$ K. For a fully ionized gas in which $N_{He}/N_H = 10^{-1}$, $N_e N_H \simeq 0.23 N^2$ (N is the total coronal number density) and the average mass per particle is $\mu \simeq 0.61\, M_H$. Adopting $N \simeq 4 \times 10^8$ cm^{-3} and $P_R(T_c) \simeq 10^{-22}$ erg cm^3 s^{-1} (Raymond *et al.*, 1976), it follows that $L \simeq 10^{27}$ erg s^{-1}; approximately one-third of this amount is radiated at wavelengths in the soft X-ray band between $\lambda = 43.6$ Å and $\lambda = 80$ Å (i.e. in the energy range 0.15–0.28 keV) with a smaller fraction emitted at shorter wavelengths (Raymond *et al.*, 1976; Raymond and Smith, 1977). If such an X-ray emitting star were located at a distance $d = 10$ pc, the energy flux incident at Earth (neglecting interstellar absorption) would be $F = L/4\pi d^2 \simeq 8 \times 10^{-14}$ erg cm^{-2} s^{-1}. For comparison, note that the data depicted in Figure 2.4 indicate that, for the Sun, the chromospheric Mg II k line (width ~ several Å) surface flux is approximately F(Mg II k) $\simeq 10^{-5}\ \sigma T_{\rm eff}^4 \simeq 6.28 \times 10^5$ erg cm^{-2} s^{-1} (for $T_{\rm eff,\odot}$ = 5770 K), corresponding to a luminosity L(Mg II k) $\simeq 3.83 \times 10^{28}$ erg s^{-1}.

Early searches for coronal soft X-ray emission were conducted with instruments having sensitivities below that required to positively detect a point source such as the star in the preceding example. Hence flux measurements were obtained for only a few of the brightest ($L_X \gtrsim 10^{30}$ erg s^{-1}) nearby stellar sources while for most late-type stars, upper limits to the X-ray luminosity were established. For example, the Copernicus observations performed by Margon *et al.* (1974) placed upper limits in the range $5–26 \times 10^{31}$ erg s^{-1} on the soft X-ray luminosities of the stars ϵ Sco (K2 III), γ Dra (K5 III), α Boo (K2 III), and θ Cen (KO IV). Cruddace *et al.* (1975) subsequently used a rocket-borne detector to observe the stars α Boo and α CMi (F5 IV). Their inability to detect a flux from either star led to luminosity limits of 10^{30} erg s^{-1} for α Boo and 10^{28} erg s^{-1} for α CMi. Similarly, the extensive soft X-ray survey carried out by Vanderhill *et al.* (1975) as part of the Skylab mission yielded upper limits typically in the range $10^{28}–10^{32}$ erg s^{-1} for the luminosities of over 50 nearby stars having a variety of spectral types. The first positive detection of coronal soft X-ray emission was reported by Catura *et al.* (1975) for the star α Aur (G6 III + F9 III), their flux measurements indicated an X-ray luminosity $L_X \sim 10^{31}$ erg s^{-1} for this star. Further observations by Mewe *et al.* (1975, 1976) using the detector on board the Astronomical Netherlands Satellite (ANS) confirmed the identification of α Aur as an X-ray source and resulted in a luminosity estimate L_X few times 10^{29} erg s^{-1}. These authors also detected X-ray emission at a rate $L_X \simeq 10^{28}$ erg s^{-2} from α CMa (A1 V + DA) and set upper limits on the luminosities of 26 other stars.

As experimental sensitivity increased, so did the number of late-type stars detected as soft X-ray sources. Using the HEAO-1 satellite, Cash *et al.* (1978) found evidence for X-ray line emission from the corona of α Aur, while Nugent and Garmire (1978) detected emission from α Cen (G2 V + K IV) at a rate $L_X \simeq 3 \times 10^{27}$ erg s^{-1}. Observational searches for coronal emission from late-type stars were carried out by Ayres *et al.* (1979) and Walter *et al.* (1980b) with the HEAO-1 satellite, and many of the brighter RS CVn binary systems were identified as strong ($L_X \sim 10^{30}–10^{31}$ erg s^{-1}) X-ray sources by Walter *et al.* (1978a, b) (see also Walter *et al.*, 1980a). In addition, Topka *et al.* (1979)

used a rocket-borne imaging X-ray telescope to observe the stars α Lyr (A0 V) and η Boo (G0 IV); the results of their observations implied luminosities of $L_X \simeq 3 \times 10^{28}$ erg s^{-1} for α Lyr and $L_X \simeq 10^{29}$ erg s^{-1} for η Boo. Finally, X-ray flares from several nearby dMe stars were detected by Heise *et al.* (1975) and Kahn *et al.* (1979).

With the launch of the Einstein (HEAO-2) Observatory (Giacconi *et al.*, 1979), it has become clear that stars in general constitute a distinct class of low-luminosity galactic X-ray sources (cf. Figure 1.1). In the following discussion, we briefly summarize some preliminary results from the first Einstein stellar survey (Vaiana *et al.*, 1981; see also Ayres *et al.*, 1981; Linsky, 1981a, b) with particular emphasis placed upon the characteristics of coronal soft X-ray emission from late-type stars. Among dwarf stars of spectral type F, the measured luminosity L_X is typically $\sim 10^{28}-10^{30}$ erg s^{-1}, with L_X/L_{bol} (i.e. the ratio of the observed soft X-ray luminosity to the stellar bolometric luminosity) of order $10^{-7}-10^{-4}$. For main sequence stars of spectral types G and K, the observed levels of X-ray emission correspond to luminosities in the range L_X $10^{26}-10^{30}$ erg s^{-1} with $L_X/L_{bol} \sim 10^{-7.5}-10^{-4.5}$. The Einstein results for dwarf M stars exhibit a similar broad range of emission levels, with $L_X \sim 10^{26}-10^{29}$ erg s^{-1} and $L_X/L_{bol} \sim 10^{-5}-10^{-3}$.

Of particular interest with regard to the Einstein observations of late-type dwarfs is (i) the wide range of luminosities which occur for stars of a given spectral type and (ii) the relatively small variation in the median X-ray luminosity as one proceeds along the main sequence from spectral type G to late M. Since the detection of soft X-ray emission implies the presence of hot ($T \gtrsim 10^6$ K) gas in the stellar atmosphere, these results have significant consequences for the nature of mechanism(s) responsible for coronal heating (see Section 2.4 below). Specifically, they suggest that the physical processes which give rise to X-ray emitting coronae in late-type stars depend upon one or more stellar parameters in addition to the surface gravity and effective temperature. Along these lines, analyses of the Einstein data have revealed some evidence for correlations between the level of coronal activity (as measured by the X-ray luminosity L_X) and the stellar rotation rate (Vaiana, 1980; Ayres and Linsky, 1980a, b; Vaiana *et al.*, 1981; Walter and Bowyer, 1981; Walter, 1981; Pallavicini *et al.*, 1981), multiplicity (Vaiana *et al.*, 1981), and the strength of chromospheric Ca II emission (Mewe and Zwaan, 1980).

Among late-type evolved stars, the Einstein pointed survey has resulted in the detection of one F supergiant ($L_X \sim 10^{30}$ erg s^{-1}), four G giants (L_X $10^{28}-10^{30}$ erg s^{-1}), and the two early K giants α Ser (K2 III) and ϵ Sco (K2 III) (L_X 10^{28} erg s^{-1}, $L_X/L_{bol} \sim 10^{-7}$) (Vaiana *et al.*, 1981; Ayres *et al.*, 1981). About the only portion of the H–R diagram in which coronal X-ray emission has not been positively detected is that region occupied by the coolest late-type giants and supergiants. For example, only upper limits to the magnitude of L_X are available for stars such as the K giants α Boo (K2 III) and α Tau (K5 III), the G supergiants β Aqr (G0 Ib), α Aqu (G2 Ib), and ϵ Gem (G8 Ib), and the M supergiants α Ori (M2 Iab) and α Sco (M1 Ib) (Vaiana *et al.*, 1981; Ayres *et al.*, 1981). In the specific case of the M supergiants α Ori and α Sco, the derived upper limits for the quantity L_X/L_{bol} are $10^{-8.5}$ and $10^{-8.7}$, respectively, less than the value $L_X/L_{bol} \sim 10^{-7}$ characteristic of solar coronal hole regions.

It is important to note that these results do not constitute conclusive proof for the absence of coronal regions in the atmospheres of late-type low-gravity stars. In particular, the X-ray flux from coronae having small emission measures or gas temperatures $< 10^6$ K could be too low for the instruments on board Einstein to detect. Alternatively, X-rays emitted near the base of a cool, massive wind of the type known to emanate from late-type giants and supergiants (see Section 2.5 below) could be attenuated by the outflowing material. However, a number of the stars which have not been detected as soft X-ray sources also have ultraviolet spectra which contain only chromospheric emission lines (e.g. α Boo, ϵ Gem, α Ori; see Linsky and Haisch, 1979). A possible explanation for the absence of both coronal and transition region emission from such stars is given by the following conjecture: As one proceeds in the cool portion of the H–R diagram from the main sequence to the giant and supergiant branches, a fundamental change in the thermal (and dynamical) structure of the outer stellar atmosphere occurs. The nature of this change is such that stars with high surface gravities possess hot coronae while those with low surface gravities have atmospheres in which T attains only chromospheric values. Similar considerations have led Ayres *et al.* (1981) to propose a coronal dividing line in the H–R diagram, analogous to the transition region dividing line discussed in Section 2.3.1. This boundary (depicted as the line labelled 'X' in Fig. 2.5) separates stars for which the detection of X-ray emission implies the presence of a corona from stars having no detectable X-ray emission. The proximity of the suggested coronal dividing line to the transition region dividing line (labelled 'T' in Figure 2.5) lends support to the hypothesis that the atmospheres of cool giants and supergiants contain primarily gas at chromospheric temperatures.

2.4. CHROMOSPHERIC AND CORONAL HEATING MECHANISMS

2.4.1. *Overview*

A complete theoretical description of the thermal and dynamical structure of late-type stellar atmospheres requires the specification of a mechanism for depositing energy at a rate sufficient to balance the observed losses due to radiation and mass motions. In the present section, we enumerate some potential nonradiative heating processes and, where possible, compare theoretical predictions with observational results. The discussion contained in the following paragraphs is brief and qualitative in nature, and is not intended as a detailed review of the many heating mechanisms which have been proposed to account for the presence of gas with $T > T_{\text{eff}}$ in the atmospheres of the Sun and late-type stars. Chromospheric and coronal heating has been treated quantitatively and at length in the review papers by Jordan (1973), Stein and Leibacher (1974, 1980), Athay (1976), Withbroe and Noyes (1977), Vaiana and Rosner (1978), Ulmschneider (1979), and Mewe (1979).

2.4.2. *Acoustic Wave Heating*

It was first suggested by Biermann (1946) and Schwarzschild (1948) that the solar corona could be heated by the dissipation of a mechanical energy flux in the form of propagating acoustic waves. A schematic description of the way in which sound wave dissipation can provide the energy needed to produce a nonradiative equilibrium temperature rise in a late-type stellar atmosphere is as follows. Small amplitude compressional disturbances are generated as a result of the turbulent motions in a subphotospheric convection zone.

Such waves will travel upward in the gravitationally stratified atmosphere provided that the wave frequency exceeds the acoustic cutoff frequency $\omega_{ac} = g/2a$, where a is the sound speed. For conditions characteristic of the temperature minimum region is the solar atmosphere, the criterion $\omega > \omega_{ac}$ corresponds to waves with periods $P \lesssim 200$–300 s. The energy flux carried by upwardly propagating waves of amplitude δV is given by $F_{ac} \simeq \rho\, \delta V^2 a$, and represents only a small fraction of the radiative energy flux at the stellar surface. If the gas temperature can be regarded as being approximately constant throughout the atmosphere, then conservation of the wave energy flux requires that $\delta V \propto \rho^{-1/2}$, and the amplitude of the wave increases as it propagates upward into regions of lower density. For a finite amplitude wave ($\delta V \sim a$), the phase velocity is greater in compressions than in rarefactions (see, e.g., Landau and Lifshitz, 1959). As a result, the wave profile steepens, and an initially sinusoidal disturbance is changed into a sawtooth wave with a shock at the front of each compression. The wave mechanical energy flux is then dissipated at a rate $\sim F_{ac}/L$ where, for a weak sawtooth shock wave of period P and Mach number M, $L \simeq (aP/4)\ [M^3/(M-1)]$ is the dissipation length (Kuperus, 1965; de Loore, 1970). The dependence of L on P suggests that short-period waves are dissipated low in the atmosphere, while the dissipation of long-period waves occurs at greater heights (see, e.g., Stein and Leibacher, 1974, for a more complete discussion of acoustic shock formation and dissipation).

The construction of shock-heated model chromospheres and coronae for late-type stars stars requires the solution of a number of formidable problems. Among other things, it is first necessary to compute the amplitude and frequency distribution of the waves generated by turbulent motions in a stellar convection zone. Once the initial acoustic wave energy flux is known, the propagation (including the effects of reflection and refraction), nonlinear evolution, and dissipation of the waves must be calculated. Finally, the overall energy balance between wave heating and losses due to conduction and radiation (including nonlocal radiative transfer effects) must be treated. As a result of these and other complexities, only simplified versions of the complete problem have been considered to date (see Ulmschneider (1979) for a review of recent theoretical calculations). However, Linsky (1980a) has noted that despite its many uncertainties, some aspects of the acoustic heating theory can still be compared with observational results. For a given star, let the acoustic wave energy flux available for atmospheric heating be denoted F_{ac}, and let F_{rad} be the observed total chromospheric and coronal radiative flux. If the outer atmospheres of late-type stars are heated solely by the dissipation of sound waves as described above, then F_{ac} must be $\gtrsim F_{rad}$. Moreover, both F_{ac} and F_{rad} should exhibit similar dependences on fundamental stellar parameters such as effective temperature and surface gravity.

To conduct these tests, it is necessary to estimate both the magnitude of F_{ac} and its dependences on T_{eff} and g. Although a quantitively reliable theory for acoustic wave generation by turbulent convection motions does not exist at present, for the purpose of illustration we adopt the treatment of Lighthill (1952) and Proudman (1952). These authors have shown that for a homogeneous layer containing isotropic turbulence, the rate at which acoustic energy is generated per unit volume due to the action of turbulent Reynolds stresses is approximately $P_{ac} \sim (\rho a^3/l)\ (u/a)^8$, where u (assumed to be $\ll a$) is the velocity and l length scale of the turbulence. The wave energy flux which emanates from the top of the convection zone is then of order $F_{ac} \sim P_{ac} l \sim \rho a^3 (u/a)^8$. To ascertain the dependence of F_{ac} on stellar properties, we follow Renzini *et al.* (1977) and use the mixing-length theory to describe the structure of the convection zone (see also Cram and Ulmschneider, 1978; Stein and Leibacher, 1980; Stein, 1981). The convective energy flux is $F_c \simeq \rho C_p u \, \Delta T$, where C_p is the specific heat at constant pressure and ΔT is the temperature difference between a convective element and its surroundings (see, e.g., Schwarzschild, 1958; Cox and Giuli, 1968). If g is the stellar gravity and l is the mixing length, the velocity u of the element is given by $u \simeq (gl \, \Delta T/T)^{1/2} = (F_c lg/\rho C_p T)^{1/3}$. Assuming that the mixing length can be expressed as a constant fraction α of the pressure scale height $H (= P/\rho g)$, it follows that $l \propto \alpha T/g$. For stars in which convective energy transport is efficient, F_c is approximately equal to the surface radiative flux σT_{eff}^4. Thus, near the top of the convection zone, $T \simeq T_{\text{eff}}$, $u \propto (\alpha T_{\text{eff}}^4/\rho)^{1/3}$, and $F_{ac} \simeq \alpha^{8/3} \rho^{-5/3} \cdot T_{\text{eff}}^{49/6}$. To estimate the density ρ, note that for an atmosphere in hydrostatic equilibrium, $p = mg$, where m is the mass column density above the level in question. If κ is the absorption coefficient of the gas, the optical depth τ corresponding to the level m is $\tau \simeq \int_0^m dm\kappa$. Assume that κ can be expressed in the form $\kappa = \kappa_0 p^c T^d$ with c and d constants; then $\tau \simeq [p^{1+c}/(1+c)] \, (\kappa_0 T^d/g)$. If the top of the convection zone occurs where $\tau \simeq 1$ and $T \simeq T_{\text{eff}}$, it follows that

$$\rho \propto g^{\frac{1}{1+c}} \, T_{\text{eff}}^{-\left(\frac{1+c+d}{1+c}\right)} \quad \text{and } F_{ac} \propto \alpha^{8/3} \, g^{-\frac{5}{3(1+c)}} \, T_{\text{eff}}^{-\frac{59(1+c)+10d}{6(1+c)}}$$

With H^- bound–free absorption taken to be the dominant source of continuous opacity, $c \simeq 0.7$, $d \simeq 5.0$ (Renzini *et al.*, 1977; Stein, 1981), and the dependence of the normalized acoustic wave energy flux on the quantities g and T_{eff} is approximately $F_{ac}/\sigma T_{\text{eff}}^4 \propto \alpha^{8/3} g^{-1} (T_{\text{eff}})^{11}$.

For the Sun, calculations using the Lighthill–Proudman generation mechanism and the mixing-length theory yield values in the range 10^7–10^8 erg cm^{-2} s^{-1} for the acoustic wave energy flux produced by turbulence in the convection zone (see, e.g., Osterbrock, 1961; Stein, 1967, 1968). While the form of the emitted wave spectrum is uncertain, the results of Stein (1968) and Milkey (1970) suggest that the maximum emission occurs at frequencies $\omega \sim a/l$ (i.e. frequencies $> \omega_{ac}$), corresponding to waves with periods in the range 20–60 s. Although radiative damping of small amplitude waves in the photosphere can reduce F_{ac} by as much as 90% (Stein and Leibacher, 1974), the remaining wave energy flux would appear to be adequate to account for the combined radiative losses from at least the low and middle chromosphere ($\sim 4 \times 10^6$ erg cm^{-2} s^{-1}; cf. Section 2.1.2). However, recent observational studies of longitudinal oscillations in the middle and upper chromosphere (Athay and White, 1978, 1979a, b; White and Athay, 1979a, b) indicate that the sound wave energy flux at heights above $h \simeq 1200$ km (relative to the level at which $\tau(\lambda = 5000 \text{ Å}) = 1$) is approximately 10^4 erg cm^{-2} s^{-1}, a value too low to account for the observed radiative losses from the transition region and corona ($\geq 3 \times 10^5$ erg cm^{-2} s^{-1}; cf. Section 2.1.2) (see also Bruner, 1978; Bruner and McWhirter, 1979). This deficiency is further exacerbated by the steep gradients in density and temperature which characterize the upper chromosphere and transition region. Sound wave reflection in such regions is expected to further reduce the mechanical energy flux available for heating the corona. These and other results (see, e.g., Withbroe and Noyes,

1977; Vaiana and Rosner, 1978) suggest that while acoustic wave dissipation may be effective in heating the low solar chromosphere, alternative mechanisms are probably required to account for the observed energy losses for the upper chromosphere and corona.

Model convection zones and acoustic wave fluxes for late-type stars with a variety of values for g and $T_{\rm eff}$ have been calculated by de Loore (1970) and Renzini *et al.* (1977) among others. As in the case of the Sun, the results of such computations are subject to considerable uncertainty since they depend sensitively upon the validity of the Lighthill–Proudman theory of sound wave generation and the mixing-length theory of convection. Not surprisingly, for stars in which convective energy transport is efficient, the calculated acoustic wave energy fluxes exhibit dependences on g and $T_{\rm eff}$ in reasonable agreement with those derived above. Hence, for the purpose of comparing the theoretical g and $T_{\rm eff}$ dependences of $F_{\rm ac}$ with those inferred from the observational results cited in preceding sections, we assume that $(F_{\rm ac}/\sigma T_{\rm eff}^4) \propto g^{-1}\, T_{\rm eff}^{11}$. This assumption is made with the following caveat: namely, that a more detailed treatment of acoustic wave generation by turbulent convective motions may lead to a different dependence of $F_{\rm ac}$ on stellar parameters. Moreover, we caution the reader that the observationally derived $T_{\rm eff}$ dependence of the quantities F(Ca II)$/\sigma T_{\rm eff}^4$, F(Mg II)$/\sigma T_{\rm eff}^4$, and $F_{\rm X}/\sigma T_{\rm eff}^4$ (see below) are at present highly uncertain.

With this in mind, we note that Linsky and Ayres (1978) and Basri and Linsky (1979) have measured surface fluxes in the Mg II h and k lines for a large number of late-type stars (see also Weiler and Oegerle, 1979). While the data analyzed by these authors contain only a small number of the coolest main sequence stars, their results suggest that the ratio F(Mg II)$/\sigma T_{\rm eff}^4$ is essentially independent of g (cf. Figure 3.4 and the discussion thereof). A smilar conclusion regarding the gravity dependence of the ratio F(Ca II)$/\sigma T_{\rm eff}^4$ (F(Ca II) is the measured surface flux in the Ca II H and K lines) has been reached by Linsky *et al.* (1979b). The results of a more recent and extensive survey of Mg II emission in late-type stars by Stencel *et al.* (1980) indicate that at a given $T_{\rm eff}(\leq 5000$ K), the value of F(Mg II)$/\sigma T_{\rm eff}^4$ for supergiants is about 3–4 times larger than that for giants, corresponding to an approximate gravity dependence, F(Mg II)$/\sigma T_{\rm eff}^4 \propto g^{-0.25}$. Moreover, for stars belonging to luminosity classes I–III, Basri and Linsky (1979) and Stencel *et al.* (1980) find that the normalized surface Mg II flux may decrease with decreasing $T_{\rm eff}$ for $T_{\rm eff} \leq 5000$ K. Although there is appreciable scatter in the data, the trend suggested by these results is approximately F(Mg II)$/\sigma T_{\rm eff}^4 \propto T_{\rm eff}^4$(cf. Figure 2.4). The normalized Ca II fluxes given by Linsky *et al.* (1979) also appear to decrease with decreasing $T_{\rm eff}$. This trend is strongest among late-type supergiants, for which the dependence on $T_{\rm eff}$ is approximately F(Ca II)$/\sigma T_{\rm eff}^4 \propto T_{\rm eff}^7$. For late-type dwarf stars, the results of the Einstein stellar survey (Vaiana *et al.*, 1981) indicate that the level of coronal X-ray emission changes relatively little as one proceeds along the main sequence from spectral type G to late M (cf. Section 2.3.2) (although the M dwarf detections may be subject to some selection effects). Since for luminosity class V stars the surface area decreases with decreasing $T_{\rm eff}$, this suggests that the surface X-ray flux $F_{\rm X}$ increases for lower values of $T_{\rm eff}$. For simplicity, assume that the mean X-ray luminosity $L_{\rm X}$ remains constant for dwarfs with spectral types later than G0. It then follows that the surface flux $F_{\rm X}$ varies according to $F_{\rm X} \propto R^{-2}$, where R is the stellar radius. A fit to the stellar data of Allen

(1973) yields $R \propto T_{\text{eff}}^{0.94}$ for dwarfs with spectral types between G0 and M0. Thus, $F_X \propto T_{\text{eff}}^{-1.88}$ and $F_X/\sigma T_{\text{eff}}^4 \propto T_{\text{eff}}^{-5.88}$. We note that apparent absence of coronal X-ray emission from the coolest late-type giants and supergiants is not necessarily a consequence of the predicted rapid decrease of F_{ac} with decreasing T_{eff}. This is because the massive stellar winds which are known to be associated with such stars (see Section 2.5) may represent a more efficient energy loss mechanism than the emission of radiation.

The results presented above suggest that the observed chromospheric and coronal radiative fluxes possess a different dependence on g and T_{eff} than that predicted by the acoustic wave heating theory. It has also been noted previously that the measured radiative losses for stars having similar spectral types and luminosity classes can differ by more than an order of magnitude (cf. Sections 2.2 and 2.3), indicating that F_{rad} depends on one or more stellar parameters in addition to g and T_{eff}. Furthermore, measurements of X-ray (Vaiana *et al.*, 1981) and Ca II K line (Blanco *et al.*, 1974; 1976) fluxes indicate that there is a quantitative discrepancy between the magnitudes of the calculated acoustic wave energy flux and the observed chromospheric and coronae radiative fluxes, particularly among late-type dwarf stars. Given the admittedly uncertain nature of the acoustic heating theory, it is important to note that some of these deficiencies may be mitigated by the inclusion of several heretofore neglected effects. For example, modifications to the g and T_{eff} dependences of F_{ac} arising from the radiative damping of waves in the stellar photosphere have not been treated in the preceding discussion (see, e.g., Schmitz and Ulmschneider, 1980a, b, 1981). Moreover, preliminary calculations by Bohn (1982) (see also Ulmschneider, 1979; Schmitz and Ulmschneider, 1981) suggest that the inclusion of effects due to the gravitational stratification of the convection zone can result in larger acoustic wave energy fluxes for cool main sequence stars. However, as Vaiana *et al.* (1981) have noted, it is unlikely that such modifications can eliminate entirely the discrepancies between theory and observations. Thus, while the dissipation of acoustic waves presumably plays some role in heating the outer atmospheres of late-type stars, alternative energy input mechanisms are required to completely account for the observed radiative losses and their dependences on stellar parameters.

2.4.3. *Magnetic Heating Mechanisms*

Recent high-resolution studies of the outer solar atmosphere have revealed that the corona (outside of coronal holes) is highly structured, consisting of a variety of loop-like features whose shape is thought to be defined by the geometry of the local solar magnetic field (see, e.g., Withbroe and Noyes, 1977; Vaiana and Rosner, 1978). Moreover, observations imply that most of the coronal radiative flux derives from hot plasma confined within such magnetically closed loops, and that the strongest soft X-ray emission originates in regions having the greatest concentration of magnetic flux (Rosner *et al.*, 1978; Golub *et al.*, 1980, 1981). These results suggests that magnetic fields are directly involved in heating the solar corona, and by analogy, in heating the coronae of other late-type stars as well. While direct observational evidence for the presence of magnetic fields is available for only a few late-type stars other than the Sun (see, e.g., Robinson *et al.*, 1980), a number of indirect indicators exist. For example, the well-known decrease in surface rotational speed for main sequence stars with spectral types later than about F5

(Kraft, 1967) and the dependence of rotation rate on age for solar-type stars (Skumanich, 1972) have generally been attributed to the action of magnetically coupled stellar winds. Further indirect evidence for the presence of magnetic fields in late-type stars is provided by the following: observations of star spots and stellar flares (Hartmann, 1980, 1981a, b); observationally determined correlations between the level of chromospheric/coronal emission and magnetic-field strength (in the case of the Sun) and rotation rate (cf. Sections 2.2.3 and 2.3.2); the apparent occurrence of activity cycles in stars other than the Sun (cf. Section 2.2.3); X-ray observations indicating the presence of coronal plasma which is too hot to be confined by gravity alone (Holt *et al.*, 1979; Walter *et al.*, 1980a, b).

As well as restricting coronal cooling processes (e.g. through inhibition of expansion and cross-field thermal conduction), magnetic fields can act in a variety of ways to influence the rate at which the coronae of the Sun and late-type stars are heated. The stellar magnetic field is presumably maintained by dynamo action in a subphotospheric hydrogen convection zone (see, e.g., Parker, 1970, 1979). In this process, the poloidal magnetic field at the surface of the star is regenerated by the interaction between differential rotation and cyclonic fluid motions deep in the convection zone. As might be expected, the generation of waves by convective turbulence can be affected by the presence of a magnetic field (Kulsrud, 1955; Osterbrock, 1961; Parker, 1964; Stein, 1981). In particular, the magnetic field permits additional wave modes to be produced, and influences their subsequent propagation and dissipation at higher levels in the atmosphere. Specific hydromagnetic wave modes which have been studied in connection with coronal heating include slow magneto-acoustic waves (Osterbrock, 1961; Stein, 1981), fast magneto-acoustic waves (Osterbrock, 1961; Habbal *et al.*, 1976; Stein, 1981), and Alfvén waves (Osterbrock, 1961; Uchida and Kaburaki, 1974; Wentzel, 1974, 1978; Hollweg, 1981; Stein and Leibacher, 1980; Stein, 1981). Of particular interest with regard to heating via the dissipation of hydromagnetic wave modes are the results of Stein (1981). Using an analysis similar to that given above for sound waves, he has shown that the normalized energy flux of Alfvén waves produced in a strong magnetic field is far less sensitive to variations in g and T_{eff} than the corresponding acoustic wave energy flux, in better agreement with observations (cf. Section 2.4.2).

In addition to the effects described above, the observed concentration of magnetic fields on the Sun into closed loop-like configurations has led investigators to consider a number of mechanisms for heating individual magnetically confined structures. For example, several authors have noted that the discontinuity in Alfvén speed which occurs at the outer boundary of a loop permits hydromagnetic surface waves to propagate (Ionson, 1978; Wentzel, 1979a, b). The material contained within the loop can therefore be heated by the dissipation of surface Alfvén waves generated by oscillations of the loop footpoints (Ionson, 1978; Wentzel, 1979c; Lee, 1980). Moreover, the stressing and twisting of magnetic loops by turbulent motions in the convection zone and photosphere can lead to heating via either the dissipation of currents by anomalous (turbulent) resistivity (Tucker, 1973; Rosner *et al.*, 1978) or rapid reconnection at magnetic neutral points (Parker, 1975a, b, 1977, 1979; Galeev *et al.*, 1981). The conditions under which heating by neutral point reconnection can occur in the solar corona have also been investigated by Levine (1974) and Parker (1981a, b). In the latter work, Parker has shown that the dislocation of a single flux tube relative to the ambient magnetic field leads to dynamical flattening and rapid dissipation (by reconnection) of the tube. Such a dislocation can result from a physical displacement of the footpoints of the tube or by inflation of the tube due to an increase in the internal gas pressure.

2.5. Observational Evidence for Mass Loss from Late-Type Stars

2.5.1. *Main Sequence Stars*

As noted in Section 2.3.2, the apparent ubiquity of soft X-ray emission among late-type dwarf stars suggests that they possess coronae characterized by gas temperatures $\gtrsim 10^6$ K. By analogy with the Sun, it therefore seems likely that such stars also possess tenuous, hot stellar winds, driven principally by the force arising from thermal pressure gradients in the outflowing material (cf. Section 2.1.2). To estimate the expected mass loss rate due to the wind from a typical late-type main sequence star, assume for simplicity that the outflow is isothermal and spherically symmetric. Note that in making these assumptions, we neglect the potentially important role played by magnetic fields in accelerating and chaneling the flow (see the reviews by Holzer (1979, 1980) for a more complete discussion). Of the many possible solutions to the fluid equations describing the steady-state expansion of the coronae, only one satisfies the boundary conditions of subsonic flow near the star and supersonic flow with constant velocity and vanishing gas pressure at large distances (Parker, 1958, 1963). This particular solution passes smoothly through a critical point, occurring where the flow velocity V becomes equal to the local speed of sound in the gas a. For isothermal flow from a star of mass M_* and radius R_*, the critical point location Z_s (in units of R_*) is given by $Z_s = (GM_*/2a^2R_*)$ where, for a gas of temperature T and mean mass per particle μ, $a = (kT/\mu)^{1/2}$. Specifying the location of a reference level $r = r_0(\simeq R_*)$ in the stellar corona, conservation of energy for the wind material can be used to derive an approximate expression for the intital velocity V_0 (i.e. V at $r = r_0$) of the flow, assuming $V_0 \ll a$, the result of this procedure is $V_0 \simeq aZ_s^2 \exp(\frac{3}{2} - 2Z_s)$ (cf. Parker, 1963). By conservation of mass, the mass loss rate $\dot{M}$ is then $\dot{M} = 4\pi r_0^2 \mu N_0 V_0$, where N_0 is the total number density at $r = r_0$. Adopting the values $M_* = M_\odot$, $r_0 = R_* = R_\odot$, $T = 1.5 \times 10^6$ K, and $\mu = 0.61\ m_H$ for illustration, it follows that $Z_s = 4.69$ and $V_0 = 8.31 \times 10^{-3}\ a = 1.19 \times 10^5$ cm s^{-1}. Thus, for a coronal density $N_0 = 10^8$ cm^{-3}, $\dot{M} = 7.35 \times 10^{11}$ g s^{-1} = $1.16 \times 10^{-14}\ M_\odot$ yr^{-1}.

For flow speeds like that of the average quiet solar wind (terminal velocity $\simeq$ 400–500 km s^{-1}), the wind column density implied by the mass loss rate estimate given above is far too low to produce any observable spectroscopic effects. Hence, the winds of cool, late-type dwarf stars are not directly detectable, and indirect evidence for their existence must be obtained. For $M_* \simeq M_\odot$, the time-scale for significant mass loss to occur is $\tau_M = M_*/\dot{M} \sim 10^{14}$ yr, a value too long for evolutionary effects to be noticeable. However, the dynamical consequences of thermally driven mass loss from main sequence stars can take place over much shorter time-scales. For a star which rotates rigidly with angular frequency Ω, the wind reduces the stellar angular momentum J at a rate $\dot{J} \simeq \frac{2}{3}\Omega R_*^2 \dot{M}$. When magnetic fields are included in the treatment of coronal expansion, magnetic stresses act to enforce approximate corotation of the flow with the star out to a distance $r \simeq r_a (> R_*)$, where r_a (the Alfvén radius) is the distance at which $\frac{1}{2}\rho V^2 = B^2/8\pi$ (Weber and Davis, 1967). The amount by which r_a exceeds R_* depends on both Ω and the strength of the magnetic field (Belcher and MacGregor, 1976), and for the solar wind,

$r_a \simeq 20–25\, R_\odot$. Since $r_a > R_*$, the torque exerted by the magnetically coupled wind causes the star to lose angular momentum at the enhanced rate $J \simeq \frac{2}{3}\Omega r_a^2 \dot{M}$. Wind calculations for stars having rotation rates and field strengths between 1 and 20 times the solar values yield e-folding times for angular momentum loss $\tau_J = J/\dot{J}$ in the range $10^8 – 10^{10}$ yr (Belcher and MacGregor, 1976). Since these values are quite comparable to evolutionary time-scales, it is expected that the angular momentum of a typical late-type dwarf star should be noticeably reduced during its main sequence lifetime. As noted previously (cf. Section 2.4.3), observations indicate that the average surface rotation speed of main sequence stars (i) decreases sharply for spectral types later than F5, and (ii), is a decreasing function of age for stars like the Sun (see also Soderblom, 1983, and references therein; Vaughn *et al.*, 1981). In view of the preceding discussion, such observations strongly suggest that late-type dwarf stars lose both mass and angular momentum via magnetically coupled stellar winds originating in hot stellar coronae.

2.5.2. *Circumstellar Absorption Lines*

Blue-shifted absorption lines due to a variety of neutral and singly ionized metals are frequently observed in the optical and ultraviolet spectra of cool giants and supergiants (see the reviews of Reimers, 1975; Goldberg, 1979; Hagen 1980; Castor, 1981; Hartmann, 1981). Such features result from the absorption and scattering of photospheric radiation by atoms or ions in an outwardly expanding circumstellar envelope (see Section 3 for a detailed discussion of line formation in a moving medium). Since the column density of wind material must be large in order to produce noticeable circumstellar absorption, these results suggest that late-type low-gravity stars undergo mass loss at rates significantly in excess of those for the Sun and dwarf stars.

Shortward displaced absorption cores were first detected in the spectra of several late-type giants and supergiants by Adams and McCormack (1935). The expansion velocities deduced from their measurements were typically ~ 10 km s^{-1}, a value substantially smaller than the surface escape speeds ($V_{esc} \sim 100$ km s^{-1}) of the particular stars observed. Subsequently, in his classic analysis of the visual binary system α Her (M5 II + G0 III), Deutsch (1956, 1960) noted that many of the blue-shifted absorption lines present in the spectrum of the M5 primary also appeared in the spectrum of its distant (separation $\simeq 360\, R_*$) G0 companion (itself a spectroscopic binary). Since none of the lines in question had ever been observed in the spectrum of a G star, Deutsch concluded that the secondary was contained within the circumstellar envelope of the primary. Besides providing an estimate of the size of the envelope, this result also established that mass was in fact being lost by α^1 Her since the observed expansion velocity exceeded the gravitational escape speed at the orbit of the secondary. More recently, similar methodology using improved spectroscopic techniques has been applied to the analysis of mass loss from α Her (Reimers, 1977c, 1978) and α Sco (M1 Ib + B2.5 V) (Kudritzki and Reimers, 1978; Van der Hucht *et al.*, 1980). It has also become possible to directly map the spatial extent of the circumstellar gas shells about some late-type supergiants. In the case of the M supergiant α Ori (M2 Iab), a number of investigators have measured the intensity of emission in the λ7699 line of K I at different angular displacements from the star (Bernat and Lambert, 1975, 1976b; Lynds *et al.*, 1977; Bernat *et al.*, 1978; Honeycutt *et al.*, 1980). Such emission results from the resonance line scattering of

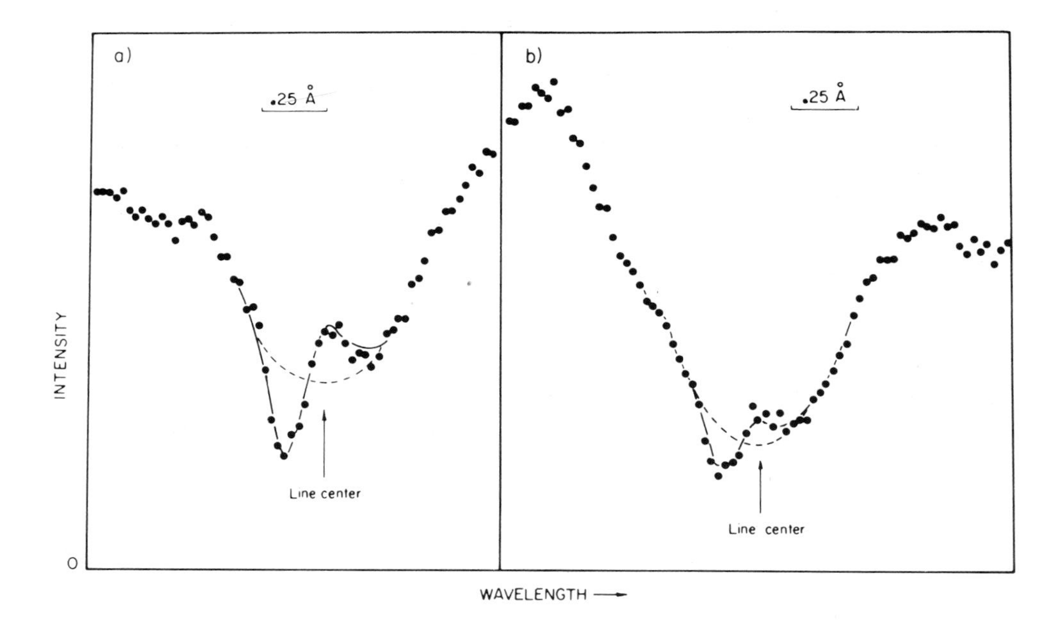

Fig. 2.6. Blue-shifted circumstellar absorption features (Mn I λλ4030.8, 4033.1) in the spectrum of α Ori (M2 Iab). The predicted shell and photospheric components of the line profile are shown by solid and dashed lines, respectively (cf. Section 2.5.2). (From Bernat and Lambert, 1976a.)

photospheric radiation by neutral potassium in the envelope surrounding the star. For the circumstellar shell of α Ori, a radius >1000 R_* is inferred from the observed spatial distribution of K I resonance line emission. In addition, there is some evidence for the existence of at least two distinct gas shells about α Ori, indicating that mass loss from this star may not occur continuously (Goldberg *et al.*, 1975; Bernat *et al.*, 1979; Bernat, 1981; Goldberg, 1981).

For a given star, an estimate of the mass loss rate $\dot{M}$ can be obtained from a detailed analysis of the observed circumstellar absorption features (such as the Mn I λλ4030, 4033 lines of α Ori depicted in Figure 2.6). The procedure for doing this is basically as follows. First, the line profile due to scattering in the circumstellar envelope is determined by removing the underlying photospheric component from the observed absorption core. It is then assumed that the line profile so derived is formed in a gas shell surrounding the star, with inner radius $R_{\min} \geq R_*$ and outer radius $\gg R_{\min}$. The flow throughout the shell is taken to be spherically symmetric and steady, and is characterized by a constant expansion velocity V (obtained from the measured line blue-shifts). With these assumptions, radiative transfer techniques can be used to fit the shell line profiles, thereby determining the column densities of the line-producing atoms or ions. Note, however, that most of the commonly observed circumstellar lines are due to optical transitions of neutral metal atoms. Since most of the metallic elements present in the flow are expected to be singly ionized, the degree of ionization of the gas in the shell must be calculated in order to determine the total column density of a given element. Because considerable error can occur in determining the ionization state of the gas, it is often preferable to work with lines due to singly ionized metals, for which the derived column density should be essentially the total column density of the element (neglecting depletion onto grains). Assuming cosmic abundances for the elements with observed lines, the total column density $\langle N \rangle$ of gaseous material in the shell can be obtained; from conservation of mass in a flow with the properties assumed above, it then follows that $\dot{M} = 4\pi R_{\mathrm{min}} \mu \langle N \rangle V$.

The procedure outlined (or a variation of it) has been applied in the determination of mass loss rates for numerous late-type giants and supergiants (see, e.g., Weymann, 1962b; Sanner, 1976a, b; Bernat, 1977; Reimers, 1977a; Hagen, 1978). Major sources of disagreement and uncertainty in these analyses include (i) treatment of the photospheric line profile, (ii) incomplete understanding of the physical processes affecting the ionization balance in the envelope, (iii) the adopted value for R_{min}, (iv) the inclusion of grains in estimating $\dot{M}$, and (v) lack of knowledge concerning the detailed geometrical and dynamical structure of the outflow. These uncertainties are reflected in the wide range of estimated mass loss rates for a given star. For example, in the case of α Ori (M2 Iab) different investigators have derived $\dot{M} = 1.5 \times 10^{-7}$ (Hagen, 1978), 3.4×10^{-5} (Bernat, 1977), and 1.7×10^{-7} $M_\odot$ yr^{-1} (Sanner, 1976a, b), while for μ Cep (M2 Ia), the values $\dot{M} = 4.9 \times 10^{-7}$ (Hagen, 1978), 4.2×10^{-4} (Bernat, 1977), and 1.0×10^{-6} $M_\odot$ yr^{-1} (Sanner, 1976a, b) have been obtained. Hence, although observations of circumstellar absorption lines indicate that late-type low-gravity stars possess massive stellar winds, estimates of $\dot{M}$ for particular stars are at present not quantitatively reliable.

We note finally that the presence of violet-displaced circumstellar absorption lines in the spectra of late-type stars can be used to delineate those regions in the cool portion of the H–R diagram where massive stellar winds occur. In an extensive survey to determine the extent of mass loss in the H–R diagram, Reimers (1977b) found that circumstellar Ca II absorption features appeared in the spectra of all giants and supergiants cooler and more luminous than the line defined by (K5, $M_v = 0.0$), (K4, -1), (K2, -1.8), and (G5, -4). This boundary is depicted in Figure 2.5 as the line labelled 'CS'; stars above and to the right of the line (that is, the coolest and most luminous stars) generally have the highest rates of mass loss. Reimers (1977b) also noted that the measured wind expansion velocities for stars in his sample were in the range 5–100 km s^{-1}, with a tendency for stars with the lowest surface gravities to have the lowest wind velocities.

2.5.3. *Chromospheric Emission Line Asymmetries*

In Section 2.2.1, it was noted that the emission cores of strong, collision-dominated spectral lines (e.g., Ca II H and K, Mg II h and k) have the characteristic double-peaked shape depicted in Figure 2.2(a). The schematic line profile of that figure is somewhat unrealistic, however, since intensity tracings of emission cores in actual stellar spectra are rarely symmetric about the line center. That is, if I_v and I_R are the intensities at the K_2 (or k_2) emission peaks (see Figure 2.2(a)), then it is generally the case the $I_v/I_R \neq 1$. Several examples of asymmetric emission lines are presented in Figure 2.7, which shows Mg II h and k line profiles for the three supergiants β Aqr (G0 Ib), α Aqr (G2 Ib), and λ Vel (K5 Ib); as is apparent from the figure, in each case $I_v/I_R < 1$. Asymmetries with the opposite sense (i.e. $I_v/I_R > 1$) can occur as well; particular examples of this behavior are the mean quiet Sun profiles of Ca II K, Mg II h and k, and Lα (see Linsky, 1980a, and references therein).

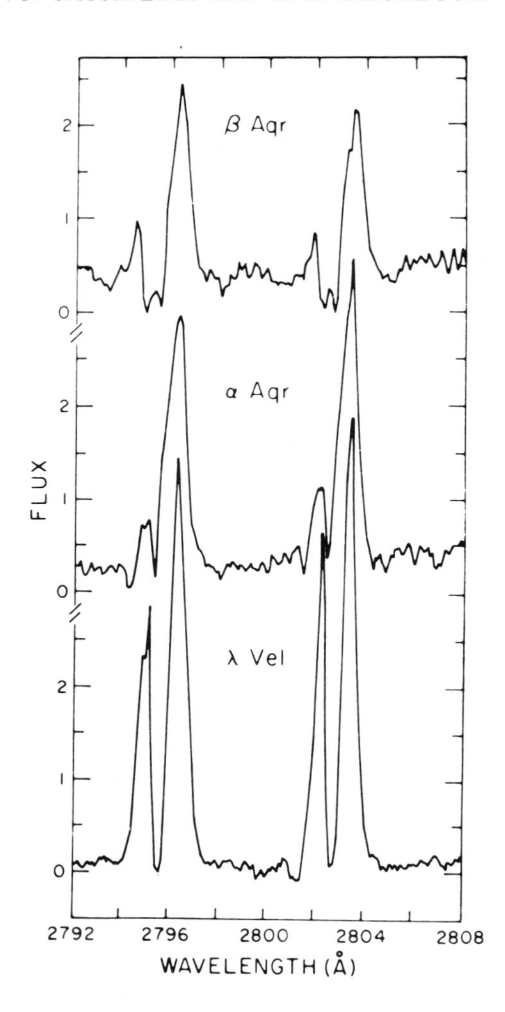

Fig. 2.7. Asymmetric Mg II h and k lines for three supergiant stars, as discussed in the text. (From Hartmann *et al.*, 1980.)

Hummer and Rybicki (1968) have interpreted asymmetries with the sense depicted in Figure 2.7 (i.e. $I_v < I_R$) as being the result of differential expansion of the stellar atmosphere. This can be seen in the following way. As discussed in Section 2.2.1, the central (K_3) absorption reversal is formed in the uppermost layers of the stellar chromosphere. If the chromospheric gas is accelerated outward (along the observer's line of sight), then the motion of these higher layers causes the K_3 feature to be blue-shifted relative to the wavelengths of the K_2 emission peaks. As a result, the intensity of the K_{2v} peak is diminished relative to that of the K_{2R} peak. This analysis therefore suggests that the observation of emission cores having $I_v < I_R$ is an indication of the collective outward expansion of the stellar chromosphere in the form of a wind. It is important, however, to qualify this statement in two respects. First, asymmetric line profiles such as those shown in Figure 2.7 may reflect chromospheric mass motions, but do not (by themselves) establish that material is physically escaping from the star. And second, the interpretation of the observed asymmetries presented above may not be unique. For example, in the case of the Sun, Athay (1970a, b, 1976) has emphasized that the observed I_v/I_R ratio for the Ca II K line can be produced by either upward or downward motions. Moreover, a number of authors (Cram, 1974, 1976; Heasley, 1975; Shine, 1975) have shown that the solar emission line asymmetries can be produced by vertically propagating acoustic waves or pulses (see Linsky, 1980a, for a more complete discussion of these effects).

With these caveats in mind, we note that emission line asymmetries as indicators of chromospheric flows and possible mass loss have been studied by Dupree (1976), Stencel (1978), Weiler and Oegerle (1979), Linsky *et al.* (1979), and Stencel and Mullan (1980), among others (see the references contained in Section 2.2.2); calculations of asymmetric Ca II K and Mg II k line profiles have been carried out by Chiu *et al.* (1977). In their survey of Ca II H and K line emission in 43 late-type stars, Linsky *et al.* (1979) found that the dwarf stars with measurable K_3 absorption features, $I(K_{2v})/I(K_{2R}) \geq 1$. As discussed above, an analogous asymmetry is seen in the mean solar Ca II K line, and is interpreted as being the result of either downflows in bright network regions or upwardly propagating acoustic waves. Among giants and supergiants, Linsky *et al.* (1979) found a wider range of values for the ratio $I(K_{2v})/I(K_{2R})$, with some stars showing $I(K_{2v})/I(K_{2R}) > 1$ and others showing $I(K_{2v})/I(K_{2R}) < 1$. Additional insight into the behavior of emission line asymmetries can be gained from the work of Stencel (1978). His observations indicate that the value of $I(K_{2v})/I(K_{2R})$ changes from > 1 for G and early K giants to < 1 for giants with spectral types later than about K3. More recently, Stencel and Mullan (1980) have measured fluxes in the Mg II k_{2v} and k_{2R} emission peaks for 47 cool giants and supergiants (spectral types G–M) and have found a similar transition in the value of $I(k_{2v})/I(k_{2R})$. These authors have delinated boundary lines in the H–R diagram across which the ratios I_v/I_R for the Ca II K and Mg II k lines change from >1 to <1. These lines (labelled 'C' and 'M', respectively, in Figure 2.5) are located in the same portion of the H–R diagram as the lines marking the apparent cessation of coronal and transition region emission (see Section 2.3). Moreover, their proximity to the boundary marking the onset of circumstellar absorption features in the spectra of late-type stars (see Section 2.5.2) suggests that the appearance of asymmetric emission cores having $I_v/I_R < 1$ is related to the occurrence of massive stellar winds. Note also that the vertical portion of the Mg II asymmetry line lies at somewhat earlier spectral types than does the corresponding portion of the Ca II asymmetry line. One possible interpretation of this behavior is as follows. In Section 2.5.2 it was noted that the coolest and most luminous stars appear to have the highest rates of mass loss. If all stars to the right of the Mg II asymmetry line in Figure 2.5 possess strong winds, this result suggests that $\dot{M}$ increases as one proceeds from the line labelled 'M' towards the line labelled 'C' in the diagram. Since the abundance of Mg is greater than that of Ca, a lower value of $\dot{M}$ is required to produce a wind column density (i.e. a line optical depth) sufficient to make the asymmetry visible in the Mg II lines. On this basis, the Mg II asymmetry should be visible at earlier spectral types (e.g. lower mass loss rates) than the corresponding Ca II asymmetry.

2.5.4. *Circumstellar Dust Shells*

Cool giants and supergiants are frequently observed to emit more radiation at infrared wavelengths than expected from a black body radiating at the stellar effective

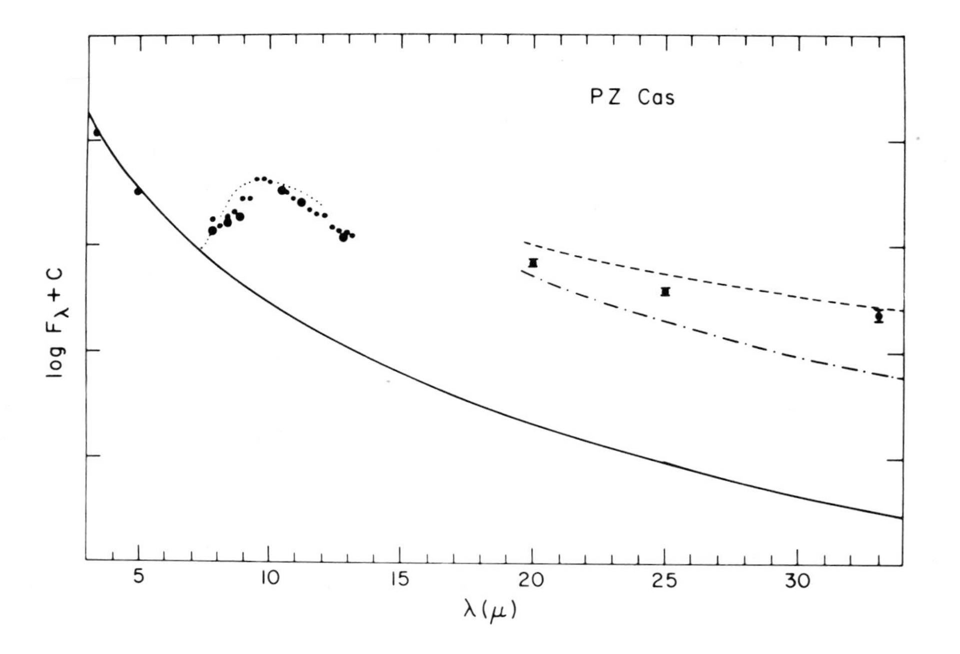

Fig. 2.8. The observed infrared flux distribution for the M supergiant PZ Cas. The solid line is the black-body curve extrapolated from the measured flux at wavelengths shortward of the emission feature at $\lambda \simeq 10$ μm. (From Hagen, 1980.)

temperature. In addition, infrared photometry often reveals the presence of a characteristic emission peak superimposed on the background continuum, with approximate central wavelength $\lambda \simeq 10$ μm. An example of these features is given in Figure 2.8, which depicts the observed infrared flux distribution for the M supergiant PZ Cas (M3 Ia). Infrared excesses and 10 μm emission were first detected by Woolf and Ney (1969) (see also Gillett *et al.*, 1968) in long wavelength spectra of the M supergiants α Ori (M2 Iab) and μ Cep (M2 Ia), and the Mira variables χ Cyg and *o* Cet. They attributed the observed infrared emission to dust grains distributed throughout a circumstellar envelope. Depending upon their composition, such grains are heated by the absorption of radiation from the stellar photosphere, and re-emit at infrared wavelengths causing the observed excess. Molecular equilibrium calculations by Gilman (1969) suggested that the grains surrounding oxygen-rich ($C/O < 1$) stars would be composed primarily of silicates (such as Mg_2SiO_4), while the grains around carbon stars ($C/O > 1$) would be largely graphite (or SiC; see Draine, 1981). Laboratory experiments indicate that optically thin samples of silicate-containing minerals have a broad band absorption feature near $\lambda = 10$ μm (see, e.g., Reimers, 1975). This result is in good agreement with the picture described above since oxygen-rich stars exhibit 10 μm emission features while carbon stars do not.

Subsequent observations have shown that excess infrared emission is present in the spectra of all stars cooler and more luminous than the line in the H–R diagram defined by M6 III, M5 II, M1 Iab, G8 Ia, and G0 la (Reimers, 1975) (for a more complete discussion of dust in circumstellar envelopes, see the reviews by Merrill, 1978, and Zuckerman,

1980). Since the region so defined lies within that portion of the H–R diagram where circumstellar absorption lines are observed (see Section 2.5.2), infrared measurements can provide additional information concerning mass loss rates for late-type giants and supergiants. In particular, because the height of the 10 μm emission peak depends upon the dust optical depth through the circumstellar shell, detailed analysis of this feature can yield an estimate for the dust column density. This can be done in the following way (Hagen, 1978; Castor, 1981). Assume that the dust is distributed throughout a circumstellar shell having the same geometrical and dynamical properties as the model adopted in Section 2.5.2. For specified grain parameters (i.e. composition, optical properties) radiative transfer techniques (see, e.g., Dyck and Simon, 1975; Apruzese, 1976; Jones and Merrill, 1976, and references therein) can then be used to fit the observed 10 μm emission feature and determine the dust optical depth $\tau_d(\nu)$ of the shell. The column density $\langle N_d \rangle$ of dust grains then follows from $\langle N_d \rangle = \tau_d(\nu)/(\pi a^2 Q_{abs}(\nu))$, where a is the grain radius and $Q_{abs}(\nu)$ is the absorption efficiency. For silicate grains, the total column density of circumstellar material $\langle N \rangle$ is obtained by multiplying $\langle N_d \rangle$ with the number of silicon atoms per grain and dividing by the fractional abundance of Si. With the wind velocity determined from measurements of circumstellar absorption lines, the mass loss rate follows from $\dot{M} = 4\pi R_{min}\, \mu \langle N \rangle V$ as before (see Section 2.5.2).

The procedure outlined above, together with a detailed analysis of circumstellar absorption features has been used by Hagen (1978) to derive mass loss rates for a number of M giants and supergiants (see Section 2.5.2). A somewhat different approach has been taken by Gehrz and Woolf (1971) who assume that the outflow is driven by the momentum imparted to the circumstellar gas due to collisions with radiatively accelerated grains. The mass loss rates derived by these authors for the M supergiants α Ori and μ Cep are 7×10^{-7} and 1×10^{-5} $M_\odot$ yr^{-1}, respectively. As was true in the case of mass loss rate estimates based on the analysis of circumstellar absorption lines, the results of methods ultilizing dust shell observations are also subject to considerable uncertainty. Among other things, the composition of the grains and their properties (i.e. size, temperature, absorption efficiency) are either not well known or difficult to determine. Moreover, the physical processes and conditions leading to grain formation are poorly understood (Salpeter, 1974a). In particular, it is not yet clear how close to the stellar surface stable grain condensation can take place (see, e.g., Weymann, 1978; Draine, 1981). In this regard, infrared observations of α Ori using the technique of heterodyne interferometry indicate that less than 20% of the measured dust emission originates at distances within 12 R_* of the star (Sutton *et al.*, 1977).

2.5.5. *Summary*

The results discussed in this and preceding sections can be summarized with the following qualitative picture of the thermal and dynamical structure of late-type stellar atmospheres (see Figures 2.9 and 2.10). Main sequence stars possess chromospheres, transition regions, and coronae whose structure and energy balance (but not always the level of activity) are probably quite similar to those of the analogous regions in the solar atmosphere. Like the Sun, these stars undergo mass loss in the form of hot, tenuous, and fairly high-velocity stellar winds. As one proceeds from the main sequence towards the giant and supergiant

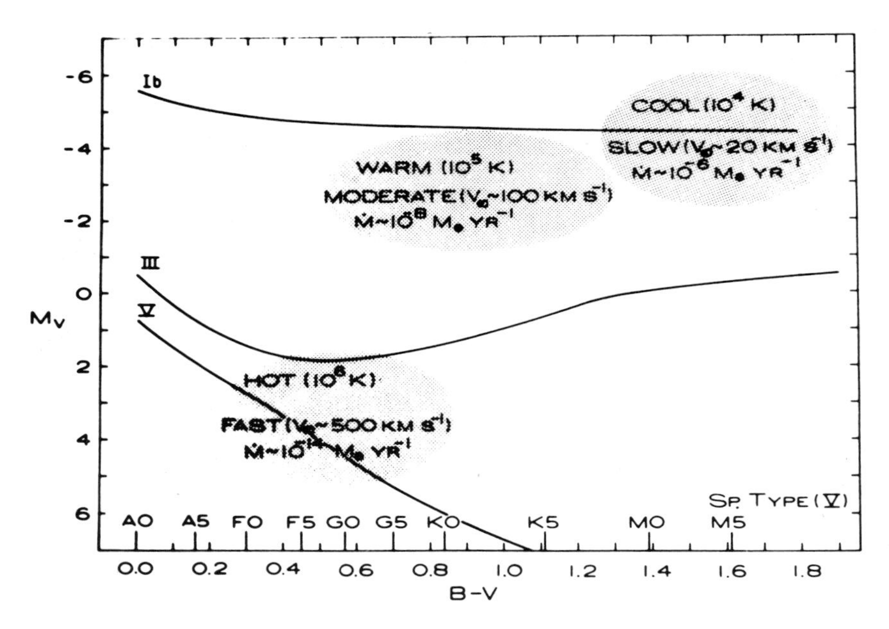

Fig. 2.9. Schematic representation of the differences in wind thermal and dynamical structure for stars in the cool portion of the H–R diagram. (From Dupree, 1981.)

branches in the cool portion of the H–R diagram, both the thermal and dynamical structure of the atmospheres under consideration change. Gas temperatures in the outer atmospheres of the coolest and most luminous supergiants never exceed values characteristic of the solar chromosphere or do so only over regions of limited spatial extent. These stars possess cool, massive winds with terminal velocities well below the gravitational escape speed at the stellar surface. Intermediate between the main sequence and supergiant stars are the so-called hybrid stars (see Section 2.3.1). The atmospheres of these stars contain gas at transition region temperatures, and observations of blue-shifted Mg II absorption features imply the existence of relatively high-velocity winds with mass loss rates substantially in excess of those appropriate to main sequence stars. The presence of broad emission lines due to Si III, C III, and C IV in spectra of the hybrid star α TrA (K4 II) has been interpreted by Hartmann *et al.* (1981) as evidence for an outflow in which $T \sim 10^5$ K throughout an extended region about the star. Alternatively, Linsky (1982) has proposed that the high-temperature line emission exhibited by hybrid stars originates in magnetically confined gas at the base of an otherwise cool outflow. Further observations are required to ascertain the spatial extent of transition regions in the atmospheres of hybrid stars.

2.6. Mass Loss Mechanisms for Late-Type Giants and Supergiants

2.6.1. *Overview*

A complete understanding of the dynamical structure of late-type stellar atmospheres requires the identification of the physical processes responsible for the observed wind

the solar wind (see, e.g., Jacques, 1977, 1978). Of particular importance in the present application is the fact that undamped Alfvén waves can drive a wind from stars having atmospheres too cool to undergo thermally driven mass loss (Belcher, 1971; Belcher and Olbert, 1975).

For isothermal flow from a star possessing a radially directed magnetic field, it is found that for wave energy fluxes in the range $10^5 - 10^6$ erg cm^{-2} s^1, the force due to Alfvén waves is capable of initiating mass loss at rates comparable to those inferred from observations of late-type giants and supergiants (cf. Hartmann and MacGregor, 1980). However, unless the waves are damped with a characteristic dissipation length L of the order of the stellar radius, the calculated wind terminal velocities are significantly higher than the observed values. This behavior is illustrated in Figure 2.11, which depicts wind solutions corresponding to the damping lengths $\lambda(\equiv L/R*) = 1$ and ∞ for a star having $M* = 16\,M_\odot$ and $R* = 400\,R_\odot$. For each solution, the values adopted for the total number density and magnetic field strength at the base of the wind are $N_0 = 10^{11}$ cm^{-3} and $B_0 =$ 10 G, respectively, and $T = 10^4$ K throughout the flow. With initial wave energy flux $F_0 = 3.4 \times 10^6$ erg cm^{-2} s^{-1} for both models, $\dot{M} = 5.5 \times 10^{-7}\,M_\odot$ yr^{-1} and $v = 408$ km s^{-1} in the case of undamped waves ($\lambda = \infty$), while for damped waves ($\lambda = 1$), $\dot{M} = 4.5 \times 10^{-7}\,M_\odot$ yr^{-1} and $v_\infty = 51$ km s^{-1}.

Although there are several plausible mechanisms for damping Alfvén waves in late-type winds (cf. Hartman and MacGregor, 1980), quantitative determination of the dissipation length L is precluded by the absence of detailed information regarding the wave periods and propagation directions, and the strength and geometry of the magnetic field. However, the wave dissipation rates required to explain the observed wind properties necessarily imply that the flow is heated as well as accelerated. For the supergiant model discussed above, a qualitative temperature estimate obtained by balancing wave heating and radiative cooling indicates that $T \sim 10^4$ K within a distance $\sim 2\,R*$ from the stellar surface. Because (for fixed F_0 and L) the density at a given location in the wind decreases with increasing surface gravity, the heating due to Alfvén wave dissipation results in higher temperatures for higher gravities. For $\log g > 2$, it is found that $T \gtrsim 10^5$ K within $\sim 2\,R*$ of the base of the wind, a value sufficiently high that thermal pressure is dynamically important. The model therefore predicts the presence of extended warm chromospheres around late-type supergiants, and suggests that the transition from tenous, coronal winds to cool, massive outflows in the H–R diagram (cf. Section 2.5.5) is a consequence of the gravity dependence of the wave momentum and energy deposition processes.

3. The Winds and Coronae of Early-Type Stars

3.1. Introduction

Rocket and satellite observations have provided a series of surprises over the past 15 years concerning the structure of the outer atmospheres of hot stars. Rocket ultraviolet spectra obtained by Morton (1967) and his collaborators showed that OB supergiants have massive high-speed winds (Morton *et al.*, 1968, 1969). The ultraviolet spectra obtained with the *Copernicus* spectra showed the presence of strong lines of unexpectedly high ionization stages, indicating that the winds are much hotter than had been expected

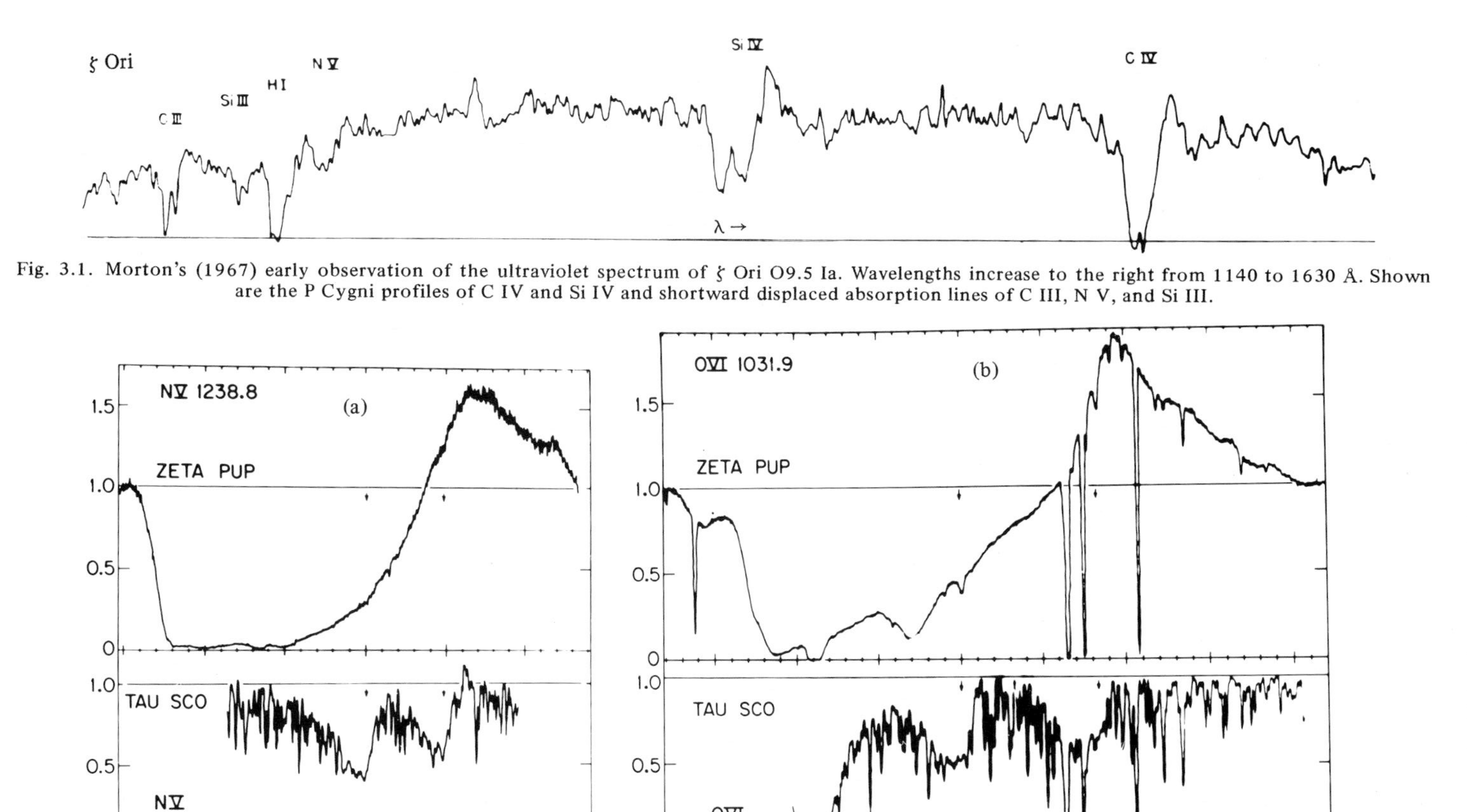

Fig. 3.1. Morton's (1967) early observation of the ultraviolet spectrum of ζ Ori O9.5 Ia. Wavelengths increase to the right from 1140 to 1630 Å. Shown are the P Cygni profiles of C IV and Si IV and shortward displaced absorption lines of C III, N V, and Si III.

Fig. 3.2. The spectra of ζ Pup O4*f* and τ Sco B0 V in the region of the resonance doublets of O VI and N V. The horizontal axes give the velocity (km s^{-1}) in the frame of the star. The arrows indicate the laboratory wavelength of the lines. In the O VI spectral region, the strong line at -1900 km s^{-1} is Ly-β. The sharp lines in the spectrum of ζ Pup are interstellar lines. (Adapted from Lamer[illegible] 976.)

(Lamers and Morton, 1976; Lamers and Rogerson, 1978). Most recently, the stars have been found to be X-ray sources, giving firm proof that coronal gas exists somewhere in the winds. All of this is occurring in stars that have no outer convection zone and had therefore commonly been thought to have no mechanism for producing acoustic or mechanical energy and the coronae that would arise as a consequence.

The rocket observations by Morton (1967) provided the first conclusive evidence for mass loss. Figure 3.1 shows a tracing of one of the early spectra of the supergiant ζ Ori (O9.5 Ia). The lines of C IV and Si IV show profiles with emission longward of line center and absorption extending far shortward of line center. Such profiles are called 'P Cygni' profiles after the prototype star P Cyg which shows prominant lines of this shape in the visible part of the spectrum. The lines indicate the presence of outward-moving gas. The broad ultraviolet P Cygni lines (see Section 3.2.1 for a discussion of the formation of these lines) indicate that the material is expanding from the stars with speeds of $\sim$2000 km s^{-1}. This is far above the escape speed (of 600 km s^{-1}) and is therefore strong proof that the stars are losing mass. Morton *et al.* (1968) estimated a mass loss rate of about 10^{-6} $M_\odot$ yr^{-1} on the basis of the strength of the lines. Thus the winds are quite extreme as compared to that of the Sun; the speeds are larger by a factor of 5 to 10; and the mass loss rates are larger by a factor of $\sim 10^8$ or more.

The relatively low stages of ionization that the observed in the flows and the high terminal speeds indicate that the Parker (1958) solar wind mechanism cannot be invoked to explain the winds. Ions such as Si^{+3} and C^{+3} should be destroyed by collisional ionzation at temperatures above a few times 10^5 K. This temperature limit is not consistent with a simple application of the solar wind mechanism, since it requires $T \geq 4 \times 10^7$ K to explain the observed wind speeds. Having ruled out hot coronal winds, Lucy and Solomon (1970) argued that matter is being driven away from the luminous stars by the gradient of radiation pressure that results from the formation of the observed P Cygni lines (see Section 3.2.1). They found that just the lines observed in the rocket ultraviolet could drive winds with speeds of 3000 km s^{-1} with mass loss rates of $\sim 10^{-8}$ $M_\odot$ yr^{-1}. Castor *et al.* (1975, 1976) extended the line driven wind model to account for the effects of many additional lines in the EUV and lines of less abundant elements. They found that it is possible for the lines to drive mass loss rates larger than 10^{-6} $M_\odot$ yr^{-1}, as is observed. It was assumed that the flow was in near radiative equilibrium, having a temperature of ~ 2–4×10^4 K. No extra heating by mechanical energy deposition was accounted for, nor was it thought necessary to explain the observed winds.

The Copernicus satellite, launched in 1973, extended the observation to a large number of hat stars and expanded the spectral region observed down to $\simeq 950$ Å (Snow and Morton, 1976; Snow and Jenkins, 1977). The survey showed that all early-type stars with $M_{\rm bol} < -6$ (i.e. $L > 2 \times 10^4$ $L_\odot$) are losing mass at a rate high enough to be detected ($M \sim 10^{-9}$–10^{-10} $M_\odot$ yr^{-1}), and that stars of lower luminosity lose mass only if their rotational speed is sufficiently large (v sin $i > 200$ km s^{-1}) (Snow and Marlborough, 1976; Marlborough and Snow, 1980). The survey also yielded an unexpected result. The O VI doublet at 1032, 1037 Å is much stronger than expected and is seen in essentially all the O stars and early B stars (earlier than B0.5). The ionization potential of O^{+4} to O^{+5} (at 114 eV), is in the far UV He II continuum and the stellar radiation field should be negligible, and therefore the O^{+5} ion could not form unless the wind temperature is higher than had been expected from line driven wind models (Lamers and Morton, 1976;

Lamers and Rogerson, 1978). The N V and O VI lines observed in ζ Pup (O4) and τ Sco (B0 V) are shown in Figure 3.2.

The discovery of the anomalously high ionization stages, or 'superionization stages', led to the realization that mechanical energy is being deposited in the winds (Cassinelli *et al.*, 1978). This could lead to an overall heating of the wind, as had been postulated by Lamers and Morton, or to the heating of a small coronal zone at the base of the wind. The radiation from a coronal zone could produce O^{+5} in the cool wind because of the Auger ionization by coronal X-ray (Cassinelli and Olson, 1979).

On the first exposures of the detectors on the Einstein X-ray Observatory in 1979, X-rays were detected from O stars (Harnden *et al.*, 1979; Seward *et al.*, 1979). Thus it is now clear that *very* hot gas exists somewhere in the outer atmosphere of early-type stars. The observational data have been coming in at a rapid rate over the past few years, and the theories to explain the presence of the hot gas have not kept step. We choose to discuss primarily the methods that are being used to derive the structure of the winds and coronae from observations made at X-ray, ultraviolet and radio wavelengths. Then we discuss some of the major mechanism that must be important in the wind dynamics, such as line radiation pressure gradient and the instabilities that arise because of this force.

3.2. THE VELOCITY AND MASS LOSS RATES DERIVED FROM LINE AND CONTINUUM OBSERVATIONS

There is evidence for outflow in several classes of early-type stars (shown in Figure 1.2); O and B main sequence stars, O*f* and OB supergiants, Be stars, Wolf-Rayet stars and central stars of planetary nebulae. This list covers a wide range of evolutionary status. A wide range of outflow properties is also seen, from the barely discernible winds of the Be stars to the massive flows from Wolf-Rayet and very luminous O*f* stars. The strong winds of the O*f* stars, OB supergiants and Wolf-Rayet stars have received the most attention thus far. This is partly because the winds are so readily studied through their strong P Cygni lines. Also, the mass loss rates are so large that the process can affect the evolution of the stars, and the kinetic energy of the massive fast winds is large enough to affect the interstellar environment around the stars. The study of these stellar winds and their effects has become one of the most active areas of astrophysical research.

As there is, as yet, no fully satisfactory theoretical explanation of the winds and coronae, we shall start with a discussion of the methods used to deduce the structure from observational data.

We would like to know the run of density, velocity, and temperature through the expanding atmospheres. It is usually assumed, for the sake of simplicity, that the flow is radial and the envelope is spherically symmetric so that ρ, v, and T are functions of radius, r, alone. In this case the velocity and density distributions are related by conservation of mass

$$\dot{M} = 4\pi\rho(r)v(r)r^2$$

so we would like to deduce the mass loss rate, $\dot{M}$, and $v(r)$ [*or* $\dot{M}$ and $\rho(r)$].

3.2.1. *The Fromation of P Cygni Profiles*

The most important diagnostic of the structure of the winds are the **P** Cygni profiles. The formation of the lines also plays an important role in the acceleration of the winds to high speeds, as it provides a mechanism for transferring momentum from the photospheric radiation field to the outflowing gas.

The formation of a P Cygni profile of a strong resonance line which has a rest wavelength, λ_0, as illustrated in Figure 3.3. Assume that the velocity $v(r)$ increases steadily in the outward direction and asymptotically reaches terminal velocity, v_∞. Assume, also, that the photosphere emits a smooth continuum of photons which must propagate through a

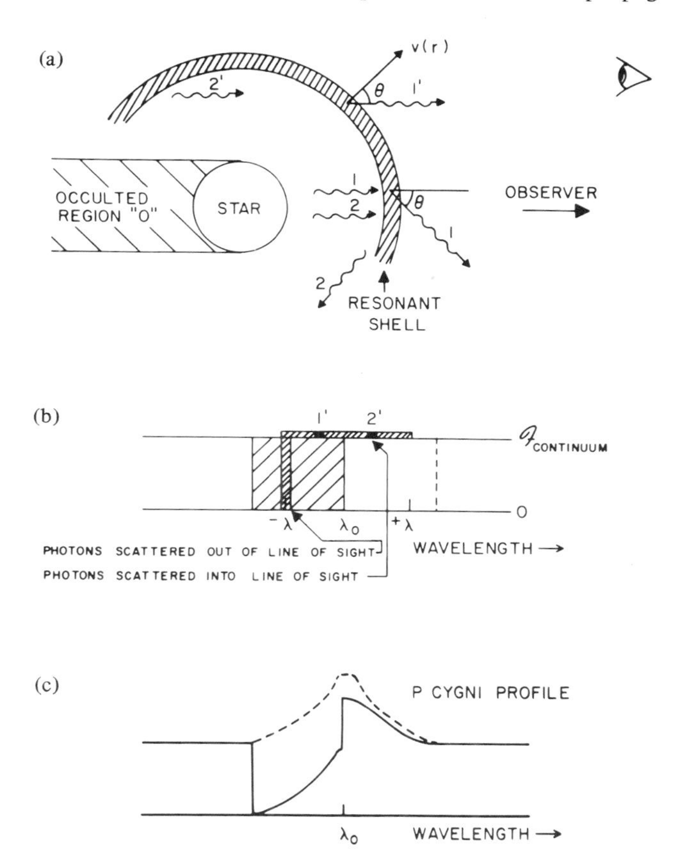

Fig. 3.3. Illustrates the formation of a P Cygni profile a resonance line. Figure (a) shows the scattering of two photons 1 and 2 *out of* the line of sight. In a spherically symmetric system, two other photons 1′ and 2′ will be scattered *into* the line of sight. As shown in (b), these will be red-shifted because of the expansion of the resonant shell. So photons are scattered from a region $-\lambda$ to $+\lambda$. This will produce the P Cygni profile shown in (c). The dashed line shows the distribution of re-emitted photons. The region labelled O corresponds to photons not seen by the observer because of the occultation of the star.

stellar wind that is capable of scattering photons whose wavelength corresponds to the resonance transition. A photon emitted radially from the photosphere at wavelength $\lambda < \lambda_0$ may be scattered at a shell in the wind at which the velocity has reached the value $v(r) = c(\lambda_0 - \lambda/\lambda_0)$. Because of thermal motions, the line absorption profile is not perfectly sharp. The transfer of the photon through the shell is described by the Sobolev escape probability method (Sobolev, 1960; Castor, 1970; Mihalas, 1978). The photon may scatter many times within the resonance shell and then escape in a direction, θ, relative to the radial direction. The wavelength of the scattered photon as seen by an outside observer will be red-shifted to λ' from its original wavelength λ by an amount determined by the angle θ; $\lambda' = \lambda + \lambda\ [v(1 - \cos\theta)/c]$. Thus (as designated in Figure 3.3), a photon 2′, which was last scattered from the back side of the envelope, is seen at a wavelength longer than λ_0, and a photon, 1′, which was last scattered from the front side, will be seen shortward of line center. If we assume that the ion is very abundant in the wind, all the photospheric photons shortward of line center, from λ_0 to $\lambda_0(1 - v_\infty/c)$, will be scattered out of the direct line of sight to the observer. Those which are scattered back to the photosphere are lost (or 'occulted' from the observer by the star, as illustrated in Figure 3.3(a)). The rest of the photons are redistributed, by the Doppler effect, into the band $\lambda_0(1 - v_\infty/c)$ to almost $\lambda_0(1 + v_\infty/c)$. The net effect is a depletion of photons at wavelengths shortward of line center and an increase in the photon flux longward of line center, thus producing the characteristic P Cygni shape.

We also see from this discussion that if a star has P Cygni lines with a sharp shortward edge, it is easy to determine the terminal velocity of the wind.

Figure 3.4 shows results of the study of terminal velocities for O and B stars. It is a plot of the ratio of v_∞ to the escape speed v_{esc} $[=(2GM(1 - \Gamma)/R)^{1/2}]$ versus spectral type. [Γ is the ratio of the outward acceleration of radiation to gravity, thus $(1 - \Gamma)$ is the effective deduction of gravity.] The data for the O stars is from Abbott (1978), who found that $v_\infty \simeq 3v_{esc}$, in agreement with a prediction of the Castor *et al.* (1975) theory for a reasonable value of a parameter, α, in the theory. The data plotted for the B supergiants is primarily from the International Ultraviolet Explorer (IUE) satellite survey of Cassinelli and Abbott (1981). It shows that v_∞/v_{esc} steadily decreases from 3 to values near 1 for late B supergiants. This could be due to a decreasing number of lines accelerating the flow. Panagia and Macchetto (1981) were able to explain the run of v_∞/v_{esc} by considering the effects of *multiple* scattering in the winds of O stars. In discussing Figure 3.3 we assumed photons to have just a single effective scattering. That is because once a photon escapes from the resonance shell it is never again resonant with the same transition. However, the photon can scatter again by, say, a different ion, in a line that lies somewhat longward of the original line (at λ_0). Panagia and Macchetto note that there is a particularly dense cluster of resonance lines in the spectral range 200–500 Å of ions that should be abundant in stars with $T_{eff} \geq 30\,000$ K. For these hotter stars there is a large photospheric flux at the 200–500 Å band. A photon scattered backward from the resonance shell for one line can find, at the other side of the envelope, another line that is suitably tuned for absorbing it again. Therefore multiple scattering can be an efficient accelerating process for stars with $T_{eff} > 30\,000$ K. We shall see later that multiple scattering must be invoked for Wolf-Rayet stars if their large mass loss rates are to be driven by line radiation forces.

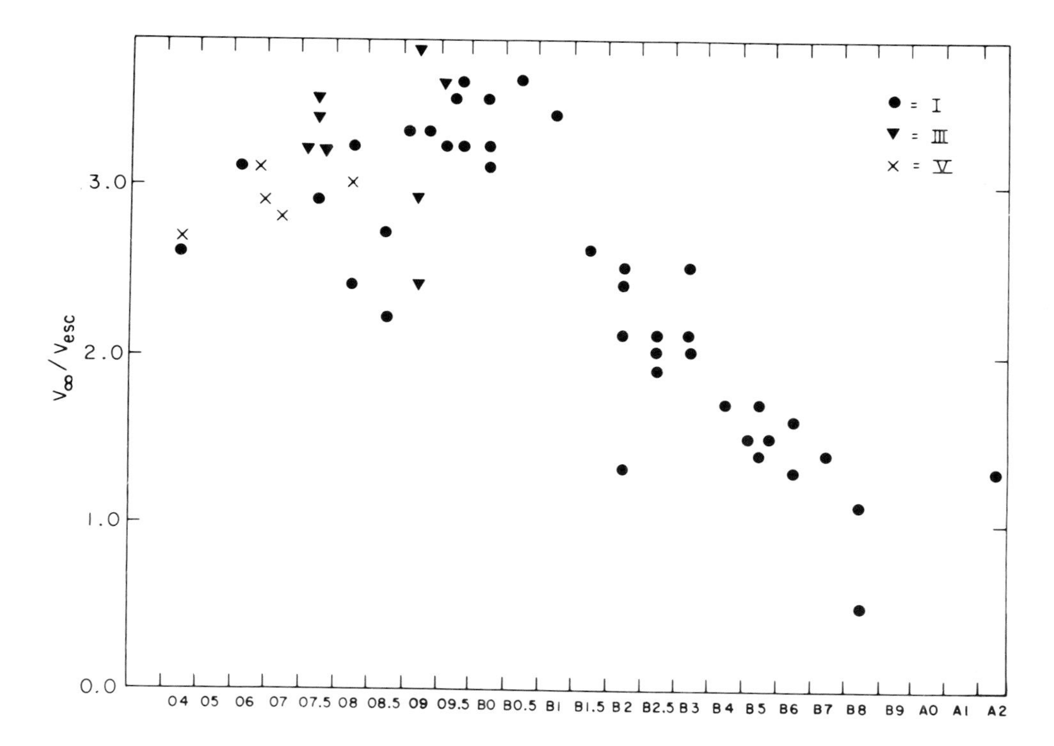

Fig. 3.4. The ratio of terminal velocity to escape speed as a function of spectral type from O4 to A2. The ratio is about 3 for the O stars (Abbott, 1978) and decreases towards unity for the later supergiants (Cassinelli and Abbott, 1981).

Thus far we have considered very strong lines. If the lines are not completely saturated, the determinations of the wind terminal speed is made more difficult, but other properties of the stellar wind can be deduced from unsaturated lines. The absorption at wavelengths, λ, shortward of line center depends on the optical depth of the resonant shell. Assuming the line has a Doppler velocity width, Δv_{D}, the effective geometrical thickness of the shell is $\Delta v_{\mathrm{D}}/(\mathrm{d}v/\mathrm{d}r)$ and the monochromatic optical depth is

$$\tau_\lambda = \kappa_{\lambda,L}\,\rho(\mathrm{r})\,\frac{\Delta v_{\mathrm{D}}}{\mathrm{d}v/\mathrm{d}r},$$

where ρ and $\mathrm{d}v/\mathrm{d}r$ are the density and velocity gradient at the resonant shell radius, and $\kappa_{\lambda,L}$ is the line opacity ($\mathrm{cm}^2\ \mathrm{g}^{-1}$). Let $g_i(r)$ be the fractional abundance of an ion stage of an element, which has an abundance, A, relative to hydrogen, then

$$\tau_\lambda = \left(\frac{\pi e^2}{mc}\right) f\,n_{\mathrm{H}} A g_i\,\frac{\Delta v_{\mathrm{D}}}{\mathrm{d}v/\mathrm{d}r},$$

where $(\pi e^2/mc)f$ is the cross-section of the transition. As an example, let's consider the

C IV $\lambda 1550$ resonance line, assume $r \simeq 1.4\, R_*$, $v = \frac{1}{2}v_\infty$, $dv/dr = \frac{1}{2}v_\infty/R_*$. Then we can relate τ_λ to the total electron scattering optical depth of the wind, τ_{es},

$$\tau_{\mathrm{C\,IV}} \simeq 10^5\, g_{\mathrm{C\,IV}} \tau_{es}$$

where τ_{es} is typically around 0.1 for O*f* stars and OB supergiants. So, if $g_i \geq 10^{-4}$, a moderately strong displaced absorption will be seen. The overall shape of the line will then depend on the optical depths through the sequence of resonance shells. Following Olson (1978), let us assume that the wind velocity distribution is known, $v^2 = v_0^2 + v_1^2(1 - R_*/r)$, then the profile shape depends only on the run of $g_i(r)$. If $g_i(r) \propto r^\beta$, we get quite different profiles depending on the index β, as is shown in Figure 3.5. The

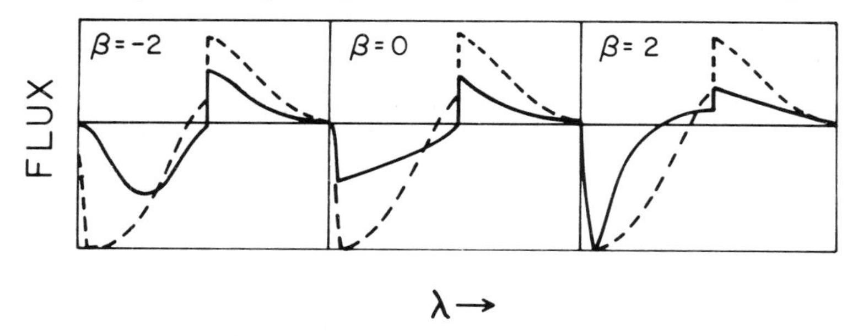

Fig. 3.5. The dependence of P Cygni profiles on the ionization abundance of a line producing ions in the wind. The ion abundance is assumed to vary with radius as $g_i \propto r^\beta$. The results for three different values of β are shown. If an ion decreases in abundance with radius, the absorption is at maximum depth for small velocity displacement. If an ion increases in abundance with r, the line may show little emission but a strong displaced absorption. The dashed lines show the results for a saturated line. Note these are all identical independent of β. (Adapted from Olson, 1978.)

profiles illustrate several interesting features. Note that if the line is saturated the profile is independent of β. As mentioned above, saturated lines are especially useful for deriving v_∞. If the line is not saturated, and if the ion has greatest abundance near the star (i.e. β is negative), the absorption occurs primarily at small velocity displacements. This is somewhat like the case of the Hα P Cygni lines seen in the optical portion of the spectrum. The line is sensitive to density because of recombination or collisional excitation effects, so it provides information concerning the velocity and temperature in the lower regions of the wind, and relatively little information about the terminal velocity and condition in the low-density outer regions. In the UV, profiles of this sort tend to occur in subordinate lines and lines of ion stages below the dominant stage. If, on the other hand, the index β is positive, the ion is concentrated in the higher velocity regions of the envelope and gives rise to a displaced absorption near v_∞. A common situation in which an ion can increase in abundance in the outward direction is that of an ion which is at a higher stage of ionization than the dominant stage. Consider the $i + 1$ stage in an atmosphere in which radiative ionizations are balanced by radiative recombinations.

$$\frac{n_{i+1}}{n_i} = \frac{W}{n_e \alpha} \int \frac{4\pi}{h\nu} a_\nu F_\nu \, d\nu,$$

n_e is the electron density, α is the recombination coefficient, a_ν is the ionization cross-section; πF_ν is the flux at the surface of the star; and W is the 'dilution factor' or the fractional solid angle at r subtended by the star. To a first approximation $W = \frac{1}{4}(R/r)^2$, and $n_e \propto 1/vr^2$, so

$$\frac{g_{i+1}}{g_i} = \frac{n_{i+1}}{n_i} \propto v(r)$$

and thus the ion increases in the outward direction in proportion to $v(r)$.

These simple examples illustrate that the profiles can be useful for deriving the ionization conditions in the wind. By comparing a number of profiles observed in a star with the theoretical profiles, it is possible to deduce the velocity and temperature structure of a wind. Significant progress has been made along these lines in the past few years. Families of theoretical profiles of ultraviolet resonance lines have been computed by Castor and Lamers (1979), Olson (1978), Olson and Castor (1981), Rumpl (1980), and others. Klein and Castor (1978) and Olson and Ebbets (1981) have calculated profiles for the interpretation of Hα lines. Olson (1981) has given special attention to the interpreation of subordinate lines. Hamaan (1980) derived a velocity and structure for the wind of ζ Pup from line profile data, and Lamers (1981) and Garmany *et al.* (1981) discuss estimates of mass loss rates from line profile fits.

3.2.2. *The Free–Free Continuum Energy Distribution of Hot Stars*

For the stars with massive winds, the infrared and radio continua are formed in the outflowing envelope. Observations in these spectral ranges have provided useful information concerning the density structure in the winds and on the mass loss rates of the stars. Continuum free–free opacity increases with wavelength ($\propto \lambda^2$), so at sufficiently long wavelength optical depth unity occurs within the wind itself. Thus we can say that the 'effective radius' of the star R_λ increases with increasing wavelength. This gives rise to an 'excess' of continuum radiation above that which would be expected from an extrapolation along black-body curve from the visual regions. The free–free continuum theory and observations of O stars, OB supergiants, and Wolf-Rayet stars are reviewed by Barlow (1979).

The basic features of the problem can be understood using the simple Eddington–Barbier relation as presented by Cassinelli and Hartmann (1977). The luminosity at any wavelength L_λ is approximately equal to the thermal emission from radial optical depth unity:

$$L_\lambda = 4\pi R_\lambda^2\, \pi B_\lambda\,(1) = 4\pi R_\lambda^2\,(2\pi c k T/\lambda^4), \tag{4}$$

where R_λ is the 'effective radius' (at $\tau_\lambda = 1$) and the temperature of the star is assumed to be hot enough that the Rayleigh–Jeans approximation of B_λ can be used. Given the density distribution, $\rho(r)$, and the temperature distribution, $T(r)$, the effective radius can be found from the optical depth integral:

$$\tau_\lambda = 1 = \int_{R_\lambda}^{\infty} \kappa_\lambda \rho\, \mathrm{d}r, \tag{5}$$

where

$$\kappa_\lambda = \kappa_0\, \rho T^{-3/2}\, \lambda^2\ (\mathrm{cm}^2\ \mathrm{g}^{-1}).$$

Let's first consider the application of this transfer model to the continuum at radio wavelengths. At these long wavelengths optical depth unity occurs typically at tens of stellar radii where the flow has reached terminal speed, and the temperature has reached an asymptotic value. From conservation of mass $\rho(r) = \dot{M}/(4\pi v_\infty r^2)$ and (5) can be integrated to give

$$R_\lambda^2 = \frac{1}{T} \kappa_0^{1/3} \lambda^{4/3} (\dot{M}/4\pi v_\infty)^{2/3} \tag{6}$$

and we see from Equation (4) that T cancels out (!). So we can write

$$\nu L_\nu = \lambda L_\lambda = \text{const} \times \lambda^{-5/3} (\dot{M}/v_\infty)^{4/3}. \tag{7}$$

This shows that observations of the continuum flux ($S_\nu = L_\nu/4\pi D^2$) of a star at distance D gives a direct measurement of $\dot{M}/v_\infty$ and, since v_∞ is well determined from the UV profiles, the radio flux gives a good measurement of $\dot{M}$ alone. The rates derived from radio flux observations are considered to be the most reliable because they are not sensitive to the assumptions about the wind model. In particular, the mass loss rates depend very little on the ionization in the wind, the temperature structure, or the velocity law $v(r)$ (Wright and Barlow, 1975). The rates derived from the UV and Hα lines depend strongly on these assumptions, as indicated earlier. Observations at 6 cm using the Very Large Array have been made of luminous O stars and OB supergiants by Abbott *et al.* (1980), and of the very luminous stars in the Cyg OB2 (VI Cyg) association by Abbott *et al.* (1981). Some of the results are shown in Figure 1.2.

At somewhat shorter wavelengths the effective radius occurs in the acceleration regions of the wind and $\rho(r)$ is not locally proportional to $1/r^2$. If we assume $\rho(r) \propto r^{-n}$ and $T \propto r^{-m}$, then

$$R_\lambda \propto \lambda^{-2/(-2n+3m/2+1)} \tag{8}$$

and

$$\lambda L_\lambda \propto \lambda^s, \tag{9}$$

where

$$s = \frac{6n - \frac{5}{2}m - 7}{-2n + \frac{3}{2}m + 1}.$$

Cassinelli and Hartmann (1977) have applied the results to study the effects of coronal regions on the infrared continua (Hartmann and Cassinelli, 1977).

The effects of the winds on the infrared and radio continuum are illustrated in Figure 3.6. An O*f* star should have a nearly plane-parallel photospheric region (Mihalas and Hummer, 1974), and the wind gives rise to a gradually increasing excess with longer wavelengths, as illustrated in Figure 3.6(a) (data from Cassinelli and Hartmann, 1977). Wolf-Rayet stars appear to have extended atmospheres on the basis of their flat energy

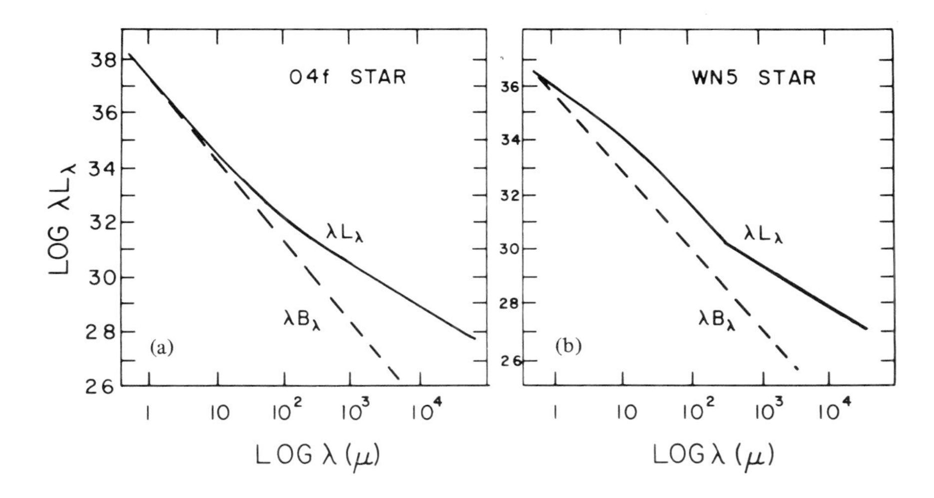

Fig. 3.6. The effects of a massive wind on the free–free flux from an O4 star and a Wolf-Rayet star. In absence of the wind the flux from the O4 star would resemble the Planck distribution B_λ. The flux in the radio region (6 cm) yields a good measure of the mass loss rate. The infrared 1–100 μ is sensitive to the density structure near the base of the wind. The data for the O4*f* star is from Cassinelli and Hartmann (1977) and that for the WN5 stars is from Van der Hucht *et al.* (1979).

distribution from thw UV though to the IR. In these stars, therefore, there is an excess over the black-body distribution even in the near IR (Figure 3.6(b); Van der Hucht *et al.*, 1979a, b) and the IR can be used to determine the run of $\rho(r)$. This has been done for Wolf-Rayet stars by Hartmann and Cassinelli (1977) and Hartmann (1978). Barlow and Cohen (1977) derived mass loss rates for a large numbers of OB stars from IR observations.

Figure 1.2 shows the mass loss rates for a variety of early-type stars. Most of the data on O stars and OB supergiants comes from the recent critical review of Lamers (1981). He used the radio observations of Abbott *et al.* (1981) to calibrate the rates obtained from the IR (Barlow and Cohen, 1977), UV (Gathier *et al.*, 1981), and H (Conti and Frost, 1977). The results for B main sequence and Be stars is from Snow (1982). The VI Cygni stars showing large mass loss rates were measured in the radio by Abbott *er al.* (1981).

3.3. Coronal Gas in Early-Type Stars

3.3.1. *Superionization of the Winds*

One of the startling discoveries of the Copernicus satellite survey was the presence of wide resonance lines of O VI in the spectrum of the B0 main sequence star τ Sco, that are shown in Figure 3.2 (Rogerson and Lamers, 1975). This indicated the presence of expanding gas in the envelope which is at a temperature far above the effective temperature of the star. The lines were also found with strong P Cygni profiles in OB supergiants as late as B0.5 I (Snow and Morton, 1976; Morton, 1979). Lamers and Snow (1978) showed that anomalously high stages of ionization such as Si^{+3} and C^{+3} are present in the envelopes of B supergiants as late as B8.

The first detailed analyses of the superionization in the expanding atmospheres of early-type stars were carried out by Lamers and Morton (1976) on the O4*f* star ζ Pup and by Lamers and Rogerson (1978) on τ Sco. A wide range of ionization is seen in the spectra from C III to O VI, and the 'warm wind model' was proposed to explain the observed ionization conditions (Lamers, 1979). In the warm wind model some type of mechanical energy propagates up from the photosphere and heats the gas to a temperature of $\sim 2 \times 10^5$ K in the O and early B stars and to temperatures of $\sim 8 \times 10^4$ K in the later B supergiants (Lamers and Snow, 1978). For τ Soc, the heating to temperatures of 2×10^5 K could initiate an outward acceleration because of radiation pressure on the O VI and N V lines. These ions do not have strong photospheric lines and the outward force would be very strong when the relative abundances, g_i, of these ions reached about 10%. Lamers and Rogerson proposed that the flow would then cool and radiative acceleration of other ions would drive to flow to terminal speed of 2×10^3 km s^{-1}. In the case of ζ Pup, Lamers and Morton found that the O VI P Cygni line extends to terminal velocity. This requires that the envelope be warm up to large distances from the star.

Cassinelli and Olson (1979) found objection to this geometrically extended heating because a mechanical luminosity of 2×10^{38} erg s^{-1} would be required to keep the wind at temperatures near 2×10^5 K. They proposed the 'corona plus cool wind model' as an alternative. In this model it is assumed that mechanical flux deposition heats up a thin coronal zone ($T > 10^6$ K) at the base of the wind. Hearn (1975) had previously suggested that the winds of hot stars could be initiated in a hot coronal region. He pointed out that in contrast with the solar case, the corona should be small in extent because in OB supergiants the density is 10^2-10^3 times as high. Thus the radiative energy losses, which are proportional to density squared, are 10^4-10^6 times greater than in the solar corona. Cassinelli and Olson (1979) assumed that the wind above the slab corona should have a temperature of $\sim$0.8 T_{eff} as would be appropriate for a gas in radiative equilibrium. X-ray from the corona should produce O^{+5} in the cool wind by the Auger ionization of O^{+3}. The model explains why the O VI line is seen as late in spectral type as B0.5 Ia and why N V and C IV are seen as late as B2.5 I and B6 I, respectively (Cassinelli and Abbott, 1981; Odegard and Cassinelli, 1982). It is because the Auger effect can produce a large number of ions two stages above the dominant stage of C, N, and O. Thus the O^{+3}, which is a dominant stage until B0.5, produces O^{+5} and so on (see Figure 1.1).

Figure 3.7 shows the ionization balance expected in the wind of ζ Pup based on the warm wind model and the corona plus cool wind model. A trace abundance of O^{+5} (i.e. $g_i \simeq 10^{-3.5}$) is sufficient to produce the strong line shown in Figure 3.2. In the warm wind model, the adjustable parameter is the electron temperature of the wind. Note in Figure 3.7(a) that the proper amount of the extreme ionization stages in the spectra, i.e. C^{+2} and O^{+5}, are produced at temperatures near log T_e = 5.25. For the coronal model, the abundance of high ions is determined by the X-ray luminosity beyond 0.6 keV. Cassinelli and Olson assumed a temperature of 5×10^6 K for the corona and adjusted the X-ray luminosity through the volume emission measure ($EM_c = \langle n_e{}^2 V \rangle$ of the corona. The effects on the ionization equilibrium in the wind are illustrated in Figure 3.7(b). The X-rays drastically increase the abundance of the high ion stages such as O^{+5} but have negligible effect on the lower stages. Therefore the observed abundance of a superionized

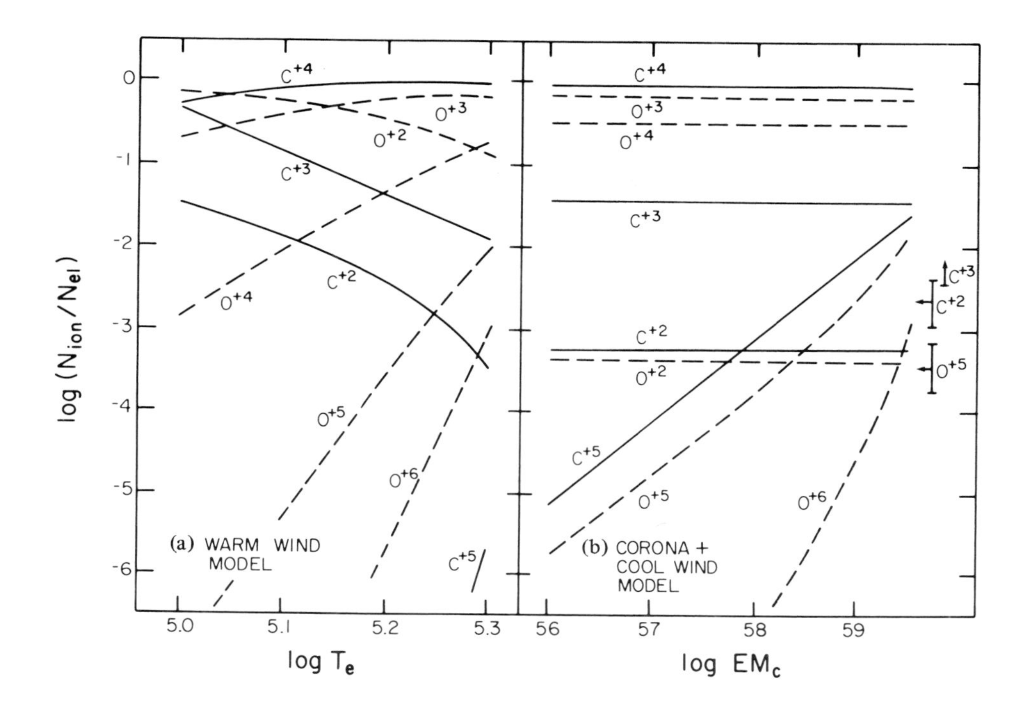

Fig. 3.7. The ionization fractions for oxygen and carbon at $2 R_*$ in the wind of ζ Pup (a) for an optically thin plasma of the warm wind model as a function of T_e and (b) for ionization due to coronal X-ray radiation that penetrates into a cool wind, as a function of the coronal emission measure. The observed ionization fractions of C^{+2}, C^{+3}, and O^{+5} are indicated to the right of (b). Either model produces about the right ionization abundance, but the corona plus cool wind model also predicts an observable X-ray flux (Cassinelli *et al.*, 1978).

state is a good diagnostic of coronal conditions, while the lower ion stages are adequately determined from the radiative equilibrium assumption. The model was used to predict that O stars and OB supergiants should be detectable with the Einstein X-ray satellite at a X-ray luminosity of $\sim 10^{32}$ erg s^{-1}. The cool wind should be optically thick between 0.05 and a few keV and thus the model also predicted that little X-ray flux would be seen below 1 keV.

3.3.2. *X-Ray Observation of Early-Type Stars*

The detection of X-rays is of course the clearest proof for the existence of very hot gas. The X-ray detectors on the ANS and HEAO-1 satellites failed to detect O and B stars, but the upper limits were consistent with the 10^{32} erg s^{-1} X-ray luminosity expected on the basis of the corona plus cool wind model. The major breakthrough came in 1979 when the Einstein Observatory (HEAO-2) was pointed at various points of the sky and discovered X-rays from a number of bright stars (Harnden *et al.*, 1979, Seward *et al.*, 1979). Figure 3.8 shows the results of the observations near η Carina using the High Resolution Imager (HRI) detector (Seward *et al.*, 1979). Apart from the peculiar object η

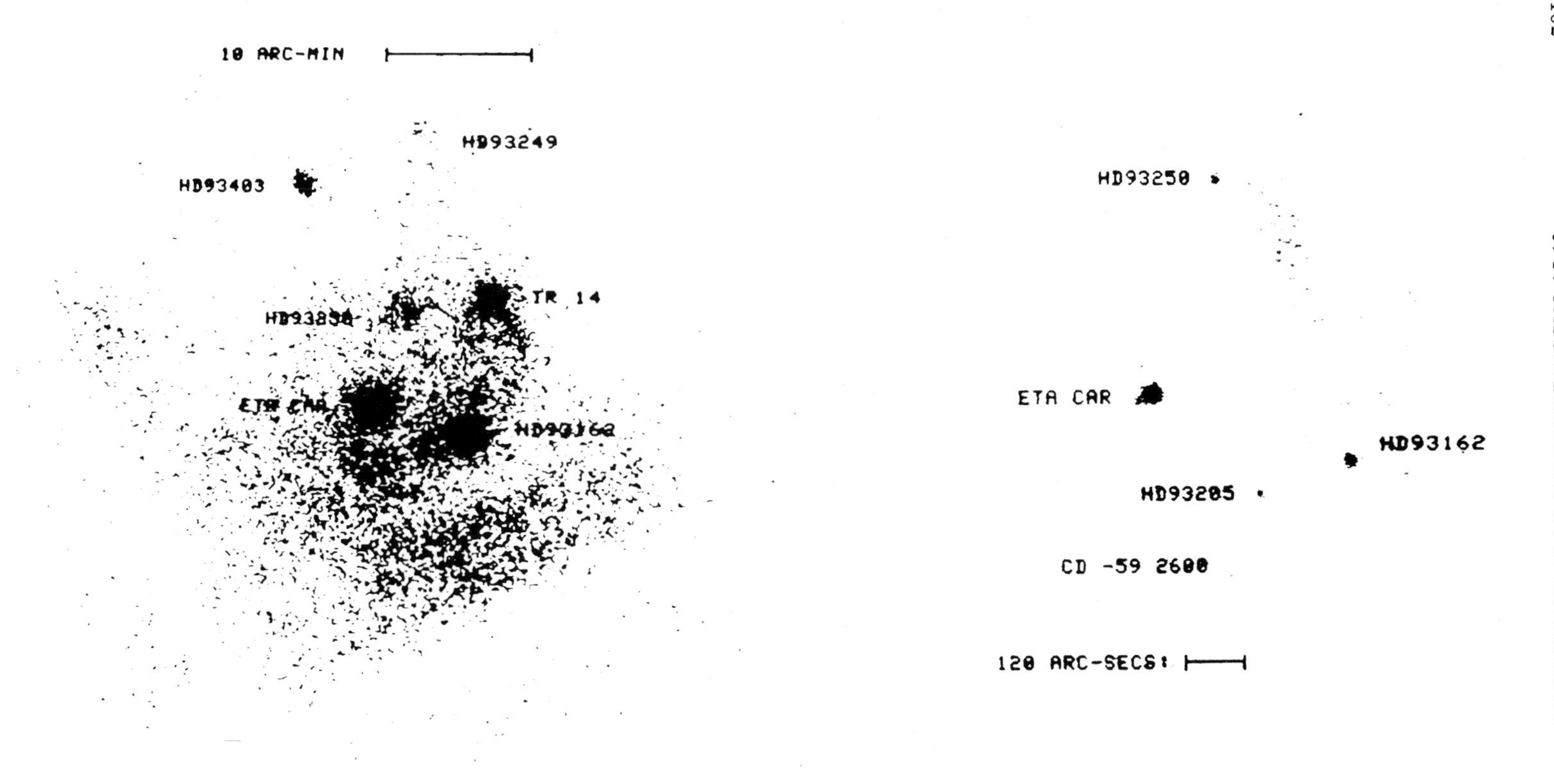

Fig. 3.8. The η Carinae region as observed by the Einstein Observatory. At the left is the low spatial resolution observation made with the IPC detector. The instrument window support structure shadows part of the field of view. At the right is shown the higher resolution observation made with the HRI. A number of bright O stars are here seen to be X-ray sources (Seward *et al.*, 1979).

Carina itself, the sources could be identified with a cluster of O stars, a Wolf-Rayet star and several individual O stars (that are also plotted on Figure 1.1).

The IPC observations provide some X-ray spectral information ($\Delta E/E \sim 1$ at 1 keV). The observations of OB supergiants and O*f* stars have indicated that the X-ray luminosity is about as expected from the corona plus cool wind model (Long and White, 1980; Cassinelli *et al.*, 1981; Stewart and Fabian, 1981), $L_X \simeq 10^{-7} L_{bol}$, but the energy distribution differs from what was expected. The slab coronal model predicted that there would be a severe attenuation of coronal X-rays at energies below 1 keV, because of absorption by C, N, and O in the cool stellar wind. The K shell ionization of oxygen is the dominant absorption process for $E > 0.6$ keV. This opacity decreases roughly as $1/E^3$ and the wind should become optically thin only at energies above 1–2 keV. Long and White (1980) and Cassinelli *et al.* (1981) have found, however, that there is a significant flux at energies below 1 keV. This indicates that the geometry proposed in the slab coronal model is inadequate. Three explanations have been proposed: (1) there must be X-rays throughout the flow, presumably in hot wisps (Long and White, 1980; Lucy and White, 1980; Lucy, 1982); (2) the base coronal region is somewhat more extended than was presumed in the slab coronal picture, and thus the attenuation is reduced (Waldron, 1980); (3) the source of X-rays is spatially confined and geometrically separated from much of the overlying attenuation (Rosner and Vaiana, 1980) (a model proposed from analogy with solar magnetic regions and solar coronal hole mass outflow). The strongest evidence that the slab coronal geometry is inadequate is shown in Figure 3.9. This shows results for one of the hot stars that were observed using the Solid State

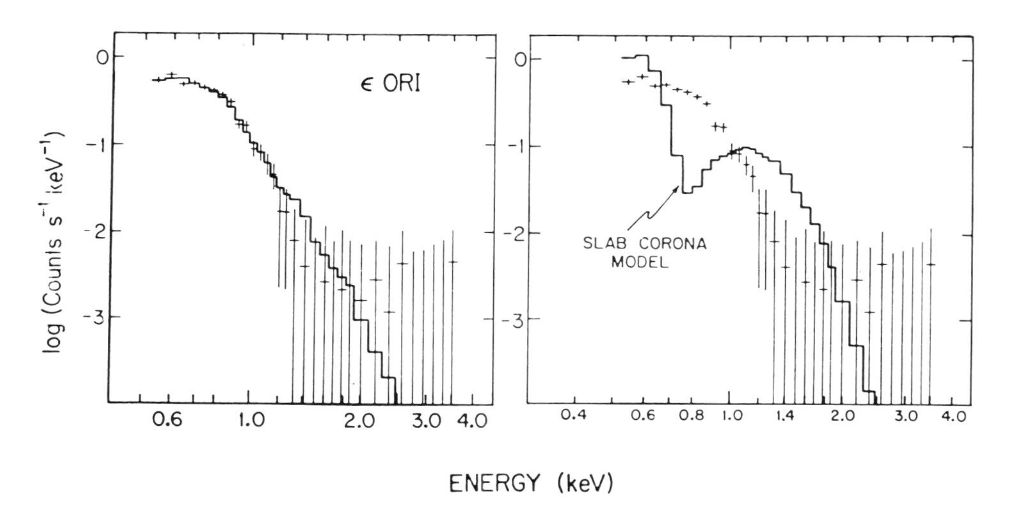

Fig. 3.9. X-ray spectral observations of ϵ Ori made with the Solid State Spectrometer on the Einstein satellite. The resolution is 160 eV and the error bars are indicated at each energy bin. The solid line shows the fit to the observed spectrum by embedded coronal source models. In (a) is shown the best fit model which corresponds to a source temperature $T = 3.3 \times 10^6$ K at an attenuation column density $N = 2.3 \times 10^{21}$ cm^{-2}. In (b) is shown the poor fit provided by a model with the source at the base of the cool wind, i.e. at $N = 2.7 \times 10^{22}$ cm^{-2}, with $T = 1.7 \times 10^6$ K. The additional soft X-ray flux in the observed spectrum indicates that X-rays are emitted from regions well above the base of the wind.

Spectrometer (SSS) on the Einstein satellite (Cassinelli and Swank, 1983). The observed photon fluxes are indicated with the crosses, the height of which corresponds to 1 σ uncertainties. Frame A is the best fit single component model. This corresponds to a hot zone with $T = 3.3 \times 10^6$ K, $EM = 2 \times 10^{55}$ cm^{-3} and an intervening absorbing column of only 2.3×10^{21} cm^{-3}. This column density is only about 10% that to the base of the wind – appropriate for a base coronal model. Figure 3.9(b) shows the completely unacceptable fit that results if the coronal radiation is constrained to the base of the wind, $N_H = 2.3 \times 10^{22}$ cm^{-2}. Obviously a great deal of the X-ray emission is arising from regions above or geometrically separated from the cool wind absorbers.

3.4. Wind Dynamics

For hot stars, the photospheric radiative flux is large in the ultraviolet portion of the spectrum, where there are many strong resonance lines. If the line opacity can be Doppler displaced into the strong continuum near the lines, the outward acceleration on the atmosphere can greatly exceed the inward acceleration of gravity. It is generally agreed that radiation forces on line opacity are responsible for accelerating the flows of the luminous early-type stars, from supersonic to the large observed terminal speeds. Now that there is good evidence for the presence of gas at coronal temperatures, it is not clear what causes the flow to be driven up to supersonic velocities, so that the radiation forces can produce the final acceleration.

Lucy and Solomon (1970) and Castor, Abbott, and Klein (CAK, 1976) argued that the Parker solar wind theory could not be responsible for the high terminal speeds, and they proceeded to develop fully radiatively driven wind models. Hearn (1975) showed that their arguments did not rule out a 'hybrid' model, however. Hearn proposed that the mass loss is initially a Parker coronal wind, heated by mechanical energy deposition at the base of the wind. To explain the large mass loss rates the sonic point of the Parker solution must be near the star. This means the temperature of the coronae must be in the range $5–9 \times 10^6$ K. Hearn also pointed out that a corona would be small in extent because of the radiative cooling that would occur in the high-density winds. In the recombination region at the top of the hot corona, strong outward acceleration by radiation forces could drive the flows to high speeds.

The discovery of hot X-ray emitting gas does not necessarily mean that the observed outflows are coronally driven, or even that there is a strong flux of acoustic or mechanical waves from the photosphere. The hot gas could be confined to magnetic loops, and the wind be driven by radiation forces in the 'hole' regions between the magnetic structures. Also, several studies have indicated that line driven winds are unstable. Thus, presumably, even if the star were acoustically quiet, the instabilities could grow and produce hot has in shocks in the line driven wind (Lucy and White, 1980; Lucy, 1982)

Recent theoretical studies have proceeded along three main lines: (1) Purely radiatively driven wind models have been developed more fully by Abbott (1977, 1980), Castor (1979), and Weber (1981). (2) Hybrid models with coronae and outer line driven wind zones have been calculated by Hearn and Vardavas (1981) and Waldron (1980, 1981). (3) Instabilities in line driven winds have been studied by Hearn (1972), Nelson and Hearn (1978), MacGregor *et al.* (1979) and Carlberg (1980). Mechanisms for the production of X-ray producing shocks from these instabilities have been studied by Lucy and

White (1980) and Lucy (1982). In addition to these models, all of which rely on radiative acceleration, there is recent work by Nerney (1980) on the possibility that some early-type stars (late B supergiants and Be stars) could have magnetically driven winds.

In this section some of the basic assumptions and theoretical predictions will be explored. We start with the line driven wind theory, then discuss the various instabilities and their consequences.

3.4.1. *Radiation Forces on Line Opacity: Momentum Deposition Considerations*

The process that leads to the production of the P Cygni lines also leads to a transfer of momentum from the stellar radiation field to the outflowing gas. How large a mass loss rate can be driven by this process? An often quoted limit is the 'single scattering limit', $\dot{M}_1$. To derive this limit, it assumed that every photon from the photosphere is scattered by one resonance shell. Assuming the radiation to be scattered isotropically from a shell, the net momentum of $h\nu/c$ is transferred to the gas. Now equating the total momentum from the photosphere L/c to the final momentum loss in the wind $\dot{M}v_\infty$, we get the single scattering upper limit

$$\dot{M}_1 = \frac{L}{v_\infty c} \, .$$

This is in the range 10^{-6} to 10^{-5} $M_\odot$ yr^{-1} for the luminous O and B stars.

Of course, the photons are not destroyed by the scattering process, but only slightly red-shifted and sent off in different directions. The photons can be scattered again and again, and it is possible, in principle, that a mass loss rate several times larger then the 'limit' $\dot{M}_1$ can be produced by the outward diffusion of photons. A particularly perplexing problem for radiation driven wind theory concerns the Wolf-Rayet stars which have mass loss rates about an order of magnitude larger than $\dot{M}_1$. It is not yet clear whether these could explained by accounting for multiple scattering.

Another important constraint concerning the deposition of radiative momentum has recently been discussed by Abbott (1980). The outward flowing radiative momentum must not only eject matter from the star but must also provide support for the extended envelope against the inward pull of gravity. A star with a slower outward increase in velocity $v(r)$ uses a larger fraction of its radiative momentum for this support than does a star with a rapidly accelerating wind. P Cygni, ζ^1 Sco and Wolf-Rayet stars appear to have slowly accelerating winds (Van Blerkom, 1978; Rumpl, 1980). To study the efficiency of the radiation field in ejecting mass from a star, we integrate the momentum equation over the mass of the outflowing gas ($\mathrm{d}m = 4\pi r^2 \rho \, \mathrm{d}r$) from the photosphere, at R_p, through the sonic radius, R_s, to infinity.

$$\int_{R_p}^{\infty} 4\pi r^2 \rho v \frac{\mathrm{d}v}{\mathrm{d}r} \, \mathrm{d}r + \int_{R_p}^{R_s} \left[\frac{1}{\rho} \frac{\mathrm{d}P}{\mathrm{d}r} + \frac{GM}{r^2} (1 - \Gamma) \right] \mathrm{d}m + \int_{R_s}^{\infty} \frac{1}{\rho} \frac{\mathrm{d}P}{\mathrm{d}r} \, \mathrm{d}m$$

$$+ \int_{R_s}^{\infty} \frac{GM(1 - \Gamma)}{r^2} \rho 4\pi r^2 \, \mathrm{d}r = \int_{R_p}^{\infty} \rho g_L \cdot 4\pi r^2 \, \mathrm{d}r. \qquad (10)$$

The first integral gives the rate at which momentum is ejected from the star $\dot{M}v_\infty$; the second integral is nearly zero since the subsonic structure is nearly hydrostatic; the third integral is nearly zero because beyond the sonic point the gas pressure is negligible. Using $\Gamma = \sigma_{es} L/4\pi cGM$ and $\tau_h = \sigma_{es} \int_{R_s}^{\infty} \rho \, dr$ (the optical depth of the supersonic 'halo') in the fourth integral, (10) becomes

$$\dot{M}v_\infty + \frac{L}{c}\left(\frac{1-\Gamma}{\Gamma}\right)\tau_h = B\frac{L}{c}, \tag{11}$$

where B is the 'fraction' of the photospheric luminosity intercepted by lines. Abbott (1980) and Panagia and Macchetto (1981) find that, when the support of the envelope is accounted for (i.e. the second term in (11)), the single scattering limit, $B = 1$, is exceeded for O4f stars and P Cygni. Therefore, multiple scattering must be occurring if the winds are indeed radiatively driven. The observational estimates of B using the observable ultraviolet portions of the spectrum does not exceed 20–30% (Panagia and Macchetto, 1981). Therefore the multiple scattering must occur in the EUV portions of the spectrum. Figure 3.10 shows the blocking of the EUV and UV continuum by lines in the wind of a 4×10^4 K star (CAK, 1976). The figure illustrates that there is a very large number of strong lines in the EUV and it is quite plausible that a photon could be scattered several times before escaping from the wind. The effects of the backscattering on the photospheric flux because of the stellar winds has not been accounted for in stellar atmosphere models. Work along those lines is now underway (Hummer, private communication).

3.4.2. *Radiative Acceleration*

The outward acceleration of a line driven wind is proportional to the flux at the line wavelengths. Lucy and Solomon (1970) considered the acceleration on one line

$$g_L = \frac{\kappa_L}{c}\int \pi F_\nu \phi(\nu)\, d\nu, \tag{12}$$

where $\phi(\nu)$ is the line profile function, πF_ν is the flux at frequency ν, and κ_L is the line opacity. To get a rough estimate of the magnitude of the line acceleration, Lucy and Solomon assumed the flux to be the unattenuated photospheric flux (i.e. $F_\nu = F_\nu^c$), and considered the opacity of one strong line (C IV λ1550) they found that the outward acceleration, g_L, exceeded the inward acceleration of gravity, g, by a factor of ~100.

Marlborough and Roy (1970) showed that if such a strong outward acceleration were to be 'turned-on' at some height in the subsonic flow, it would not drive the flow to large supersonic speeds, as might be expected, but would lead to a compression and deceleration of the flow! Lucy and Solomon (1970) and Castor (1974) showed that the line force does not simply turn on suddenly. The resonance line opacity also is present in the photosphere and is saturated ($F_\nu \ll F_\nu^c$), therefore the acceleration does not become large until the shortward wing of the line desaturates because of the gradual acceleration of the atmosphere.

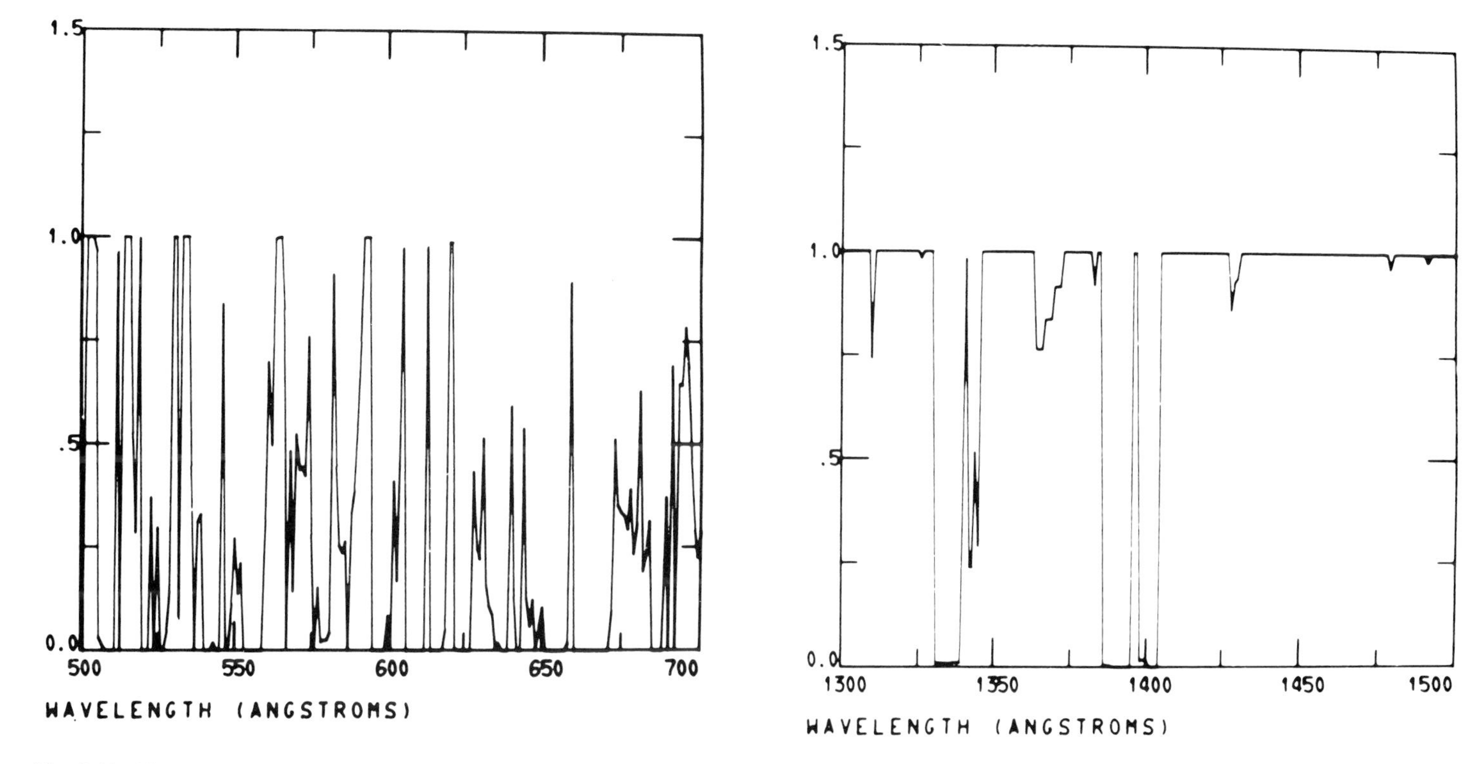

Fig. 3.10. Illustrates the possible blocking of stellar radiation by lines in the EUV and UV regions, in the wind of an O4 star with a mass loss rate of $7 \times 10^{-6} M_{\odot}\,\mathrm{yr}^{-1}$. Re-emission of the radiation is not accounted for, so the flux shown illustrates the density and strength of the lines in these two spectral regions (Castor *et al.*, 1976).

Castor (1974) developed an expression for the acceleration of the line that is valid in the supersonic regions

$$g_L = \kappa_L \frac{\pi F_\nu^c}{c} \Delta\nu_D \frac{1 - \exp(-\tau_L)}{\tau_L}$$

$$\approx \kappa_L \frac{\pi F_\nu^c}{c} \Delta\nu_D \min\left[1, \frac{1}{\tau_L}\right], \tag{13}$$

where F_ν^c is the continuum flux at line center frequency, $\Delta\nu_D$ is the Doppler width of the line, and τ_L is the optical depth of the shell through which the velocity change equals the thermal velocity, i.e.

$$\tau_L = \kappa_L \rho \frac{v_D}{dv/dr}. \tag{14}$$

Thus we see that for strong lines g_L is proportional to the velocity gradient dv/dr, while for weak lines ($\tau_L < 1.$), g_L is independent of dv/dr. The fundamental approximation of CAK was to write the acceleration due to all of the lines, some of which are optically thin and some of which are optically thick, as a power of the quantity $1/\rho(dv/dr)$. That is, for the sum of all lines

$$g_L = \sum g_{L,i}$$

CAK used

$$g_L = \frac{\sigma L}{c(4\pi r^2)} k \left(\frac{1}{\sigma\rho v_D} \frac{dv}{dr}\right)^\alpha.$$

where σ is the electron scattering opacity per gram. CAK and Abbott (1977) used extensive lists of line strengths to find values for the fitting parameters k and α. ($k \simeq 1/30$, $\alpha \simeq 0.7$ for the opacities in an O4f star.) In this representation for the line acceleration the momentum equation, assuming isothermal conditions for simplicity, becomes

$$\frac{1}{v}\frac{dv}{dr}(v^2 - a^2) + \frac{GM}{r^2}(1 - \Gamma) - \frac{2a^2}{r} - \frac{\sigma L k}{4\pi c}\left[\frac{4\pi}{\sigma v_D \dot{M}}\right]^\alpha \left(r^2 v \frac{dv}{dr}\right)^\alpha = 0. \tag{15}$$

In contrast with the Parker solar wind equation, the CAK momentum equation is not linear in dv/dr. This leads to several interesting new features. The equation does not have a singularity at the sonic point where $v = a$. The solutions of the equations do not have the X-type topology familiar from solar wind theory. There is, however, a unique solution corresponding to the transition from low-speed flow near the star to rapid flow far from the star. This solution passes through a 'critical point' that is discussed in some detail by Abbott (1977, 1980) and Cassinelli (1979). The critical point is at a radius r_c that is farther from the star than the sonic point, and the velocity, v_c, at the critical point tends to be much large than the sonic velocity. CAK derive expressions for two auxiliary expressions that must be satisfied at the critical point, analogous to the vanishing of the numerator and denominator in Parker's solar wind equations, and asymptotic expressions for the flow structure are derived from expansions about the critical solution. The asymptotic velocity structure satisfies the expression

$$r^2 v \frac{dv}{dr} = \frac{\alpha}{1-\alpha} GM(1-\Gamma), \tag{16}$$

giving, on integrating from the sonic radius, r_S, outward

$$v^2 = v_S^2 + \frac{\alpha}{1-\alpha} \frac{2GM(1-\Gamma)}{r_S} \left(1 - \frac{r_S}{r}\right). \tag{17}$$

so that

$$v_\infty = \sqrt{\frac{\alpha}{1-\alpha}}\, v_{esc}. \tag{18}$$

This equals the empirically derived expression $v = 3v_{esc}$ (Abbott, 1978) for O stars (Figure 3.5) if $\alpha = 0.9$, and can decrease to the result $v = v_{esc}$ for the later B supergiants if α decreases to 0.5.

The mass loss rate is given by the CAK theory as a function of k, α, M, and L. For an Of star with $T_{eff} = 5 \times 10^4$ K, and using the CAK values $k = 1/30$, $\alpha = 0.7$, we obtain $\dot{M} = 7.9 \times 10^{-6}\ M_\odot\ yr^{-1}$, in good agreement with observationally estimated mass loss rates. The first extensive survey of mass loss rates for luminous early-type stars was that of Barlow and Cohen (1977) from IR observations of 44 O, B, and A stars. The rates were found to corrolate with luminosity as $\dot{M} \sim L^{1.15}$. These observations were considered a strong argument for the CAK theory which predicts $\dot{M} \sim L^{1/\alpha}$ ($\sim L^{1.15}$ if $\alpha = 0.80$) and a weak dependence on gravity. Recent IUE observations suggest, however, that there is a strong dependence on gravity for stars on and near the main sequence (Lamers, 1981), so the simplest version of the CAK theory appears to be in need of improvement, at least.

Weber (1981) has shown that the CAK treatment of the radiation force is not adequate for the low-velocity portions of the winds, where the velocity and velocity gradients are low and the Sobolov approximation is not valid. Weber has calculated line driven wind models using the line transfer technique of Mihalas *et al.* (1975). This is a computational method for solving the radiation transfer equation in the fluid frame of a spherically symmetric expanding atmosphere, and it is accurate for both subsonic and supersonic flows. Weber found that the flow velocity increases more slowly with radius than had been the case in the CAK models. Perhaps a more important conclusion is that massive stellar winds could indeed be initiated by line radiation forces. This question of the initiation of the wind had been a source of contention since the line driven theory was first developed. The CAK argument was not fully convincing since it was based on a transfer technique that is invalid in the low-velocity regions. Of course, Weber's result does not mean that the winds of hot stars actually are driven by the line forces but only that, in principle, they could be.

Other characteristics of radiation driven winds have been studied by Abbott (1977, 1980) and Castor (1979). Abbott (1980) shows that the run of $v(r)$ deduced from CAK theory is very sensitive to rotation, overlapping lines, and nonradial radiation transfer. These all tend to make the increase of $v(r)$ less steep than expected from the CAK models. Abbott (1980) also has stressed the similarities of CAK solutions to the Parker and nozzle flow solutions. For example, he shows that the critical point velocity equals the speed of radiatively driven acoustic wave modes. Thus the critical point is the point farthest downstream that can communicate to all parts of the flow. This result has cleared up a conceptual problem with the CAK theory, which showed the solution was controlled by conditions at the critical point, yet that point, being in the supersonic region, could not communicate upstream.

3.4.3. *Instability of Line Driven Winds and the Consequences*

Lucy and Solomon (1970) pointed out a possible instability in their newly developed line driven wind theory. An outward velocity perturbation on a parcel of the gas would be amplified because of the Doppler shifting of the line opacity into frequencies at which there would be a higher flux. They suggested that this 'line shape instability' (Carlberg, 1980) could lead to the production of the high ion stage N^{+4} that had been seen in the rocket spectra of OB stars. Another heating mechanism was proposed by Hearn (1972, 1973). He suggested that heating of the upper photospheric regions (subsonic regions) could occur in hot stars by a mechanical flux originating from sound waves amplified by continuum radiation forces. This 'radiation driven' sound instability arises because bound–free continuous opacity increases with density, and energy can be fed into the wave by photospheric radiation during compression phases. Hearn (1973) proposed that these waves would grow to become a train of shock waves.

Several observations brought on an increased interest in the study of the heating of winds by radiation driven instabilities. The discovery of the strong P Cygni lines of O VI seen in the Copernicus spectra indicated heating was certainly occurring in some form (Lamers and Morton, 1976). Short time-scale variability (1 day) was discovered in line profile observations (Brucato, 1971; York *et al.*, 1977, Snow and Hayes, 1978; Wegner and Snow, 1978). These suggested the presence of small-scale inhomogeneities in the winds. Finally, the discovery of X-ray emission in early-type stars (Harnden *et al.*, 1979; Seward *et al.*, 1979) proved conclusively that very hot gas is present in the winds. The excess of detected soft X-rays in the 0.5–1 keV bound (Long and White, 1980; Cassinelli *et al.*, 1981; Cassinelli and Swank, 1981) suggested that some of the hot gas is in the outer regions (supersonic regions) of the winds.

Carlberg (1980) summarizes the recent work on studies of the stability of line driven winds by Nelson and Hearn (1978), MacGregor *et al.* (1979), Martens (1979), and others. Carlberg (1980) has carried out the most complete analysis to date. He applied a completely linearized analysis to a realistic unperturbed CAK model of a radiation driven stellar wind and identified three instabilities: the two mentioned above and a 'density gradient instability'. In this instability, an increased density perturbation feels an increased radiation force in the radial direction and develops an excess pressure which pushes the gas sideways as well as radially. The radiative sound wave and the gradient instabilities will tend to cause the flow to break up into clumps or slabs of typical size $10^9 - 10^{11}$ cm. This may lead to a serious departure from the smooth flow condition implicit in the CAK prediction of stellar mass loss rates. Carlberg also suggests that collisions between the clumps could lead to the production of detectable 0.5–1 keV X-rays.

Lucy and White (1980) had already considered the possibility that clumpy outflow could explain the observed soft X-ray flux from O stars (Long and White, 1980). In the Lucy and White model the instabilities lead to a two-component wind structure. There is a population of clumps or 'blobs' that are radiatively driven through an ambient nonradiatively driven gas. An element of the ambient gas in not radiatively driven because it is shadowed by blobs closer to the source of radiation. The drag force between the blobs and the ambient gas confines the blobs by ram pressure, and drags the ambient gas outward. Using the de Young and Axford (1967) treatment of ram pressure confined plasma clouds, they determine the structure of the blobs. Then, by considering the heat generated at the front bow shocks, an X-ray spectrum is deduced. For ζ Pup the blobs tend to have masses of $\sim 10^{15}$ g, size $\sim 10^{10}$ cm. The maximum relative velocity between the blobs is ~ 360 km s^{-1}. This occurs where $v(r) \simeq \frac{1}{2} v_\infty$ and produces a hot X-ray producing shock with $T \simeq 10^{6.2}$ K. The density of the gas in the wind of ζ Pup is large enough that ambient gas can cool to low temperatures between collisions with blobs, and the blobs can remain cool enough to continue to be radiatively driven.

This is not the case for stars with lower mass loss rates, such as τ Soc (B0 V). Lucy and White find that the cooling time-scales are longer because of the lower density. Therefore, the temperature of both the blob and ambient gas phases are quickly increased to values such that the UV line opacity is lost and the proposed acceleration and shielding do not occur. The accelerated blob model therefore fails to predict the X-ray emission at the level observed in B stars.

Lucy (1982) has modified the Lucy and White model to better account for the decay of shocks with radial distance. It is assumed that at any given point in the wind, shocks traveling radially pass at a fixed frequency. In contrast with the Lucy and White model, which has two free parameters that are adjustable to match the observations, the Lucy model has only one, and there are plausible physical arguments to roughly fix that parameter. The dimensionless parameter, ν, in the theory determines the frequency and strength of the shocks. Its value can be estimated from Lucy's new condition that shocks survive until they are shadowed by more recently formed shocks upstream. The shocks persist farther out in the flow than had been the case in the blob model, and it predicts that there should be only very small absorption discontinuities at the K shell edges in the X-ray flux. The model in its simplest form, with ν fixed by plausibility arguments, does not give rise to enough X-ray emission (down by a factor 1/20) and the characteristic temperatures of the X-ray emission is too low to match the observations. Lucy suggests that the model can be made to match the observed X-ray luminosities and spectra if there is distribution of shock strengths such that, occasionally, a markedly stronger shock is produced. If this is the case, one should see a sporadic variability in the X-ray flux. Such variability was not predicted in the Lucy and White blob model. Snow *et al.* (1981) have recently reported detecting variability in the X-ray flux of several O stars, but the time-scales are longer (1 year for OB supergiants) and the increased X-ray flux indicates a contribution from a hotter emitting region (10^7 K) than would be expected in Lucy's shock model. Variability was not detected in the simultaneous UV and X-ray observations of two OB supergiants by Cassinelli *et al.* (1981).

3.4.4. *Hybrid Models with a Base Coronal Zone*

We have discussed several plausible instabilities that could plausibly produce hot gas at almost any radius in a radiatively driven wind. Stewart and Fabian (1981) point out that considerations of 'radiative efficiency' argue for the presence of hot gas primarily at the base of the flows. To be an efficient X-ray producing region, the cooling time, t_r, due to line and bremsstrahlung emission, must be shorter than the adiabatic expansion time $(r/v(r))$ or the Compton cooling time, t_c, (i.e. time to transfer energy from the hot gas to UV photons via electron scattering). On the basis of these considerations, Stewart and Fabian argue that the hot gas, even in the wind of shocked O stars, should be within a few tenths of a stellar radius above the photosphere.

In addition to the various flow instabilities for producing base coronae (Hearn, 1975; Nelson and Hearn, 1978; Cannon and Thomas, 1977), the low lying coronae could be produced by several other mechanisms. Maeder (1980) suggests that the outer envelopes of even early-type supergiants could be convective, and this could lead to the generation of an acoustic flux. Underhill (1980) suggests that hot gas is present in magnetic loops. The early-type stars should have moderately strong magnetic fields even in the absence of a dynamo mechanism because the ages are less than the characteristic decay time for the initial magnetic field (Stencel and Ionson, 1980). The rapid rotation that is characteristic of early-type stars is another possible source of mechanical heating. Only an infinitesimal fraction of the rotational kinetic energy would need to be tapped to provide a substantial base coronae (Stothers, 1980; Nerney, 1980).

The primary argument against base coronae as being the source of the observed X-ray emission in O*f* stars and OB supergiants has been the unexpectedly strong flux of 0.5–1.0 keV X-ray seen with the IPC and SSS detectors on Einstein, as discussed earlier (Lucy and White, 1980; Cassinelli *et al.*, 1981; Cassinelli and Swank, 1981). Cassinelli and Swank (1983) noted that the SSS spectra could not rule out the presence of even a rather massive base corona ($EM > 10^{57}$) if $T_c \simeq 2 \times 10^6$ K. Also in ϵ Ori, the SSS spectra showed some indication of a hot component ($T \simeq 15 \times 10^6$ K; $EM = 10^{55}$) that could give rise to Si XIII, S XV X-ray line emission. From the models of Lucy (1982) it appears that very hot gas cannot be produced by shocks in the radiatively driven wind. So the 15×10^6 K gas would presumably be in a base coronal region, perhaps one that is magnetically confined.

There may well be base coronal zones *and* hot shocked regions far out in the winds of early-type stars. However, it is desirable to isolate the effects of the base coronal zone and see what observations can be explained without invoking the outer shocked regions.

This was the approach taken by Stewart and Fabian (1981) and Waldron (1980, 1984). Stewart and Fabian noted that the absorption of soft X-ray is a strong function of the ionization conditions in the cool wind region, and of the stellar mass loss rate. They got a reasonably good fit to the IPC spectra of ζ Pup O4*f* by reducing the mass loss rate from $3.7 \times 10^{-6}\ M_\odot\ \mathrm{yr}^{-1}$ (as measured in the radio region) to $2 \times 10^{-6}\ M_\odot\ \mathrm{yr}^{-1}$. They also argued that the winds may be clumpy, and the measured mass loss rates may therefore be uncertain by this amount.

Waldron considered the temperature distribution in the recombination region at the base corona–cool wind interface. The wind momentum and energy equations were integrated out from a starting point in the base corona where the temperature is T_c. When the temperature decreases to about 2×10^5 K, the line radiation forces become very strong and so Waldron assumed that there is a transition to a CAK type of wind. The initial starting velocity was chosen to permit a transition though the CAK critical point. The emission measure, EM_c of the base corona below his starting point was iteratively adjusted so that the total X-ray flux matched that observed for the O stars and B supergiants that he was studying.

An interesting result from Waldron's calculations is that the starting velocity in the corona, v_c, must be supersonic if the star is to have a large mass loss rate ($>10^{-9}$ $M_\odot$ yr^{-1}). This appears to agree with Hearn's (1975) contention that the mass loss in a hybrid flow is determined by the coronal condition (i.e. v is already very large in the corona) and the line acceleration mechanism merely provides the final source of momentum deposition necessary to achieve high velocity. (See Leer and Holzer, 1980, for a discussion of the effects of momentum deposition in supersonic flows.)

Waldron (1984) also accounted for the effects of the coronal radiation on the ionization on the ionization conditions in the flow. He found that the extension of the recombination region into the flow and the reduction of the attenuation column density to soft X-ray give good fits to the IPC spectra of both O4*f* stars with massive winds such as (9 Sgr and B supergiants with less massive winds such as ϵ Ori. However, Cassinelli and Swank (1983) show that the model has too large a jump at 0.6 keV to fit the higher resolution SSS data.

3.4.5. *Magnetically Driven Winds and Magnetically Dominated Coronae*

Mihalas and Conti (1980) suggested that the peculiar Hα profiles seen in Oe stars indicate that there is significant corotation of the expanding atmosphere, and proposed that this is due to magnetic fields. Their interpretation of the Hα profile was not supported by the more detailed calculations of Barker *et al.* (1981), yet the paper has aroused interest in the possible effects of magnetic fields on early-type stellar winds. Weber and Davis (1967) developed the equations for the solar wind, including a magnetic field which is drawn out into an Archimedes spiral about the rotating Sun. The equations were used to study angular momentum loss from the Sun. Michel (1969) and Belcher and MacGregor (1976) applied the equations to rapidly rotating pulsars and late-type main sequence stars, respectively. Recently Nerney (1980) has applied this fast magnetic rotator theory to estimate the magnetic fields in early-type stars.

Because of the magnetic field, a rapidly rotating star can transfer angular momentum from the star to the outer envelope. In the process a wind with a maximum speed near the 'Michel' speed (Belcher and MacGregor, 1976) can be driven. This is given by

$$v_M = \left\{ \frac{\Omega^2 \, [r_A{}^2 B_R(r_A)]^2}{\dot{M}} \right\}^{1/3}, \tag{19}$$

where Ω is the angular velocity of the star, r_A is the Alfvén radius, B_R is the radial component of the magnetic field at the Alfvén radius. To derive an upper limit on B, Nerney assumes the winds are magnetically driven and sets $v_M = v_\infty$, $r_A = R_*$, uses the rotation rate Ω estimated from line profile observations, and uses the observationally determined mass loss rates.

As we might expect, since there is good reason to think that line forces play an important role in accelerating O star winds, Nerney deduces large magnetic field upper limits for these stars ($B < 2800$ gauss). More interesting results are derived from the consideration of other early-type stars. He finds quite reasonable estimates for the magnetic fields of Be stars and late B supergiants. For example, for ϕ Per (B2 Ve), which has a rotation speed at the equator estimated to be 550 km s^{-1} and a mass loss rate $\sim 5 \times 10^{-11}$ $M_\odot$ yr^{-1}, Nerney finds that a magnetic field like that of the Sun ($\lesssim 1$ gauss) could drive the winds to the observed terminal speed of 350 km s^{-1}.

Nerney also suggests that the process occurs in the late B supergiants, for which fields $\lesssim 100$ gauss could drive the winds. These stars have been considered as likely candidates for a nonradiatively driven wind mechanism because of the Hα profile observations of Rosendahl (1973). Rosendahl found that there is a change in the outflow velocity structure that sets in for B supergiants of type later than B5. These later B supergiants show evidence for a slower increase in $v(r)$ as might be expected if a different acceleration process is operating. Nerney finds a good inverse correlation between the efficiency of the magnetically driven wind mechanism and the radiation driven wind mechanism ($\dot{M}/\dot{M}_1$) for late B supergiants. Hartmann and Cassinelli (1982) have investigated the process as applied to the Wolf-Rayet stars which have mass loss rates in excess of what can be explained with the line driven wind theory.

Magnetic fields might also have a strong effect on the coronal zones in hot stars. Vaiana *et al.* (1981) argue that the observed X-ray emission is controlled by magnetic structures at the base of the stellar winds. The arguments in favor of magnetically formed regions are:

(1) Harnden *et al.* (1979), Lucy and White (1980), and Vaiana *et al.* (1981) find that the X-ray emission is not dependent on any characteristic of the star involved, other than its luminosity. $L_X = 10^{-7} L_{bol}$ for both main sequence and supergiant stars. Therefore one cannot explain the luminosities by appealing to binary processes such as colliding winds but it must be intrinsic to the stars.

(2) The strong X-ray flux from main sequence stars is not expected from the shocked wind models because of the excessive cooling time in the winds.

(3) There is no agreement between the observed X-ray luminosities and that predicted by acoustic wave theories.

(4) There is good evidence for magnetically dominated coronae of late type stars.

(5) There is some evidence (discussed earlier) for hot regions with $T > 10^7$ at the base of the Orion supergiant winds.

The magnetic fields could control the coronal emission because the fields can channel the nonradiative flux to the corona, and thereby participate in the coronal heating process. Also, the magnetic fields allow for nongravitational confinement of very hot gas.

We cannot claim to have acceptable theories for the chromospheres, coronae, and winds of stars of any spectral type. Over the past decade we have obtained our first opportunities to study the outer layers of stellar atmospheres at wavelengths along the spectrum running from X-ray to radio wavelengths. We have learned that almost all stars are X-ray sources. We now know that we shall need much higher spectral resolution and time resolution to sort and determine the properties of the coronal regions giving rise to the X-rays. These observations will hopefully be made by the AXAF satellite. In the ultraviolet region the Space Telescope will provide information on the chromospheres and winds of fainter objects and stars in other galaxies. Orbiting infrared observations will provide thorough surveys of the effects of winds of young stars on their surrounding dense clouds. Combining the infrared data with ground-based and ultraviolet line profile observations should permit good empirical models to be developed of the temperature and velocity structures in the outflows. Along the way, perhaps, we shall develop sufficient insight from the observations to develop an appropriate theoretical understanding of the outer atmospheres.

Acknowledgements

The authors are grateful to Drs J. I. Castor, L. Hartmann, T. E. Holzer, J. L. Linsky, D. Mihalas, and E. N. Parker for comments on the manuscript. K.B.M. was supprted in part by NASA grant NGL 14-001-001 to the University of Chicago; and J.P.C. was supported in part by NSF grant AST 7912141 to the University of Wisconsin.

References

Abbott, D. C.: 1977, Ph.D. Thesis, Univ. of Colorado.
Abbott, D. C.: 1978, *Astrophys. J.* **225**, 893.
Abbott, D. C.: 1980, *Astrophys. J.* **242**, 1183.
Abbott, D. C., Bieging, J. H., Churchwell, E., and Cassinelli, J. P.: 1980, *Astrophys. J.* **238**. 196.
Abbott, D. C., Bieging, J. H., and Churchwell, E.: 1981, *Astrophys. J.* **250**, 645.
Adams, W. S. and McCormack, E.: 1935, *Astrophys. J.* **81**, 119.
Alazraki, G. and Couturier, P.: 1971, *Astron. Astrophys.* **13**, 380.
Allen, C. W.: 1973, *Astrophysical Quantities*, Athlone Press, London.
Altenhoff, W. J., Oster, L., and Wendker, H. J.: 1979, *Astron. Astrophys.* **73**, L21.
Anderson, C. M.: 1974, *Astrophys. J.* **190**, 585.
Apruzese, J. P.: 1976, *Astrophys. J.* **207**, 799.
Athay, R. G.: 1966, *Astrophys. J.* **145**, 784.
Athay, R. G.: 1970a, *Solar Phys.* **11**, 347.
Athay, R. G.: 1970b, *Solar Phys.* **12**, 175.
Athay, R. G.: 1972, *Radiation Transport in Spectral Lines*, D. Reidel, Dordrecht.
Athay, R. G.: 1976, *The Solar Chromosphere and Corona: Quiet Sun*, D. Reidel, Dordrecht.
Athay, R. G. and Skumanich, A.: 1968, *Astrophys. J.* **152**, 141.
Athay, R. G. and White, O. R.: 1978, *Astrophys. J.* **226**, 1135.
Athay, R. G. and White, O. R.: 1979a, *Astrophys. J. Suppl.* **39**, 333.
Athay, R. G. and White, O. R.: 1979b, *Astrophys. J.* **229**, 1147.
Avrett, E. H.: 1981, in R. M. Bonnet and A. K. Dupree (eds), *Solar Phenomena in Stars and Stellar Systems*, D. Reidel, Dordrecht, p. 173.
Ayres, T. R.: 1979, *Astrophys. J.* **228**, 509.
Ayres, T. R.: 1981, *Astrophys. J.* **244**, 1064.
Ayres, T. R., Linsky, J. L., Vaiana, G. S., Golub, L., and Rosner, R.: 1981, *Astrophys. J.* **250**, 293.
Ayres, T. R. and Linsky, J. L.: 1980a, *Astrophys. J.* **235**, 76.
Ayres, T. R. and Linsky, J. L.: 1980b, *Astrophys. J.* **241**, 279.
Ayres, T. R., Linsky, J. L., and Shine, R. A.: 1975, *Astrophys. J.* **195**, L121.
Ayres, T. R., Linsky, J. L., Garmire, G., and Cordova, F.: 1979, *Astrophys. J.* **232**, L117.
Baliunas, S. L. and Dupree, A. K.: 1979, *Astrophys. J.* **227**, 870.
Barker, P. K., Landstreet, J. D., Marlborough, J. M., Thompson, I., and Maza, J.: 1981, *Astrophys. J.* **250**, 300.

Barlow, M. J.: 1979 in P. S. Conti and C. deLoore (eds), *Mass Loss and Evolution of O-Type Stars*, D. Reidel, Dordrecht, p. 119.
Barlow, M. J. and Cohen, M.: 1977, *Astrophys. J.* **213**, 737.
Basri, G. S.: 1980, *Astrophys. J.* **242**, 1133.
Basri, G. S. and Linsky, J. L.: 1979, *Astrophys. J.* **234**, 1023.
Belcher, J. W.: 1971, *Astrophys. J.* **168**, 509.
Belcher, J. W. and Davis, L.: 1971, *J. Geophys. Res.* **76**, 3534.
Belcher, J. W. and Olbert, S.: 1975, *Astrophys. J.* **200**, 369.
Belcher, J. W. and MacGregor, K. B.: 1976, *Astrophys. J.* **210**, 498.
Bernat, A. P.: 1977, *Astrophys. J.* **213**, 756.
Bernat, A. P.: 1981, *Astrophys. J.* **246**, 184.
Bernat, A. P. and Lambert, D. L.: 1975, *Astrophys. J.* **201**, L153.
Bernat, A. P. and Lambert, D. L.: 1976a, *Astrophys. J.* **204**, 830.
Bernat, A. P. and Lambert, D. L.: 1976b, *Astrophys. J.* **210**, 395.
Bernat, A. P., Hall, D. N. B., Hinkle, K. H., and Ridgway, S. T.: 1979, *Astrophys. J.* **233**, L135.
Bernat, A. P., Honeycutt, R. K., Kephart, J. E., Gow, C. E., Sanford, M. T., and Lambert, D. L.: 1978, *Astrophys. J.* **219**, 532.
Bidelman, W. P.: 1954, *Astrophys. J. Suppl.* **1**, 175.
Biermann, L.: 1946, *Naturwiss.* **33**, 118.
Blanco, C. Catalano, S., Marilli, E., and Rodono, M.: 1974, *Astron. Astrophys.* **33**, 257.
Blanco, C., Catalano, S., and Marilli, E.: 1976, *Astron. Astrophys.* **48**, 19.
Bohm-Vitense, E. and Dettmann, T.: 1980, *Astrophys. J.* **236**, 560.
Bohn, H. U.: 1982, in M. S. Giampapa and L. Golub (eds), *Second Cambridge Workshop on Cool Stars, Stellar Systems and the Sun,* Smithsonian Astrophysical Obs. Special Report # 392, p. 67.
Brandt, J. C.: 1970, *Introductory to the Solar Wind*, Freeman, San Francisco.
Brucato, R. J.: 1971, *Monthly Notices Roy. Astron. Soc.* **153**, 435.
Bruner, E. C.: 1978, *Astrophys. J.* **226**, 1140.
Bruner, E. C. and McWhirter, R. W. P.: 1979, *Astrophys. J.* **231**, 557.
Cannon, C. J. and Thomas, R. N.: 1977, *Astrophys. J.* **211**, 910.
Carlberg, R. G.: 1980, *Astrophys. J.* **241**, 1131.
Carpenter, K. G. and Wing, R. F.: 1979, *Bull. Amer. Astron. Soc.* **11**, 419.
Cash, W., Bowyer, S., Charles, P. A., Lampton, M., Garmire, G., and Riegler, G.: 1978, *Astrophys. J.* **223**, L21.
Cash, W., Charles, P., Bowyer, S., Walter, F., Garmire, G., and Riegler, G.: 1980, *Astrophys. J.* **238**, L71.
Cassinelli, J. P.: 1979, *Ann. Rev. Astron. Astrophys.* **17**, 275.
Cassinelli, J. P. and Abbott, D. C.: 1981, in R. D. Chapman (ed.), *The Universe at Ultraviolet Wavelengths: The First Two Years of IUE*, NASA Conference Publ. 2171, p. 127.
Cassinelli, J. P., Castor, J. I., and Lamers, H. J. G. L. M.: 1978, *Publ. Astron. Soc. Pacific* **90**, 496.
Cassinelli, J. P. and Hartmann, L.: 1975, *Astrophys. J.* **202**, 718.
Cassinelli, J. P. and Hartmann, L.: 1977, *Astrophys. J.* **212**, 488.
Cassinelli, J. P. and Olson, G. L.: 1979, *Astrophys. J.* **229**, 304.
Cassinelli, J. P., Olson, G. L., and Stalio, R.: 1978, *Astrophys. J.* **220**, 573.
Cassinelli, J. P. and Swank, J. H.: 1983, *Astrophys. J.* **271**, 681.
Cassinelli, J. P., Waldron, W. L., Sanders, W. T., Harnden, F. R., Rosner, R., and Vaiana, G. S.: 1981, *Astrophys. J.* **250**, 677.
Castor, J. I.: 1970, *Monthly Notices Roy. Astron. Soc.* **149**, 111.
Castor, J. I.: 1972, *Astrophys. J.* **178**, 779.
Castor, J. I.: 1974, *Astrophys. J.* **189**, 273.
Castor, J. I.: 1979, *IAU Symp.* **83**, 175.
Castor, J. I.: 1981, in I. Iben Jr and A. Renzini (eds), *Physical Processes in Red Giants,* D. Reidel, Dordrecht, p. 285.
Castor, J. I. and Lamers, H. J. G. L. M.: 1979, *Astrophys. J. Suppl.* **39**, 481.
Castor, J. I., Abbott, D. C., and Klein, R. I.: 1975, *Astrophys. J.* **195**, 157.
Castor, J. I., Abbott, D. C., and Klein, R. I.: 1976, in R. Cayrel and M. Steinberg (eds), *Physique des Mouvements dans les Atmospheres Stellaires*, CNRS, Paris, p. 363.

Castor, J. I., McCary, R., and Weaver, R.: 1975, *Astrophys. J.* **200**, L107.
Catura, R. C., Acton, L. W., and Johnson, H. M.: 1975, *Astrophys J.* **196**, 247.
Chiu, H. Y., Adams, P. J., Linsky, J. L., Basri, G. S., Maran, S. P., and Hobbs, R. W.: 1977, *Astrophys. J.* **211**, 453.
Conti, P. S. and Frost, S. A.: 1977, *Astrophys. J.* **212**, 728.
Conti, P. S. and de Loore, C. (eds): 1979, *IAU Symp.* **83**.
Cox, J. P. and Giuli, R. T.: 1968, *Principles of Stellar Structure*, Gordon and Breach, New York.
Cox, D. P. and Tucker, W. H.: 1969, *Astrophys. J.* **157**, 1157.
Cram, L. E.: 1974, *Solar Phys.* **37**, 75.
Cram, L. E.: 1976, *Astron. Astrophys.* **50**, 263.
Cram, L. E. and Ulmschneider, P. 1978, *Astron. Astrophys.* **62**, 239.
Cram, L. E., Krikorian, R., and Jeffries, J. T.: 1979, *Astron. Astrophys.* **71**, 14.
Cruddace, R., Bowyer, S., Malina, R., Margon, B., and Lampton, M.: 1975, *Astrophys. J.* **202**, L9.
de Campli, W. M.: 1981, *Astrophys. J.* **244**, 124.
de Jager, C.: 1980, *The Brightest Stars*, D. Reidel, Dordrecht.
de Jager, C., Kondo, Y., Hoekstra, R., Van der Hucht, K. A., Kamperman, T. M., Lamers, H. J. G. L. M., Modisette, J. L., and Morgan, T. H.: 1979, *Astrophys. J.* **230**, 543.
de Loore, C.: 1970, *Astrophys. Space Sci.* **6**, 60.
de Loore, C.: 1980, *Space Sci. Rev.* **26**, 113.
Deutsch, A. J.: 1956, *Astrophys J.* **123**, 210.
Deutsch, A. J.: 1960, in J. L. Greenstein (ed.), *Stellar Astmospheres*, Univ. Of Chicago Press, Chicago, p. 543.
de Young, D. S. and Axford, W. I.: 1967, *Nature* **216**, 129.
Doherty, L. R.: 1972, *Astrophys. J.* **178**, 495.
Doherty, L. R.: 1973, in S. D. Jordan and E. H. Avrett (eds), *Stellar Chromospheres*, NASA SP-317, Washington, D.C., p. 99.
Doschek, G. A., Feldman, U., Mariska, J. T., and Linsky, J. L.: 1978, *Astrophys. J.* **226**, L35.
Draine, B. T.: 1981, in I. Iben Jr and A. Renzini (eds), *Physical Processes in Red Giants*, D. Reidel, p. 317.
Dupree, A. K.: 1975, *Astrophys. J.* **200**, L27.
Dupree, A. K.: 1976, in R. Cayrel and M. Steinberg (eds), *Physique des Mouvements dans les Atmospheres Stellaires*, CNRS, Paris, p. 439.
Dupree, A. K.: 1978, in D. R. Bates and B. Bederson (eds), *Advances in Atomic and Molecular Physics*, Vol. 14 Academic Press, New York, p. 393.
Dupree, A. K.: 1981, *IAU Coll.* **59**, 87.
Dupree, A. K. and Hartmann, L.: 1980, *IAU Coll.* **51**, 279.
Dyck, H. M. and Johnson, H. R.: 1969, *Astrophys J.* **156**, 389.
Dyck, H. M. and Simon, T.: 1975, *Astrophys. J.* **195**, 689.
Engvold, O. and Rygh, B. O.: 1978, *Astron. Astrophys.* **70**, 399.
Evans, R. G., Jordan, C., and Wilson, R.: 1975, *Monthly Notices Roy. Astron. Soc.* **172**, 585.
Fix, J. D. and Alexander, D. R.: 1974, *Astrophys. J.* **188**, L91.
Fosbury, R. A. E.: 1973, *Astron. Astrophys.* **27**, 129.
Galeev, A. A., Rosner, R., Serio, S., and Vaiana, G. S.: 1981, *Astrophys. J.* **243**, 301.
Garmany, C. D., Olson, G. L., Conti, P. S., and Van Steenberg, M.: 1981, *Astrophys. J.* **250**, 660.
Gary, D. E. and Linsky, J. L.: 1981, *Astrophys. J.* **250**, 284.
Gathier, R., Lamers, H. J. G. L. M., and Snow, T. P.: 1981, *Astrophys. J.* **247**, 173.
Gehrz, R. D. and Woolf, N. J.: 1971, *Astrophys. J.* **165**, 285.
Giacconi, R. *et al.*: 1979, *Astrophys. J.* **230**, 540.
Gillett, F. C., Low, F. J., and Stein, W. A.: 1968, *Astrophys. J.* **154**, 677.
Gilman, R. C.: 1969, *Astrophys. J.* **155**, L185.
Gilman, R. C.: 1972, *Astrophys. J.* **178**, 423.
Goldberg, L.. 1957, *Astrophys. J.* **126**, 318.
Goldberg, L.: 1979, *Quart. J. Roy. Astron. Soc.* **20**, 361.
Goldberg, L.: 1981, in I. Iben Jr and A. Renzini (eds), *Physical Processes in Red Giants*, D. Reidel, Dordrecht, p. 301.
Goldberg, L., Ramsey, L., Testerman, L., and Carbon, D.: 1975, *Astrophys J.* **199**, 427.

Goldreich, P. and Scoville, N.: 1976, *Astrophys. J.* **205**, 144.
Golub, L., Maxson, C., Rosner, R., Serio, S., and Vaiana, G. S.: 1980, *Astrophys J.* **238**, 343.
Golub, L. Rosner, R., Vaiana, G. S., and Weiss, N. O.: 1981, *Astrophys. J.* **243**, 309.
Habbal, S. R., Leer, E, and Holzer, T. E.: 1979, *Solar Physi.* **64**, 287.
Hagen, W.: 1978, *Astrophys J. Suppl.* **38**, 1.
Hagen, W.: 1980, in A. K. Dupree (ed.), *Cool Stars, Stellar Systems, and the Sun*, Smithsonian Astrophysical Observatory Special Rept No. 389, p. 143.
Haisch, B. M. and Linsky, J. L.: 1976, *Astrophys. J.* **205**, L39.
Haisch, B. M. and Linsky, J. L.: 1980, *Astrophys. J.* **236**, L33.
Haisch, B. M., Linsky, J. L., and Basri, G. S.: 1980, *Astrophys. J.* **235**, 519.
Hamann, W. R.: 1980, *Astron. Astrophys.* **84**, 342.
Harnden, F. R., Branduardi, G., Elvis, M., Gorenstein, P., Grindlay, J., Pye, J. P., Rosner, R., Topka, K., and Vaiana, G. S.: 1979, *Astrophys J.* **234**, L51.
Hartmann, L.: 1978, *Astrophys. J.* **224**, 520.
Hartmann, L.: 1980, in A. K. Dupree (ed.), *Cool Stars, Stellar Systems, and the Sun*, Smithsonian Astrophysical Observatory, Special Rept No. 389, p. 15.
Hartmann, L.: 1981a, in R. M. Bonnet and A. K. Dupree (eds), *Solar Phenomena in Stars and Stellar Systems*, D. Reidel, Dordrecht, p. 487.
Hartmann, L.: 1981b, in R. M. Bonnet and A. K. Dupree (eds), *Solar Phenomena in Stars and Stellar Systems*, D. Reidel, Dordrecht, p. 331.
Hartmann, L. and Cassinelli, J. P.: 1977, *Astrophys. J.* **215**, 155.
Hartmann, L. and Cassinelli, J. P.: 1982, *Bull. Amer. Astron. Soc.* **13**, 785.
Hartmann, L. and MacGregor, K. B.: 1980, *Astrophys. J.* **242**, 260.
Hartmann, L, Davis, R., Dupree, A. K., Raymond, J., Schmidtke, P. C., and Wing, R. F.: 1979, *Astrophys. J.* **232**, L69.
Hartmann, L., Dupree, A. K., and Raymond, J. C.: 1980, *Astrophys. J.* **236**, L143.
Hartmann, L., Dupree, A. K., and Raymond, J. C.: 1981, *Astrophys. J.* **246**, 193.
Hartmann, L., Dupree, A. K., and Raymond, J. C.: 1982, *Astrophys. J.* **252**, 214.
Hearn, A. G.: 1972, *Astron. Astrophys.* **19**, 417.
Hearn, A. G.: 1973, *Astron. Astrophys.* **19**, 417.
Hearn, A. G.: 1975, *Astron. Astrophys.* **40**, 277.
Hearn, A. G. and Vardavas, I. M.: 1981, *Astron. Astrophys.* **98**, 230.
Heasley, J. N.: 1975, *Solar Phys.* **44**, 275.
Heise, J., Brinkman, A. C., Schrijver, J., Mewe, R., Gronenshild, E., and den Boggende, A.: 1975, *Astrophys. J.* **202**, L73.
Hollweg, J. V.: 1973, *Astrophys. J.* **181**, 547.
Hollweg, J. V.: 1981, *Solar Phys.* **70**, 25.
Holt, S. S., White, N. E., Becker, R. H., Boldt, E. A., Mushotzky, R. F., Serlemitsos, P. J., and Smith, B. W.: 1979, *Astrophys. J.* **234**, L65.
Holzer, T. E.: 1979, in C. F. Kennel, L. J. Lanzerotti, and E. N. Parker (eds), *Solar System Plasma Physics*, North-Holland, Amsterdam, p. 101.
Holzer, T. E.: 1980, in A. K. Dupree, *Cool Stars, Stellar Systems, and the Sun*, Smithsonian Astrophysical Observatory Special Rept No. 389, p. 153.
Honeycutt, R. K., Bernat, A. P., Kephart, J. E., Gow, C. E., Sanford, M. T., and Lambert, D. L.: 1980, *Astrophys. J.* **239**, 565.
Hummer, D. G. and Rybicki, G. B.: 1968, *Astrophys. J.* **153**, L107.
Hundhausen, A. J.. 1972, *Coronal Expansion and Solar Wind*, Springer-Verlag, New York.
Iben, I.: 1967, *Ann. Rev. Astron. Astrophys.* **5**, 571.
Ionson, J. A.: 1978, *Astrophys. J.* **226**, 650.
Jacques, S. A.: 1977, *Astrophys. J.* **215**, 942.
Jacques, S. A.: 1978, *Astrophys. J.* **226**, 632.
Jeffries, J. T.. 1968, *Spectral Line Formation*, Blaisdell, Waltham.
Jeffries, J. T. and Thomas, R. N.: 1959, *Astrophys. J.* **129**, 401.
Jennings, M. C.: 1973, *Astrophys. J.* **185**, 197.
Jennings, M. C. and Dyck, H. M.: 1972, *Astrophys. J.* **177**, 427.
Jones, T. W. and Merrill, K. M.: 1976, *Astrophys. J.* **209**, 509.

Jordan, C.: 1980, in P. A. Wayman (ed.), *Highlights of Astronomy*, Vol. 5, D. Reidel, Dordrecht, p. 553.
Jordan, S. D. 1973, in S. D. Jordan and E. H. Avrett (eds), *Stellar Chromospheres*, NASA SP-317, Washington, D.C., p. 181.
Jordan, C. and Wilson, R.: 1971, in C. J. Macris (ed.), *Physics of the Solar Corona*, D. Reidel, Dordrecht, p. 219.
Kahn, S. M., Linsky, J. L., Mason, K. O., Haisch, B. M., Bowyer, C. S., White, N. E., and Pravdo, S. H.: 1979, *Astrophys. J.* **234**, L107.
Klein, R. I. and Castor, J. I.: 1978, *Astrophys. J.* **220**, 902.
Kondo, Y., Duval, J. E. Modisette, J. L., and Morgan, T. H.: 1976a, *Astrophys. J.* **210**, 713.
Kondo, Y., Giuli, R. T., Modisette, J. L., and Rydyren, A. E.: 1972, *Astrophys. J.* **176**, 153.
Kondo, Y., Morgan, T. H., and Modisette, J. L.: 1975, *Astrophys. J.* **196**, L125.
Kondo, Y., Morgan, T. H., and Modisette, J. L.: 1976b, *Astrophys. J.* **207**, 167.
Lpmdp. Y., Morgan, T. H., and Modisette, J. L.: 1976c, *Astrophys. J.* **209**, 489.
Kraft, R. P.: 1976, *Astrophys. J.* **150**, 551.
Kudritzki, R. P. and Reimers, D.: 1978, *Astron. Astrophys.* **70**, 227.
Kulsrud, R. M.: 1955, *Astrophys. J.* **121**, 461.
Kuperus, M.: 1965, *Rech. Astron. Obs. Utrecht* **17**, 1.
Kwok, S.: 1975, *Astrophys. J.* **198**, 583.
Kwok, S., Purton, C. R., and FitzGerald, P. M.: 1978, *Astrophys. J.* **219**, L125.
Lamers, H. J. G. L. M.: 1976, in R. Cayrel and M. Steinberg (eds), *Physique des Mouvements dans les Atmospheres Stellaires*, CNRS, Paris, p. 405.
Lamers, H. J. G. L. M.: 1981, *Astrophys. J.* **245**, 593.
Lamers, H. J. G. L. M. and Morton, D. C.: 1976, *Astrophys. J. Suppl.* **32**, 715.
Lamers, H. J. G. L. M. and Rogerson, J. B.: 1978, *Astron. Astrophys.* **66**, 417.
Lamers, H. J. G. L. M. and Snow, T. P.: 1978, *Astrophys. J.* **219**, 504.
Landau, L. and Lifshitz, E. M.: 1959, *Fluid Mechanics*, Pergamon, London.
Lee, M. A.: 1980, *Astrophys. J.* **240**, 693.
Leer, E. and Holzer, T. E.: 1980, *J. Geophys. Res.* 85, 4681.
Levine, R.: 1974, *Astrophys. J.* **190**, 457.
Liebert, J., Dahn, C., Gresham, M., and Strittmatter, P. A.: 1979, *Astrophys. J.* **233**, 226.
Lighthill, M. J.: 1952, *Proc, Roy. Soc. London* **A211**, 564.
Linsky, J. L.: 1977, in O. R. White (ed.), *The Solar Output and its Variation*, Colorado Assoc. Univ. Press, Boulder, p. 477.
Linsky, J. L.: 1980a, *Ann. Rev. Astron. Astrophys.* **18**, 439.
Linsky, J. L.: 1980b, in A. K. Dupree (ed.), *Cool Stars, Stellar Systems, and the Sun*, Smithsonian Astrophysical Observatory Special Rept No. 389, p. 217.
Linsky, J. L.: 1981a, in R. M. Bonnet and A. K. Dupree (eds), *Solar Phenomena in Stars and Stellar Systems*, D. Reidel, Dordrecht, p. 99.
Linsky, J. L.: 1981b, in I. Iben Jr and A. Renzini (eds), *Physical Processes in Red Giants*, D. Reidel, Dordrecht, p. 247.
Linsky, J. L.: 1981c, *IAU Coll.* **59**, 187.
Linsky, J. L.: 1982, in Y. Kondo, J. M. Mead, and R. D. Chapman (eds), *Advances in Ultraviolet Astronomy: Four Years of IUE Research,* NASA Conference Pub. 2238, p. 17.
Linsky, J. L. *et al.*: 1978, *Nature* **275**, 389.
Linsky, J. L. and Ayres, T. R.: 1978, *Astrophys. J.* **220**, 619.
Linsky, J. L. and Haisch, B. M.: 1979, *Astrophys. J.* **229**, L27.
Linsky, J. L., Hunton, D. M., Sowell, R., Glackin, D. L., and Kelch, W. L.: 1979a, *Astrophys. J. Suppl.* **41**, 481.
Linsky, J. L., Worden, S. P., McClintock, W., and Robertson, R. M.: 1979b, *Astrophys. J. Suppl.* **41**, 47.
Long, K. S. and White, R. L.: 1980, *Astrophys. J.* **239**, L65.
Lucy, L. B.: 1976, *Astrophys. J.* **205**, 482.
Lucy, L. B.: 1982, *Astrophys. J.* **255**, 286.
Lucy, L. B. and Solomon, P. M.: 1970, *Astrophys. J.* **159**, 879.
Lucy, L. B. and White, R. L.: 1980, *Astrophys. J.* **241**, 300.

Lutz, T. E., Furenlid, I., and Lutz, J. H.: 1973, *Astrophys. J.* **184**, 787.
Lynds, C. R., Harvey, J., and Goldberg, L.: 1977, *Bull. Amer. Astron. Soc.* **9**, 345.
MacGregor, K. B., Hartmann, L., and Raymond, J. C.: 1979, *Astrophys. J.* **231**, 514.
Maciel, W. J.: 1976, *Astron. Astrophys.* **48**, 27.
Maciel, W. J.: 1977, *Astron, Astrophys.* **57**, 273.
Maeder, A.: 1980, *Astron. Astrophys.* **90**, 311.
Margon, B., Mason, K. O., and Sanford, P. W.: 1974, *Astrophys. J.* **194**, L75.
Marlborough, J. M. and Snow, T. P.: 1980, *Astrophys. J.* **235**, 85.
Marlborough, J. M. and Roy, J. R.: 1980, *Astrophys. J.* **160**, 221.
Martens, P. C. H.: 1979, *Astron. Astrophys.* **75**, L7.
McClintock, W., Linsky, J. L., Henry, R. C., Moos, H. W., and Gerola, H.: 1975a, *Astrophys. J.* **202**, 165.
McClintock, W., Henry, R. C., Moos, H. W., and Linsky, J. L.: 1975b, *Astrophys. J.* **202**, 733.
McClintock, W., Moos, H. W., Henry, R. C., Linsky, J. L., and Barker, E. S.: 1978, *Astrophys. J. Suppl.* **37**, 223.
McWhirter, R. W. P., Thonemann, P. C., and Wilson, R.: 1975, *Astron. Astrophys.* **40**, 63.
Menietti, J. D. and Fix, J. D.: 1978, *Astrophys. J.* **224**, 961.
Merrill, K. M.: 1978, *IAU Coll.* **42**, 446.
Mewe, R.: 1979, *Space Sci. Rev.* **24**, 101.
Mewe, R. and Zwaan, C.: 1980, in A. K. Dupree (ed.), *Cool Stars, Stellar Systems, and the Sun*, Smithsonian Astrophysical Observatory Special Rept No. 389, p. 123.
Mewe, R., Heise, J., Gronenschild, E. H. B. M., Brinkman, A. C., Schrijver, J., and den Boggende, A. J. F.: 1975, *Astrophys. J.* **202**, L67.
Mewe, R., Heise, J., Gronenschild, E. H. B. M., Brinkman, A. C., Schrijver, J., den Boggende, A. J. F.: 1976, *Astrophys. Space Sci.* **42**, 217.
Michel, F. C.: 1969, *Astrophys. J.* **158**, 727.
Mihalas, D.: 1978, *Stellar Atmospheres*, Freeman, San Francisco.
Mihalas, D. and Conti, P. S.: 1980, *Astrophys. J.* **235**, 515.
Mihalas, D. and Hummer, D. G.: 1974, *Astrophys J. Suppl.* **28**, 343.
Mihalas, D., Kunasz, P. B., and Hummer, D. G.: 1975, *Astrophys. J.* **202**, 465.
Milkey, R. W.: 1970, *Solar Phys.* **14**, 77.
Moos, H. W., Linsky, J. L., Henry, R. C., and McClintock, W.: 1974, *Astrophys. J.* **188**, L93.
Morton, D. C.: 1967, *Astrophys. J.* **147**, 1017.
Morton, D. C.: 1979, *Monthly Notices Roy. Astron. Soc.* **189**, 57.
Morton, D. C., Jenkins, E. B., and Bohlin, R. C.: 1968, *Astrophys. J.* **154**, 66.
Morton, D. C., Jenkins, E. B., and Brooks, N.: 1969, *Astrophys. J.* **155**, 875.
Noyes, R. W.: 1971, *Ann. Rev. Astron. Astrophys.* **9**, 209.
Nelson, G. and Hearn, A. G.: 1978, *Astron. Astrophys.* **65**, 223.
Nerney, S.: 1980, *Astrophys. J.* **242**, 723.
Nugent, J. and Garmire, G.: 1978, *Astrophys. J.* **226**, L83.
O'Brien, G. and Lambert, D. L.: 1979, *Astrophys J.* **229**, L33.
Odegard, N. and Cassinelli, J. P.: 1982, *Astrophys. J.* **256**, 568.
Olson, G. L.: 1978, *Astrophys. J.* **226**, 124.
Olson, G. L.: 1981, *Astrophys. J.* **245**, 1054.
Olson, G. L. and Castor, J. I.: 1981, *Astrophys. J.* **244**, 179.
Olson, G. L. and Ebbets, D.: 1981, *Astrophys. J.* **248**, 1021.
Osterbrock, D. E.: 1961, *Astrophys. J.* **134**, 347.
Pallavicini, R., Golub, L., Rosner, R., Vaiana, G. S., Ayres, T., and Linsky, J. L.: 1981, *Astrophys. J.* **248**, 279.
Panagia, N. and Macchetto, F.: 1981, *IAU Coll.* **59**, 173.
Parker, E. N.: 1958, *Astrophys. J.* **128**, 664.
Parker, E. N.: 1963, *Interplanetary Dynamical Processes*, Interscience, New York.
Parker, E. N.: 1964, *Astrophys. J.* **140**, 1170.
Parker, E. N.: 1970, *Ann. Rev. Astron. Astrophys.* **8**, 1.
Parker, E. N.: 1975a, *Astrophys. J.* **201**, 494.
Parker, E. N.: 1975b, *Astrophys. J.* **201**, 502.
Parker, E. N.: 1977, *Ann. Rev. Astron. Astrophys.* **15**, 45.

Parker, E. N.: 1979, *Cosmical Magnetic Fields*, Clarendon Press, Oxford.
Parker, E. N.: 1981a, *Astrophys. J.* **244**, 631.
Parker, E. N., 1981b, *Astrophys. J.* **244**, 644.
Philips, J. P.: 1979, *Astron. Astrophys.* **71**, 115.
Pneuman, G. W. and Kopp, R. A.: 1977, *Astron. Astrophys.* **55**, 305.
Pneuman, G. W. and Kopp, R. A.: 1978, *Solar Phys.* **57**, 49.
Pottasch, S. R.: 1964, *Space Sci. Rev.* **3**, 816.
Praderie, F.: 1973, in S. D. Jordan and E. H. Avrett (eds), *Stellar Chromospheres*, NASA SP-317, Washington, D. C., p. 79.
Proudman, I.: 1952, *Proc. Roy. Soc. London* **A214**, 119.
Raymond, J. C., Cox, D. P. and Smith, B. W.: 1976, *Astrophys. J.* **204**, 290.
Raymond, J. C. and Smith, B. W.: 1977, *Astrophys. J. Suppl.* **35**, 419.
Reimers, D.: 1973, *Astron. Astrophys.* **24**, 79.
Reimers, D.: 1975, in B. Baschek, W. H. Kegel, and G. Traving (eds), *Problems in Stellar Atmospheres and Envelopes*, Springer, Berlin, p. 229.
Reimers, D.: 1977a, *Astron. Astrophys.* **54**, 485.
Reimers, D.: 1977b, *Astron. Astrophys.* **57**, 395.
Reimers, D.: 1978, *Astron. Astrophys.* **67**, 161.
Renzini, A., Cacciari, C., Ulmschneider, P., and Schmitz, F.: 1977, *Astron. Astrophys.* **61**, 39.
Robinson, R. D., Worden, S. P., Harvey, J. W.: 1980, *Astrophys. J.* **236**, L155.
Rogerson, J. and Lamers, H.: 1975, *Nature* **256**, 190.
Rosendahl, J. D.: 1973, *Astrophys. J.* **186**, 909.
Rosner, R.: 1980, in A. K. Dupree (ed.), *Cool Stars, Stellar Systems, and the Sun*, Smithsonian Astrophysical Observatory, Special Rept No. 389, p. 79.
Rosner, R., Golub, L., Coppi, B., and Vaiana, G. S.: 1978a, *Astrophys. J.* **222**, 317.
Rosner, R., Tucker, W. H., and Vaiana, G. S.: 1978b, *Astrophys. J.* **220**, 643.
Rosner, R. and Vaiana G. S.: 1980, in R. Giacconi (ed.), *X-Ray Astronomy*, D. Reidel, Dordrecht, p. 129.
Rumpl, W. M.: 1980, *Astrophys. J.* **241**, 1055.
Salpeter, E. E.: 1974a, *Astrophys. J.* **193**, 579.
Salpeter, E. E.: 1974b, *Astrophys. J.* **193**, 585.
Sanner, F.: 1976a, *Astrophys. J.* **204**, L41.
Sanner, F.: 1976b, *Astrophys. J. Suppl.* **32**, 115.
Scharmer, G. B.: 1976, *Astron. Astrophys.* **53**, 341.
Schmitz, F. and Ulmschneider, P.: 1980a, *Astron, Astrophys.* **84**, 93.
Schmitz, F. and Ulmschneider, P.: 1980b, *Astron. Astrophys.* **84**, 191.
Schmitz, F. and Ulmschneider, P.: 1981, *Astron. Astrophys.* **93**, 178.
Schwarzschild, K. and Eberhard, G.: 1913, *Astrophys. J.* **38**, 292.
Schwarzschild, M.: 1948, *Astrophys. J.* **107**, 1.
Schwarzschild, M.: 1985, *Structure and Evolution of the Stars*, Princeton Univ. Press, Princeton.
Seward, F. D., Forman, W. R., Giacconi, R., Griffith, R. B., Harnden, F. R., Jones, C., and Pye, J. P.: 1979, *Astrophys. J.* **234**, L51.
Shine, R. A.: 1975, *Astrophys. J.* **202**, 543.
Skumanich, A.: 1972, *Astrophys. J.* **171**, 565.
Skumanich, A., Smythe, C., and Frazier, E. N.: 1975, *Astrophys. J.* **200**, 747.
Snow, T. P.: 1982, *IAU Symp.* **98**, 377.
Snow, T. P. and Hayes, D. P.: 1978, *Astrophys. J.* **226**, 897.
Snow, T. P. and Jenkins, E. B.: 1977, *Astrophys. J. Suppl.* **33**, 269.
Snow, T. P. and Marlborough, J. M.: 1976, *Astrophys. J.* **203**, L87.
Snow, T. P. and Morton, D. C.: 1976, *Astrophys. J. Suppl.* **33**, 269.
Snow, T. P., Cash, W., and Grady, C. A.: 1981, *Astrophys. J.* **244**, L19.
Sobolev, V. V.: 1960, *Moving Envelopes of Stars*, Harvard Univ. Press, Cambridge.
Soderblom, D. R.: 1983, *Astrophys. J. Suppl.* **53**, 1.
Stalio, R. and Chiosi, C. (eds): 1981, *IAU Coll.* **59**.
Stein, R. F.: 1967, *Solar Phys.* **2**, 385.
Stein, R. F.: 1968, *Astrophys. J.* **154**, 297.
Stein, R. F.: 1981, *Astrophys. J.* **246**, 966.

Stein, R. F. and Leibacher, J. W.: 1974, *Ann. Rev. Astron. Astrophys.* **12**, 407.
Stein, R. F. and Leibacher, J. W.: 1980, *IAU Coll.* **51**, 225.
Stencel, R. E.: 1978, *Astrophys J.* **223**, L37.
Stencel, R. E.: 1980, in A. K. Dupree (ed.), *Cool Stars, Stellar Systems, and the Sun*, Smithsonian Astrophysical Observatory Special Rept No. 389, p. 183.
Stencel, R. E. and Ionson, J. A.: 1979, *Publ. Astron. Soc. Pacific.* **91**, 452.
Stencel, R. E. and Mullan, D. J.: 1980, *Astrophys. J.* **238**, 221.
Stencel, R. E., Mullan, D. J., Linsky, J. L., Basri, G. S., and Worden, S. P.: 1980, *Astrophys. J. Suppl.* **44**, 383.
Stewart, G. C. and Fabian, A. C.: 1981, *Monthly Notices Roy. Astron. Soc.* **197**, 713.
Stothers, R.:1980, *Astrophys. J.* **242**, 756.
Stothers, R. and Chin, C.: 1976, *Astrophys. J.* **204**, 472.
Sutton, E. C., Storey, J. W. V., Betz, A. L., Townes, C. H., and Spears, D. L.: 1977, *Astrophys. J.* **217**, L97.
Thomas, R. N.: 1957, *Astrophys. J.* **125**, 260.
Thomas, R. N. and Athay, R. G.: 1961, *Physics of the Solar Chromosphere*, Interscience, New York.
Topka, K., Fabricant, D., Harnden, F. R., Gorenstein, P., and Rosner, R.: 1979, *Astrophys J.* **229**, 661.
Tucker, W. H.: 1973, *Astrophys. J.* **186**, 285.
Tucker, W. H. and Koren, M.: 1971, *Astrophys. J.* **168**, 283.
Uchida, Y. and Kaburaki, O.: 1974, *Solar Phs.* **35**, 451.
Ulmschneider, P.: 1974, *Solar Phys.* **39**, 327.
Ulmschneider, P.: 1979, *Space Sci. Rev.* **24**, 71.
Underhill, A. B.: 1980, *Astrophys. J.* **240**, L153.
Vaiana, G. S.: 1980, in A. K. Dupree (ed.), *Cool Stars Stellar Systems, and the Sun*, Smithsonian Astrophysical Observatory Special Rept No. 389, p. 195.
Vaiana, G. S. and Rosner, R.: 1978, *Ann. Rev. Astron. Astrophys.* **16**, 393.
Vaiana, G. S. *et al.*: 1981, *Astrophys. J.* **245**, 163.
Van Blerkom, D.: 1978, *Astrophys. J.* **221**, 186.
Vanderhill, M. J., Borken, R. J., Bunner, A. W., Burstein, P. H., and Kraushaar, W. L.: 1975, *Astrophys. J.* **197**, L19.
Van der Hucht, K. A., Bernat, A. P., and Kondo, Y.: 1980, *Astron. Astrophys.* **82**, 14.
Van der Hucht, K. A., Cassinelli, J. P., Wesselius, P. R., and Wu, C. C.: 1979a, *Astron. Astrophys. Suppl.* **38**, 279.
Van der Hucht, K. A., Stencel, R. E., Haisch, B. M., and Kondo, Y.: 1979b, *Astron. Astrophys. Suppl.* **36**, 377.
Vaughn, A. H., Baliunas, S. L., Middelkoop, F., Hartmann, L. W., Mihalas, D., Noyes, R. W., and Preston, G. W.: 1981, *Astrophys. J.* **250**, 276.
Vernazza, J. E., Avrett, E. H., and Loeser, R.: 1973, *Astrophys. J.* **183**, 605.
Vernazza, J. E., Avrett, E. H., and Loeser, R.: 1981, *Astrophys. J. Suppl.* **45**, 635.
Vitz, R. C., Weiser, H., Moos, H. W., Weinstein, A., and Warden, E. S.: 1976, *Astrophys. J.* **205**, L35.
Waldron, W. L.: 1980, Ph.D. Thesis, Univ. of Wisconsin.
Waldron, W. L.: 1984, *Astrophys. J.* **282**, 256.
Walter, F. M.: 1981, *Astrophys. J.* **245**, 677.
Walter, F. M. and Bowyer, S.: 1981, *Astrophys. J.* **245**, 671.
Walter, F. M., Cash, W., Charles, P. A., and Bowyer, C. S.: 1980a, *Astrophys. J.* **236**, 212.
Walter, F. M., Cash, W., Charles, P. A., and Bowyer, C. S.: 1978a, *Astrophys. J.* **225**, L119.
Walter, F., Charles, P., and Bowyer, S.: 1978b, *Astrophys. J.* **83**, 1539.
Walter, F. M., Linsky, J. L., Bowyer, S., and Garmire, G.: 1980b, *Astrophys. J.* **236**, L137.
Weaver, R., McCray, R., and Castor, J. I.: 1977, *Astrophys. J.* **218**, 377.
Weber, E. J. and Davis, L.: 1967, *Astrophys. J.* **148**, 217.
Weber, S. V.: 1981, *Astrophys. J.* **243**, 954.
Wegner, G. A. and Snow, T. P.: 1978, *Astrophys. J.* **226**, L25.
Weiler, E. J. and Oegerle, W. R.: 1979, *Astrophys. J. Suppl.* **39**, 537.
Wentzel, D. G.: 1974, *Solar Phys.* **39**, 129.
Wentzel, D. G.: 1978, *Rev. Geophys. Space Phys.* **16**, 757.

Wentzel, D. G.: 1979a, *Astrophys. J.* **227**, 319.
Wentzel, D. G.: 1979b, *Astron. Astrophys.* **76**, 20.
Wentzel, D. G.: 1979c, Astrophys. J. 233, 756.
Weymann, R.: 1962a, *Astrophys. J.* **136**, 476.
Weymann, R.: 1962b, *Astrophys. J.* **136**, 844.
Weymann, R.: 1963, *Ann. Rev. Astron. Astrophys.* **1**, 97.
Weymann, R. J.: 1978, *IAU Coll.* **42**, 577.
White, O. R. and Athay, R. G.: 1979a, *Astrophys. J. Suppl.* **39**, 317.
White, O. R. and Athay, R. G.: 1979b, *Astrophys. J. Suppl.* **39**, 347.
Wilson, O. C.: 1959a, *Astrophys. J.* **130**, 499.
Wilson, O. C.: 1959b, *Astrophys J.* **131**, 75.
Wilson, O. C.: 1963, *Astrophys. J.* **138**, 832.
Wilson, O. C.: 1966, *Science* **151**, 1487.
Wilson, O. C.: 1978, *Astrophys. J.* **226**, 379.
Wilson, O. C. and Bappu, M. K. V.: 1957, *Astrophys. J.* **125**, 661.
Wilson, O. C. and Skumanich, A.: 1964, *Astrophys. J.* **140**, 1401.
Withbroe, G. L.: 1971, in K. B. Gebbie (ed.), *The Menzel Symposium on Solar Physics, Atomic Spectra, and Gaseous Nebulae*, NBS Spec. Publ. 353, Washington, D.C., p. 127.
Withbroe, G. L. and Noyes, R. W.: 1977, *Ann. Rev. Astron. Astrophys.* **15**, 363.
Woolf, N. J. and Ney, E. P.: 1969, *Astrophys. J.* **155**, L181.
Wright, A. E. and Barlow, M. J.: 1975, *Monthly Notices Roy Astron. Soc.* **170**, 41.
York, D. G., Vidal-Madjar, A., Laurent, C., and Bonnet, R.: 1977, *Astrophys. J.* **213**, L61.
Zappella, R. R., Becklin, E. E., Matthews, K., and Neugebauer, G.: 1974, *Astrophys. J.* **192**, 109.
Zirin, H.: 1966, *The Solar Atmosphere*, Blaisdell, Waltham.
Zirin, H.: 1976, *Astrophys. J.* **208**, 414.
Zuckerman, B.: 1980, *Ann. Rev. Astron. Astrophys.* 18, 263.

J. P. Cassinelli,
Univ. of Wisconsin,
Madison, Wisconsin 53706
U.S.A.

K. B. MacGregor,
High Altitude Observatory,
National Center for Atmospheric Research,
Boulder, Colorado 80303,
U.S.A.

CHAPTER 19

SOLAR AND STELLAR MAGNETIC ACTIVITY

ROBERT W. NOYES

1. Introduction

The Sun is a magnetic star, and almost all of the solar surface phenomena which attract the interest of solar physicists are influenced or even dominated by its magnetic fields. The list of such phenomena includes sunspots, pores, plages, the photospheric and chromospheric network, spicules, prominences, coronal loops, coronal rays and streamers, coronal holes, the solar wind, and many others. Solar magnetic variability gives rise to time-varying phenomena on scales from seconds to millennia, including flares, magnetohydrodynamic waves, photospheric flux emergence, growth and decay of magnetic regions, the sunspot cycle, and long-term variations in cycle amplitude such as the Maunder Minimum. Even the solar luminosity appears to be controlled by the magnetic field.

Because the Sun is a rather ordinary star, with no obvious characteristics that distinguish it from other late-type dwarf stars, it is reasonable to suppose that magnetic fields play an equally important role in creating surface activity on other similar stars. To the extent that we can observe that activity and compare and contrast it with solar activity we shall enormously broaden the range of physical parameters for which magnetic activity is seen, and our understanding of its nature will increase accordingly.

Progress towards such a goal has long been sought, but only rather recently have there been really significant advances, to the point where the 'solar–stellar connection' has become an exciting area of solar (and stellar) research. In this chapter we review some aspects of solar–stellar research that are particularly relevant to understanding magnetic activity.

The most widely accepted explanation of solar magnetic activity is that it is rooted in a self-regenerating, reversing dynamo, created by the joint interaction of convection and differential rotation (see Chapter 5 by Gilman in this work). While dynamo models can be made to reproduce many observed properties of the activity cycle, physically self-consistent dynamo calculations do not successfully predict all observed properties (Gilman and Miller, 1981; La Bonte and Howard, 1982). Nevertheless, there is little doubt that, whatever the true nature of the dynamo mechanism, it depends on the interaction of convection and differential rotation. It is reasonable, therefore, to expect stellar magnetic activity to depend in a basic way on stellar rotation and convection. As we discuss below, this expectation is well borne out by the observations. These observations reveal a correlation between rotation and magnetic activity in convecting stars, as well

Peter A. Sturrock (ed.), Physics of the Sun, Vol. III, pp. 125–153.

as a dependence on spectral type which probably reflects the variation of convective properties with spectral type. While study of the details of stellar magnetic activity as a function of rotation rate and spectral type (or convective properties) is only just beginning, it is already clear that this study is providing both a useful regime for testing existing theories of stellar magnetic activity, and fertile ground for the growth of new theory.

One important aspect of solar magnetic activity is its activity cycle, which provides the most important of all clues to the mechanism of solar magnetic field generation. The recent detection of activity cycles on other stars has furnished a wealth of new information about stellar activity, and it may be anticipated that the characteristics (cycle period, amplitude, and growth and decay rates) of stellar activity cycles will provide stringent tests of any magnetic dynamo theory.

The observation of magnetic activity on other stars is important not only because it allows us to study different regimes of convection and rotation. Such observations permit study of a third important parameter – stellar age. It is a widely held view that stars spin down as they age, due to angular momentum loss produced by stellar winds. As stars spin down, the amplification of magnetic fields becomes weaker, and surface magnetic activity declines. The age-dependence of magnetic activity, as well as rotation, can be learned from a study of late-type stars whose ages are known independently. Of particular interest are stars in galactic clusters, which are approximately coeval, or the components of widely spaced binaries, which share a common origin but whose rotation and activity may evolve differently, depending on their individual spectral types.

A fortunate separation of the roles of age and rotation in producing magnetic activity is provided by the components of RS CVn binaries, whose rotation is maintained against spindown by tidal torques. The evidence from such stars further verifies the concept that rotation, rather than age, is the primary determinant of magnetic activity, and that the decay of stellar activity with age is simply a consequence of spindown with age.

As a byproduct of our learning in detail how the magnetic activity and rotation of late-type stars decrease with age, we may infer the life history of rotation and activity in our own Sun. It has been argued that the Sun is anomalous in having a much lower rotation rate and activity level than other G2 V stars its age; however, the data we review suggest that its departure from the mean for these properties is not large enough to be significant. It therefore seems reasonable that the Sun's magnetic activity and rotation did in fact evolve in a way similar to that of other main sequence stars of similar spectral type.

The purpose of this chapter is twofold: (a) to survey evidence for overall magnetic activity levels in stars from the perspective of our knowledge of solar magnetic activity in order that we may apply, where appropriate, our more detailed knowledge of the Sun to stellar situations and extend our description of solar magnetic phenomena to the differing physical regimes presented by other stars; (b) to extract from observed magnetic behavior of other stars, as compared with the Sun, clues about the nature and evolution of stellar magnetic activity in general, and solar activity in particular. Section 2 addresses the first point, in surveying some properties of late-type stars that appear similar to, and invite comparison with, solar magnetic activity. Section 3 discusses more specifically the relation between stellar rotation, activity, and age, and also the implications for the evolutionary history of solar and stellar magnetic activity. In Section 4 we close with some suggestions for future research.

The subject of solar and stellar magnetic activity is a very large one, and in order to keep this review sharply focused and of manageable length, a number of interesting topics are left out. Thus, we do not discuss transient stellar activity (flares), which, although extremely interesting to compare with solar flares, do not seem critical to the central issue of the basic nature and evolution of solar and stellar magnetic activity. Recent reviews of solar and stellar flares are given by Mullan (1977), Linsky (1977), and Gershberg (1975). We also do not discuss solar and stellar winds, even though they appear to play a fundamental role in determining magnetic activity evolution through spindown. Finally, we focus our discussion mainly on magnetic activity in single, main sequence stars with convective envelopes – stars that may be more or less directly compared with the Sun in various stages of its main sequence evolution. Thus we omit discussions of magnetic fields or activity in T-Tauri stars, early-type stars, close binary systems, or magnetic white dwarfs. We also omit discussing activity in late-type giants, even though magnetic field generation in such stars could involve processes very similar to the solar dynamo.

2. Solar and Stellar Magnetic Activity: A Phenomenological Comparison

2.1. SURFACE MAGNETIC FIELDS AND THEIR EFFECTS ON STELLAR RADIATIVE FLUX

In the solar photosphere, magnetic fields are invariably seen in concentrated form. The field strength in magnetic structures may range from about 1200 gauss in small elements of the quiet solar magnetic 'network' to 3000 gauss or more in sunspots, but nonzero fields less than about 1000 gauss have not been conclusively detected. The field apparently emerges through the photosphere in concentrated flux bundles (Zwaan, 1978), but then may be further concentrated by radiative cooling and convective collapse (Spruit, 1979; see also Parker, 1978). Spruit and Zweibel (1979) calculate that when the field within a flux tube reaches a value of 1350 gauss, the flux tube becomes convectively unstable. While upward flow within the tube causes it to expand dissipate, downward flow causes it to contract to a still higher field configuration. The maximum field in the Sun and other late-type stars should be of the order $(8\pi P_g)^{1/2}$, thus balancing the photospheric gas pressure P_g. Zwaan (1977) has pointed out that all photospheric magnetic elements, ranging from network elements through pores to sunspots, may be explained by a set of magnetohydrostatic models near-vertical flux tubes in the upper convection zone, within which convective heat flow is much reduced. Horizontal heat flow from the surrounding nonmagnetic photosphere into the flux tubes is also very much reduced. If the flux tube is large enough in diameter (a pore or a sunspot), the result is a dark feature. Small flux tubes in faculae or the chromospheric network, however, appear bright (especially towards the limb) due to horizontal radiative heat transfer in the photospheric surface layers where the photon mean free path becomes comparable to the width of the flux tube.

The concentration of magnetic fields into discrete structures with surface magnetic pressure $B^2/8\pi$ comparable to the photospheric gas pressure, as observed on the Sun, has been shown by Weiss and coworkers (e.g. Galloway and Weiss, 1981) to be required by the interaction of convection and magnetic fields. In other words, in stars with convection zones, homogeneous surface magnetic fields would be impossible. To the extent that magnetic fields exist at all in stars with convective zones, field concentrations, including perhaps dark starspots, should then be anticipated. On the other hand, for stars earlier than about F5, where convective zones are weak or absent, homogeneous global-scale fields might be anticipated. This dichotomy appears to explain the difference between the global, time invariant fields inferred for magnetic A_p stars and the intermittent, time-varying fields seen on the Sun and other convective stars (Zwaan, 1977, 1981a).

One body of evidence for spatially intermittent concentrated magnetic fields on other convective stars is the apparent presence of dark structures in the photospheres of many of these stars, as inferred from photometric observations of decreases in total stellar radiative flux received at the telescope. The photometric light curves are consistent with the hypothesis that dark 'starspots' are rotating across the stellar disk. However, until recently there has been no empirical evidence that physical entities like sunspots actually produce a decrease of total luminous flux integrated over the stellar surface. Sunspot umbrae, whose effective temperatures are typically about 1700 K cooler than the nonmagnetic photosphere, have an emittance in visible light less than about 20% that of the photosphere. However, there is no *a priori* reason to expect that the 'missing flux' would not reappear in other places on the disk, thus canceling the flux deficit and making the spot's presence undetectable by photometry of the disk-integrated sunlight or starlight.

Recently Willson and colleagues (Willson *et al.*, 1981) have used a space-borne precision solar radiometer to detect variations of solar irradiance, to a fractional sensitivity of 10^{-5}. Variations as large as 1.5×10^{-3} were observed. These variations were remarkably well correlated with the variation of area of the visible disk occupied by sunspots during the period of observation. The amplitude of the variation was found to agree rather closely with expectations if there were no instantaneous compensating increase of photospheric radiation away from the spots; in other words, the deficit calculated from the known area and emergent flux of the spots agreed closely with the observed deficit (Willson *et al.*, 1981).

This important result has several interesting implications for solar-stellar research. First, by demonstrating that sunspots *do* affect the total irradiance (i.e. that the 'missing flux' is at least temporarily missing and not redistributed quickly to the surrounding photosphere), the observation supports the interpretation that the very much large photometric variations seen on other stars are due to the stellar analog of magnetically formed sunspots. Second, and more fundamentally, the observation implies that the flux blocked by the sunspot is stored in the convection zone, and leaks out later at a rate so small that the irradiance increase, although inevitable (by conservation of energy), is unobservably small. It is important to extend these observations to greater sensitivity levels and longer observation times, in order to determine the characteristic rates of storage and subsequent leakage of the missing flux (see Foukal and Vernazza, 1979).

As we have noted, the total variation of luminous flux from the Sun due to sunspots is found to be only about 10^{-3}, which is close to what is calculated from areas and temperatures of spots, assuming that the 'missing energy' in spots is stored in the hydrogen convection zone. A variation of 10^{-3} (i.e. 0.001 magnitude) is at the limit of detectability of stellar photometry, so spots similar to sunspots would be almost undetectable on other stars. However, large variations – sometimes up to 30% – are frequently seen in the luminosity of late-type stars (cf. Hartmann, 1981). In spite of their large amplitude, these variation have many properties in common with what one might expect from hugely scaled-up sunspots, and are generally ascribed to 'starspots'.

Starspots occur in two principal classes of stars: (a) BY Dra stars, which are rapidly rotating (less than 4 day period) dwarf M stars, occurring sometimes singly and sometimes as members of binary pairs; and (b) RS CVn stars, which are evolved subgiant K stars, also rapidly rotating (period generally less than 15 days), and are always members of binary pairs.

The word 'starspot' is somewhat presumptuous, since the critical proof of commonality of physical origin – namely a strong umbral magnetic field – has not yet been demonstrated. Nevertheless, the above-mentioned similarities in behavior to sunspots make the case for a similar physical origin rather strong. Specifically:

(1) The luminous flux variations are quasi-periodic, as if some dark feature is modulating the light as the star rotates. For stars whose rotational velocity $v \sin i$ has also been measured, the calculated rotation period is close to the observed period of photometric modulation, implying that the dark features are corotating with the surface layers.

(2) The photometric variation is not strictly periodic, but (like the solar luminous flux variations that now may be attributed to sunspots) varies in amplitude and phase, presumably as starspots are born and die at various locations on the stellar surface. Sometimes significant variations are seen in times less than one day (Oskanyan *et al.*, 1977).

(3) The phase and amplitude coherence time of stellar photometric modulation is often months or longer; this is long compared to the convective turnover time in the stellar convection zone, and hence suggests (Hartmann and Rosner, 1979) that the convection is being drastically modified by some other influence. Magnetic fields are the only known phenomenon energetically capable of such modifications over time-scales of the order of months.

(4) Starspots appear to be cooler than the surrounding photosphere, by amounts of several hundred degrees or more. This is deduced from broad-band color changes between 'spot-covered' and 'spot-free' faces of the star (Vogt, 1975; Torres and Ferraz-Mello, 1973; Bopp and Espenak, 1977) or from the strengthening of temperature-sensitive TiO bands during the 'spotted' phase of the photometric curve (Ramsey and Nations, 1980).

(5) There is an association between minima of photometric brightness (i.e. when the 'spot' is on the side of the star facing the Earth), and stellar chromospheric phenomena similar to solar phenomena associated with sunspots. Specifically, at these times (a) there is an enhancement of chromospheric Ca II and other UV emission, as if the 'spot' were surrounded by chromospheric plages and (b) there may be a *decrease* in the rate of mass loss or stellar wind, as is found in the Sun (see Baliunas and Dupree, 1982).

(6) The amount of 'starspottedness' shows long-term changes, and occasionally

evidence for quasi-cyclical variations with decade-long time-scales (Phillips and Hartmann, 1978), in a stellar analog of the sunspot cycle. It has also been noted that stars may pass from a long-term spot-free state (possibly analogous to the Maunder Minimum) to one showing large amplitude photometric variability indicating a significantly star-spotted state (Hartmann, Londono, and Phillips, 1979). The data on these points are scantly, for sufficiently accurate photometry on decades-long time spans has been obtained for only a few stars.

(7) Spots on stars in tidally synchronized binary systems produce modulations at periods close to, but not exactly equal to the binary revolution period. In other words the spot migrates gradually relative to a coordinate frame rotating at the system's orbital period. At first glance the difference might seem due to lack of precise phase-locking of rotation rate and orbital period. However, the migration rate changes over time-scales of years, occasionally even switching from retrograde to prograde (Hall, 1981). This could be due to differential rotation with latitude, as observed on the Sun, combined with migration of sunspot zones in latitude, as is also seen on the Sun (Spörer's law, embodied in the familiar 'butterfly diagram'). Needless to say, if differential rotation combined with a stellar Spörer's law, can be established as the cause of variable star spot rotation rate (perhaps by observations over one or more 'starspot cycles'), the information will be of key importance for solar and stellar dynamo theory. This is because differential rotation, as much as rotation itself, is a fundamental parameter affecting dynamo activity.

Notwithstanding the overall similarities between starspots and sunspots, there are significant differences, which emphasize the danger of simply appropriating known characteristics of sunspots to starspots. The most striking difference is the one we have already noted: the starspots giving rise to observable luminosity variations are, relatively speaking, huge. The total decrease of luminous flux of more than 10% in many cases (rather than 0.01–0.1% as on the Sun) means that the fractional area covered is 100–1000 times as great. If the darkening is caused by a single spot, its radius must be of the order of 30% of the stellar radius. On the other hand, it is possible to match the observed light curve with a collection of many small spots (Eaton and Hall, 1979). Given the uncertainties in the data and the large number of free parameters involved, it seems premature to draw firm conclusions about the geometry of the dark areas.

In any case, if as photometric observations suggest, spots on some stars can at least in aggregate cover more than 10% of the area of the visible disk, their effects on the structure and dynamics (i.e. convective energy flow) in the stars convection zone, will be far greater than for the Sun. It has been pointed out by Hartmann and Rosner (1979) that for BY Dra stars the 'missing flux' is not simply redistributed into other spectral regions or redistributed spatially over the surface of the star; rather, it must be stored in the convection zone for times long compared to the convective energy transport time-scale, although short compared to the thermal time-scale of the envelope. The storage is effected through magnetic suppression of the efficiency of convection in the convection zone, possibly by as simple a mechanism as freezing of convective motions wherever the field in the convection zone exceeds a critical value.

Stars with very large inferred spot areas are generally of late K or M spectral type, and thus reflect an apparent tendency for fractional area coverage of spots to increase with advancing spectral type. In addition, there is some evidence that for stars of earlier (G and early K) spectral type, spot areas increase with increasing overall level of chromospheric activity and thus with increasing youth. Thus, late F to K stars in the Hyades, whose age is about 700 million years, are variable at the 1% level, probably due to starspots covering 1% or more of the stellar disk (Radick *et al.*, 1982). Also, photometric studies of the G2 V star HD 1835 (Chugainov, 1980) indicate that this star, of the same spectral type as the Sun but with age appropriate to Hyades stars (i.e. some 600 million years), has a spot area covering about 3% of the disk, or some 30 times the area of sunspot coverage at solar activity maximum.

The continuity of photometric variations from the Sun to other stars both younger in age and later in spectral type suggests that in both cases we are seeing a phenomenon very similar to sunspots, but on a greatly expanded scale.

Sunface magnetic fields may cause stellar photometric modulation in ways other than the creation of starspots. On the Sun, magnetic flux concentrations of size less than about 5×10^{20} Mx do not create dark sunspots, but rather produce complex patterns of surface magnetism on size scales from the subtelescopic up to large-scale photospheric faculae underlying active regions. These features do not affect the visible light emission from the Sun to the degree that sunspots do, although magnetic active regions are easily visible near the limb as white light faculae. Foukal and Vernazza (1979) report the marginal detection of faculae in the solar irradiance measurements of Abbott, at the level of $\Delta I/I \sim 7 \times 10^{-4}$. This suggests the possibility, at least, that some of the photometric variations seen in other late-type stars could be due to stellar analogs of magnetic faculae. Indeed, it has been suggested (Oskanyan *et al.*, 1977) that the 'spots' on the prototype spotted star BY Dra are sometimes dark and sometimes bright.

It is possible that magnetic fields in the convection zone can alter solar or stellar luminosities without producing any inhmogeneity in surface brightness. Spiegel and Weiss (1980) argue that deep-lying magnetic fields at the base of the solar convection zone can alter the properties of convection so as to cause a modulation of solar luminosity over the 11 year activity cycle at the 0.1% level. Similar arguments are presented by Dearborn and Blake (1980), but with respect to the superadiabatic region in the outer 3000 km of the hydrogen convection zone. They find that an increase in the amount of magnetic flux penetrating the superadiabatic zone causes a decrease in luminosity, with the energy deficit being stored as gravitational potential energy. A total magnetic flux of 2×10^{24} Mx (approximately equal to the total magnetic flux at the base of the convection zone; see Galloway and Weiss, 1981) is calculated to cause a decrease of 1% in solar luminosity. In either case, the result would be a modulation of solar luminosity with a period of the 11 year magnetic cycle. The theoretical calculations underlying such predictions will require considerable elaboration and, if possible, observational confirmation by techniques such as Willson's (Willson *et al.*, 1981) before they can be trusted as a likely explanation for either solar or stellar luminosity variations, in addition to those caused by spots or faculae directly.

2.2. Direct Detection of Magnetic Fields on Stars Like the Sun

The magnetic field on the Sun seen as a star averages about 0.15 gauss (Severny *et al.*, 1970; Scherer *et al.*, 1977). This is the value as measured with a Babcock-type magnetograph, an instrument which records the mean Zeeman shift of a spectral line, averaged over the entire solar disk. The value has little physical meaning, for as we have noted, in magnetic regions of the photosphere the actual field is 10^4 times larger. The great reduction in field strength, as measured by standard magnetographic techniques is due to the facts that (a) a large fraction of the photosphere is essentially field-free and (b) positive and negative fields tend to cancel each other. The cancellation occurs because most magnetic flux tends to occur in bipolar regions whose characteristic size is small compared to the solar radius, and whose opposite polarities then tend to cancel in the mean field.

Unfortunately, the most sensitive stellar spectrographs have a sensitivity to Zeeman shifts of about 20 gauss (Brown and Landstreet, 1981), so the mean field of the Sun would be below the threshold of detectability if it were at a typical stellar distance.

It is possible, however, to detect the net Zeeman broadening, rather than the Zeeman shift, due to magnetic fields on the Sun seen as a star (Robinson, 1980; see also Unno, 1959; Preston, 1971). Fields of both polarities contribute equally to Zeeman broadening, so there is no cancellation effect; the signal is however weak due to the small fractional area of the solar surface covered by nonzero fields. The technique of measuring differential Zeeman broadening between spectral lines of high and low magnetic sensitivity has been applied successfully (Robinson, Worden, and Harvey, 1980) to magnetically active stars with plages that apparently cover a significant fraction (10–20%) of the surface. It would not be successful (Marcy, 1982), for a star like the Sun in which active regions cover substantially less than 10% of the surface area.

Robinson, Worden, and Harvey detected Zeeman broadening in disk-averaged light from ξ Boo A (G8 V) and 70 Oph A (K0 V), indicating the existence of strong surface fields (2550 and 1880 gauss, respectively) and large fractional area coverage (20–45% and 10% respectively) in the two stars. Later, Marcy (1981) carried out a similar observation of ξ Boo A, and failed to detect evidence for photospheric fields in the star at that time; he suggested that, as on the Sun, the amount and detectability of magnetic activity changes with time. Thus Marcy's result does not negate the important implication of Robinson, Worden, and Harvey's work – that some other lower main sequence stars have magnetic active regions very similar in properties (except perhaps for geometrical size) to those on the Sun. It is hardly necessary to emphasize that extending this work to further studies of the same and other stars is of critical importance to our understanding of solar and stellar magnetic activity.

It would be extremely useful if magnetic fields could be detected *within* starspots. Not only would such detection immediately verify their magnetic origin, but in addition determination of the field strength would permit deriving magnetohydrostatic models of temperature and density within the spot, for comparison to sunspot models. The observational problem is very difficult because of the darkness of starspots, and remains a challenge for the future.

2.3. Ca II H and K Emission as Indicators of Stellar Magnetic Fields

The Ca II H and K line cores are probably the most thoroughly studied spectral feature in late-type stars. It is clear from the solar analogy that stellar Ca II emission reveals the presence of a stellar chromosphere. The strength of the emission gives a measure of total radiative losses, and hence heat input, to the chromosphere. In the Sun, there is a clearly established one-to-one spatial relationship between areas of detectable photospheric magnetic field and enhanced Ca II emission (Leighton, 1959) which persists to scales at least as small as 1 arcsec. There also appears to be a smooth dependence of strength of the Ca II emission upon magnetic flux integrated over the observing aperture (Skumanich *et al.*, 1975). Thus the total emission in the Ca II line cores provides a very useful, if indirect, empirical tool for studying stellar magnetic fields. Unlike Zeeman shifts, Ca II intensification occurs independently of the algebraic sign of the field, so that opposite fields on the stellar surface add instead of canceling. As a result, Ca II emission enhancements are a gross, rather than subtle, effect. Therefore, to the extent that the reality of a simple relation between magnetic flux and Ca II emission can be assumed or demonstrated, Ca II observations are the most widely applicable probe (accessible from ground-based observatories) of magnetic activity in late-type stars. For this reason it is extremely important to obtain enough direct magnetic field measures on stars showing Ca II emission to verify the extrapolation of the solar magnetic field/Ca II emission relation to other stars. In addition, the quantitative relationship between Ca II emission and surface magnetic flux may be expected to vary with T_{eff}, g, and stellar metallicity. This relationship must be determined, through direct field measurements, before Ca II emission data can be considered a quantitative indicator of stellar surface magnetism.

On the Sun seen as a star, the total Ca II emission varies significantly on several time-scales: (a) about 27 days due to rotational modulation as active regions pass onto and off the disk; (b) weeks to months due to the growth and decay of active regions; (c) up to several years due to recurring activity within a single complex of activity; and (d) 11 years, due to the activity cycle. White and Livingston (1981) showed that the intensity of the Ca II K_3 central emission increased by 30% from solar minimum in 1975 to solar maximum in 1980, while the emission within a 1 Å window centered on the K line (directly comparable to stellar data to be discussed below) increased by 18%. These increases are due almost entirely to increased area occupied by chromospheric plages, rather than changes in the brightness of the quiet chromospheric network, or in the brightness of plages, at least on the rising side of the cycle.

In comparison to the Sun, many other late-type stars show substanially higher Ca II chromospheric emission (i.e. corrected for the 'photospheric' radiative equilibrium component, and normalized to the surface area of the star). It is an interesting question whether these stars are brighter in Ca II because their plages are bigger than those on the Sun (implying a different behavior of the field-generating mechanism that causes the field to erupt over larger areas), or because their plages are brighter per unit area than the Sun's. The observed brightness of Ca II averaged over the surface of 'active chromosphere' stars such as ϵ Eri (K2 V), ξ Boo A (G8 V) or 70 Oph A (K0 V) is comparable to that of a solar active region (Hartmann *et al.*, 1979; see also Kelch, 1978; Kelch, *et al.*, 1979; and Simon, *et al.*, 1980). The simple idea that these stars are like the Sun if it were completely covered with active regions, however, cannot be correct in detail. As we shall discuss immediately below, late-type stars generally show evidence for a substantial degree of large-scale chromospheric inhomogeneity. A mean surface brightness equal to that of solar plages but also an inhomogeneous structure of course implies that the mean surface brightness of plages exceeds that of solar plages.

The magnetic field in solar active regions is highly intermittent, having a 'filling factor', or fractional area coverage, of at most 10 to 30% (Zwaan,1981b). Therefore it is possible for active regions on other Sun-like stars to have mean surface brightness considerably larger than the Sun's simply through larger filling factors. In other words, the data do not rule out the possibility that the chromospheric emission directly above a *resolved* photospheric flux tube in a plage is the same on an active chromosphere star as on the Sun. Indeed, for an active chromosphere star of the same spectral type as the Sun (and therefore with about the same photospheric gas pressure and hence surface magnetic field strength, and also with similar convective properties) such a circumstance seems plausible.

Perhaps the best indication of inhomogenous surface brightness distribution of magnetic active regions on stars comes from the recent work of Vaughan *et al.*, (1981). They observed the rotational modulation of chromospherically 'active' and 'quiet' stars, revealed by the change from night to night of the ratio of emission in 1 Å bands centered on Ca II H and K to that in 25 Å bands equally spaced on each side of the Ca II lines. Typical rotational modulation curves are shown on the right-hand side of Figure 1. While motivated principally by the desire to obtain accurate rotation periods (discussed in Section 3 below), Vaughan *et al.* also obtained some information on size, lifetimes, and contrast (relative to the quiet chromosphere) of inhomogeneous structures on young and old G and K dwarf stars. It was found that the chromospherically more 'active' (and younger, as we describe below) stars always showed large fluctuations over time-scales of several days, and that in most cases the modulation showed a strong quasi-periodic component, as if produced by rotation modulation of surface structures which had a lifetime greater than one rotational period. The patches cannot be numerous, evenly distributed small structures akin to the solar chromospheric network elements; in order to produce the observed modulation the bulk of the emission must come from patches isolated by a significant fraction of a stellar radius from other patches. By extension of solar terminology, these patches may be called stellar chromospheric plages.

Some of the less active (and older) stars in the sample of Vaughan *et al.* (1981) also showed clear rotational modulation, but the detected fraction was much smaller than for the active stars. Quite probably the smaller success rate is a selection effect, for the amplitude of fluctuation is considerably smaller for these stars (see also Wilson, 1978). This, of course, would be expected if, as we have suggested, plages on young 'active' stars are larger and have a higher surface brightness relative to the background chromosphere than plages on older stars (such as the Sun).

Vaughan *et al.* (1981) pointed out that the rotational modulation of young stars often persists essentially unchanged in amplitude or phase for an entire observing season. For several of these stars, phase coherence has been inferred over two successive observing seasons, in a follow-on study by Baliunas *et al.* (1983). In fact, analysis of the long-term data of Wilson (1978) indicates that coherence (albeit with apparent gradual drift in phase from year to year) persists for as long as 15 years in the star HD 165341, and

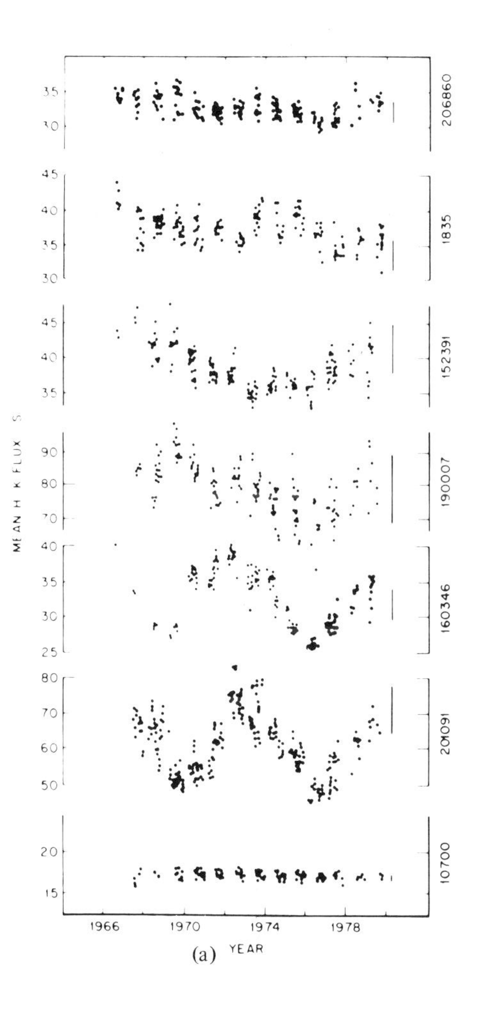

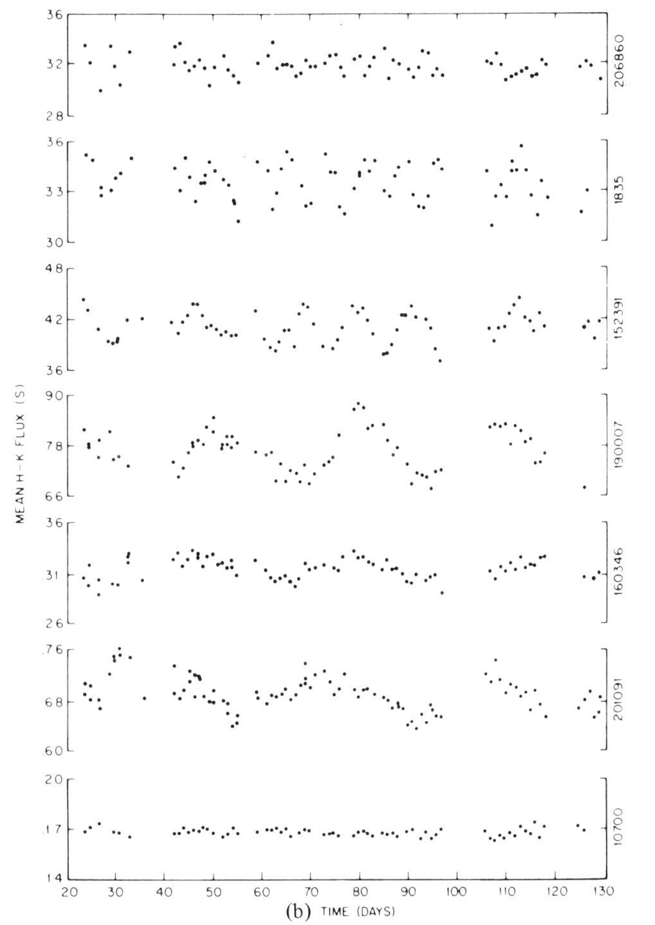

Fig. 1. (a) Mean H and K fluxes for seven of the stars monitored by O. C. Wilson, plotted *versus* time from 1966 to 1978. (b) Nightly observations of the same stars over a 130-day interval in 1980. The range of variation for each star is given by a vertical bar in figure (a). (From Vaughan *et al.*, 1981.)

possibly others (Noyes, 1982). During this period of time the overall level of chromospheric activity rose and fell, presumably as more short-lived active regions appeared and disappeared within an 'envelope' of long-term activity (see discussion on activity cycles below). The implication is that longitudes of recurrent activity ('active longitudes') are prominent on young late-type stars, and may persist for at least a decade or more. Such active longitudes have long been known on the Sun, but their chromospheric signature in integrated sunlight is probably too weak for them to be readily apparent in the Sun seen as a star. It is not yet clear whether the long-lived active longitudes on younger stars differ from solar active longitudes in physically significant ways, or whether they are simply high-visibility manifestations of rather similar behavior.

2.4. CORONAL ACTIVE REGIONS

We have discussed Ca II emission from the Sun and Sun-like stars at some length, in part because of the long history and ready availability of stellar Ca II observations. While Ca II emission presents the most obvious indicator of stellar chromospheric properties obtainable through ground-based observations, far ultraviolet and X-ray observations from space have now begun to extend our knowledge of magnetic activity in stars well beyond what can be learned from study of Ca II emission.

Spatially resolved solar X-ray and extreme ultraviolet observations reveal emission mainly in highly structured loop-like features in the corona ($T \gtrsim 10^6$ K) with a narrow transition region ($T \sim 10^5$ K) lying between the hotter coronal features and the cooler chromosphere. The coronal loops delineate closed magnetic field lines, whose 'footpoints' are anchored in the photospheric magnetic fields and hence are well correlated spatially with the structured chromospheric Ca II emission discussed above. Indeed, the transition region lines, observed in the far ultraviolet, show an emission pattern closely similar to that of the Ca II lines.

The solar coronal emission pattern differs significantly from that of the chromosphere and transition zone, in that coronal emission depends sensitively on the *geometry* of the magnetic field, as well as the magnetic flux per unit area. Magnetic fields in the corona consist of two distinct topological types: open and closed field lines. Coronal extreme ultraviolet and X-ray emission is relatively much weaker in open magnetic regions, known as coronal holes, within which the process of generation of the expanding solar wind gives rise to relatively low densities and temperature of the coronal plasma. By contrast, chromospheric or transition region emission is comparable in strength for both geometries, depending mainly on the magnetic flux per unit area and less on field line topology. In addition, in closed magnetic regions, coronal emission depends sensitively on the compactness of the magnetic loops, or equivalently on the separation of their footpoints. Compact (generally young) active region loops are far more intense emitters of coronal radiation than larger scale structures characteristic of older or decaying active regions. However, in the chromosphere the emission from closed loops is much less sensitive to the size of the loops, but depends mainly on the magnetic flux threading the emitting regions.

As a result, the ratio of coronal to chromospheric emission from compact solar active regions is much greater than from the 'quiet' Sun. It would not be surprising, therefore, if 'active chromosphere' stars showed a greater ratio of coronal to chromospheric emission than 'quiet chromosphere' stars like the Sun. This is indeed found to be the case (Mewe *et al.*, 1981); the ratio F_X/F_{HK} is found to increase appoximately as $(F_{HK})^{1.34}$, where F_X and F_{HK} are the stellar surface fluxes in soft X-rays and in the Ca II H and K lines, respectively. It has also been found (Oranje *et al.*, 1982) that the transition zone lines, even though their emission from the Sun is approximately cospatial with the H and K lines, have a mean surface flux F_{tr} varying like the coronal lines: $F_{tr}/F_{chrom} \sim (F_{chrom})^{1.4}$, where F_{chrom} is the surface flux from representative far ultraviolet chromospheric lines. This is probably due to the fact that the transition zone is heated primarily by downward thermal conduction from the corona.

Interestingly, and perhaps surprisingly, the above relations hold for stars of all luminosity classes from the main sequence (V) to luminous giants (II), irrespective of spectral type. In other words, the physics governing relative coronal and chromospheric heating rates appears to be relatively insensitive to stellar surface gravity and effective temperatures (Oranje *et al.*,1982).

Detailed studies of solar coronal loops have led to a reasonable understanding of the processes which control their density, temperature, and emissivity, and a number of attempts have been made to extrapolate this understanding to other stars. To review briefly the solar picture, it appears that the high density in magnetically confined loops is caused by magnetic field-dominated heating at coronal levels. By raising the temperatue of the interior volume of the loop, chromospheric material is caused to 'evaporate' upward into the loop, filling it to just the degree that radiative losses from the denser loop plasma just balance the heating rate (so that no excess heat is conducted down into the chromosphere to cause further 'evaporation'). If the input per unit volume rises, the density increases, and vice versa. The thermal structure of the loop is insensitive to the detailed distribution of heat input along the loop axis, for thermal conduction is highly efficient in smoothing out the temperature structure. A number of authors (e.g. Landini and Monsignori-Fossi, 1975; Rosner, Tucker, and Vaiana, 1978; and others; see Withbroe, 1981, for a review) have calculated the structure of coronal loops in static equilibrium, deriving relations between total heat input rate, temperature, and pressure for a loop of given size (the size of the loop is assumed to be determined by the surface magnetic field geometry). For example, Rosner, Tucker, and Vaiana derived the scaling relation $T_{max} \sim 1400(PL)^{1/3}$ for the relation between the temperature T_{max} at the top of a loop, its gas pressure P, and its length L. All these quantities are determinable from solar X-ray data, and it was found that the relation fits solar coronal loops of all sizes from bright point loops to large-scale coronal structures (provided the structures are static; flare loops do not fit the relation.

Such scaling laws are very appealing in that they strip extremely complex physical processes down to a few parameters, but it is possible that some important physics are also stripped away. For example, it has been argued (Raymond and Foukal, 1982) that the observed thermal structure of loops by no means has the simple conductively smoothed profile assumed in deriving the scaling laws, and that the loops may not even be in static equilibrium. It is also interesting that the scaling law quoted is insensitive to the heat input rate – either its nature or how it should vary from loop to loop on the Sun, or from star to star (see also Pallavicini *et al.*, 1981a; Withbroe, 1981). Some additional physics is needed to specify the heating rate.

Golub *et al.* (1980) have attempted to parametrize the heating of active region loops through twisting of the magnetic loop structure. They equate the heating rate to the generation of nonpotential magnetic energy, which is parametrized in terms of an azimuthal velocity v_ϕ at the photospheric loop footpoints and the ratio α of longitudinal to azimuthal field strength. By comparing with solar X-ray observations they find that the product αv_ϕ is approximately constant for active region loops of various sizes. They derive a scaling relation for the total energy content of coronal active region loops: $U_T \sim (\alpha v_\phi)^{0.9} {\Phi_T}^{1.7} L^{-0.6}$, where Φ_T is the total magnetic flux entering the loop. To the extent that such a relation is generally valid, it implies that stellar coronal loops have an energy content depending on the vigor of the surface convection (which determines the azimuthal velocity v_ϕ) as well as the characteristics of the dynamo (which determines the total magnetic flux Φ_T emerging in an active region and, through the geometry of the loop footpoints, the active region size L).

The comparison of stellar coronal data with these or other models of solar coronal structures is still at an early stage. Walter *et al.* (1980) and Holt *et al.* (1980) have observed X-ray emission from Capella and interpreted it in terms of a corona dominated by magnetically confined loops, as in the Sun. (Models in which the corona is gravitationally, rather than magnetically, bound to the star fail by several orders of magnitude to reproduce the observed X-ray luminosity.) Walter *et al.* calculated the size and number of loops on Capella and other stars using the scaling relations of Rosner *et al.* (1978), assuming for simplicity that all loops had the same size and pressure. They concluded that some stars (e.g. Capella) were only sparsely covered with loops, whereas others (e.g. UX Ari) must be essentially completely covered with active regions and loops. This conclusion should be subject to observational test from X-ray (if not Ca II) rotational modulation studies.

Swank *et al.* (1981), using data from the Solid State Spectrometer on the Einstein Observatory, were able to determine temperatures for coronal loops on several RS CVn stars. They found two regimes of temperature: (a) plasma at 4 to 8×10^6 K, not greatly in excess of the hottest solar active region loops, and (b) plasma an order of magnitude hotter, from 20 to 100×10^6 K. The lower temperature material is probably confined near to the surface, just as are solar coronal loops; however it is not yet clear whether the hotter plasma is also confined near the surface (in which case the gas pressure must exceed 100 dyn cm^{-2}) or extends far above the surface, so far that it would fill the binary orbit of the RS CVn pair. In the latter case the situation is so far removed from the Sun that detailed comparisons with solar loops seems not too useful.

2.5. Magnetic Activity Cycles

In 1966, Wilson (1978) began a pioneering long-term study of nearby lower main sequence stars, to see if it is possible to detect stellar analogues of the 11 year sunspot cycle. He monitored the long-term variations of integrated emission in 1 Å bands centered on the Ca II H and K lines, normalized to broader bands situated in the nearby 'continuum' for 91 lower main sequence stars. This work was partly inspired by indications that the solar cycle could be detected as an 11 year modulation of Ca II emission seen from the Sun as a star (Sheeley, 1967). Recent photoelectric monitoring of the current solar cycle beginning with the 1975 activity minimum bears out these indications quantitatively (White and Livingston, 1981). The amplitude of the solar 11 year modulation is such that if the Sun were at several parsecs, its cycle would be clearly detectable in

Wilson's survey. (Wilson actually detected the solar cycle, using sunlight reflected by the moon.)

It was obvious that the detection of stellar activity cycles can richly contribute to our understanding of the solar activity cycle, for we may expect to measure many properties of stellar activity cycles that can be directly compared with analogous properties for the Sun. Some basic cycle properties of interest are period, amplitude, and cycle morphology (i.e. rate of rise and fall, smoothness, etc.) If enough stars can be measured, one may learn empirically how these properties depend on fundamental stellar attributes such as mass, rotation rate, or age.

As became clear in Wilson's (1978) landmark paper, the results clearly show the hoped-for result: activity cycles on late-type stars are by no means uncommon. Of 73 stars surveyed for their activity variation (excluding an additional 18 used as standards), most were found to show significant long-term Ca II variations; many of these were cyclical in nature (Figure 1). Quite often the variations were rather similar to the solar activity cycle, in period, amplitude of Ca II variations, and waveform. This tendency has been borne out by continuing observations since Wilson's (1978) report, during which additional stars have been observed to complete Ca II activity cycles. Vaughan (1983) notes that well established cycles can be seen ranging in period from as short as 7 years to about 15 years or more (in the latter case the full waveform has not yet been seen, but the partial waveform suggests a regular cycle), and points out that the separation of observed maxima of the sunspot number also have varied from about 8 to 17 years. It is interesting, and perhaps somewhat surprising, that the period of activity cycles observed so far is so similar to that of the solar cycle, in spite of the wide range of spectral type, rotation rate (Vaughan *et al.* 1981), and age of these stars. For example Belvedere *et al.* (1980) calculate that there should be a rapid increase of cycle period towards later types, on the basis of α–ω dynamo theory. However, the obvious selection effect that periods much longer than the total survey duration cannot be reliably inferred weakens the significance of the lack of apparent trend in Wilson's data. It may be noted, by contrast, that Hartmann and colleagues (see Hartmann, 1981) have detected cycle-like 'starspot' variations with apparent periods of the order of 50 years in M dwarfs, using the extensive Harvard photographic plate collection.

The waveform of the cycles detected by Wilson varies from star to star, some stars showing a typical solar-type waveform with the rise of activity being rather steeper than the subsequent decline, but others showing the reverse. There are cases in Wilson's sample of stars (especially F7–G3 stars) showing long-term secular decreases over the period of observation; the preponderance of observed decreases over increases suggests (Wilson, 1978) that the curves may be strongly asymmetrical, with short periods of rapid recovery following long periods of show decline. There is no obvious relation between wave form and other properties (i.e. cycle period, or mean chromospheric emission level) in Wilson's data.

The amplitude of the detected stellar activity cycles, from minimum to maximum of Wilson's H and K emission index, varies from about 10% to as much as 50%. The Sun's variation of 18% in a similar 1 Å K index from minimum in 1975 to maximum in 1980 (White and Livingston, 1981) place it between the extreme amplitude ranges detected in Wilson's study, close to but not at the lower limit of detection.

Vaughan (1980) pointed out that there is an apparently significant trend in Wilson's stellar variability data: The less chromospherically active (and presumably older) stars show smoothly undulating cycles (Figure 1), whereas the most chromospherically active (youngest) stars exhibit chaotic long-term behavior, in which a cycle, if it exists at all, is almost buried beneath short-term fluctuations.

3. The Rotation/Activity/Age Connection

In the remainder of this chapter we turn to the second principal question raised in the Introduction: 'How does magnetic activity on stars change as the star ages?' Again we restrict ourselves to lower main sequence stars like the Sun, for which an abundance of new data on rotation, surface activity, and age is now becoming available.

3.1. THE AGING OF MAGNETIC ACTIVITY AND ROTATION

The fact that chromospheric activity decreases with stellar age has been known for many years, starting with the work of Delhaye (1953) who noted that the velocities perpendicular to the galactic disk of dMe stars were very much less than dM stars, and concluded that the active emission-line stars must be much younger. Wilson (1968) and Wilson and Wooley (1970) obtained extensive data on Ca II emission in nearby stars, as well as data on the space velocities of the same stars, and showed that the stars with smaller Ca II emission were older, based on kinematic properties. Skumanich (1972) put the results on a quantitative basis by comparing Ca II emission indices for the Pleiades, Ursa Major, Hyades clusters, and the Sun, all of which have reasonably well-known ages. After converting the observed Ca II emission indices to values appropriate to a common spectral type, Skumanich derived the well-known relation that Ca II emission decays as $(\mathrm{age})^{-1/2}$ (see Skumanich and Eddy, 1981). Given the apparent linear relation between excess (chromospheric) Ca II flux on the Sun and surface magnetic flux, (Leighton, 1959; Skumanich, *et al.*, 1975) it was natural to infer that the mean magnetic flux of late-type stars, presumably brought to the surface via dynamo activity, declines as $(\mathrm{age})^{-1/2}$.

At the same time the idea was current that magnetic field generation in convecting stars was due to dynamo activity induced by the interaction of rotation with convective motions (Parker, 1955; Babcock, 1961; Leighton, 1969) and it was natural to expect the younger stars, with high magnetic activity, to be the more rapidly rotating. This was shown to be the case by Kraft (1967) who found a striking correlation of rotation rate with Ca II emission in F and G main sequence stars in the field and in the Hyades and Pleiades; the mean rotation rates of Pleiades, Hyades, and field stars decreased with age along with the Ca II emission. Skumanich's (1972) analysis showed that, as with Ca II emission, the mean rotation rate $\langle v \sin i \rangle$ of Pleiades, Ursa Major, Hyades, and field stars, declined as $(\mathrm{age})^{-1/2}$.

The decrease of rotation rate with age was in gratifying agreement with theoretical suggestions of a number of researchers (Lüst and Schluter, 1958; Schatzman, 1959, 1962, 1965; Mestel, 1968; and others) that mass ejected from stellar winds or flares would also carry off angular momentum, thus de-spinning the star. Surface magnetic fields are important in the process of angular momentum loss primarily through their action in 'stiffening' the corona so that the lever arm for angular momentum loss is increased.

This hypothesis was bolstered by the observation that the solar wind does in fact carry off angular momentum from the Sun (Brandt,1966; Weber and Davis, 1967). Durney (1972; see also Durney and Stenflo, 1972) showed that under the assumptions that (a) the mean magnetic field strength of open field regions is proportional to the stellar angular velocity, (b) the Sun rotates as a solid body, (c) the solar wind velocity is constant over time, then both angular velocity and surface magnetic field strength decrease as $(\text{age})^{-1/2}$.

3.2. ROTATION AS THE FUNDAMENTAL DETERMINANT OF MAGNETIC ACTIVITY

A skeptic might argue that, as stars age, both their rotational velocity and surface magnetic activity level decrease independently, for unrelated reasons, in spite of the rather plausible picture sketched above. However, there is abundant observational evidence that rotation, rather than age, is in fact the governing parameter. The clearest evidence comes from stars that are members of binary systems whose members are close enough that tidal interactions maintain synchronism between the stellar rotation period and the orbital revolution period (de Campli and Baliunas, 1979). Angular momentum, lost to the star through winds or magnetic activity, is replaced from the much larger reservior of angular momentum contained in the system's orbital motion, ahd the star does not de-spin. It is found that stars in such systems have anomalously high rotation rates compared with single stars of their age, and also anomalously high levels of surface magnetic activity.

Middelkoop (1981) has shown that main sequence single-lined spectroscopic binaries with periods less than 9 days have orbital eccentricities less than 0.1. This implied that the rotational and orbital motions are probably synchronized by tidal friction, Middelkoop found the chromospheric emission of the binaries with short orbital period and low eccentricity (and hence probable rapid rotation) to be significantly higher than the average for main sequence stars of the same spectral type.

Hall (1981) notes that in almost all RS CVn binaries the active star is rotating in near-synchronism with the orbital period, as deduced from photometric ('starspot') modulation data (λ And is one well-established exception), and concludes that a necessary and sufficient condition for strong surface activity is rapid rotation. Ayres and Linsky (1980) studied the (G6 III + F9 III) binary system Capella with the IUE ultraviolet spectrograph, and found that by far the bulk of the chromospheric activity seems to come from the F-type secondary, in spite of its earlier type, smaller optical luminosity, and identical age. They noted that the F-type secondary was the more rapidly rotating of the pair, presumably as an 'accident of youth' and ascribed the difference in X-ray emission to a stronger level of dynamo activity.

The overall dependence of stellar activity level on rotation rate is well illustrated in Figure 2, taken from Pallavicini *et al.* (1981b). We note that for G to M dwarfs the total X-ray luminosity varies as $L_X \sim (v \sin i)^2$, independent of spectral type. This is equivalent to a relation for the surface-averaged flux $f_X = L_X/4\pi R_*{}^2$ given by $f_X \sim \Omega_p{}^2$, where $\Omega_p = v \sin i/R_*$ is the angular velocity projected against the sky. (Interestingly, the same relation also appears to hold for luminosity classes II through IV. The approximate dependence persists over nearly two orders of magnitude in rotation velocity, and four

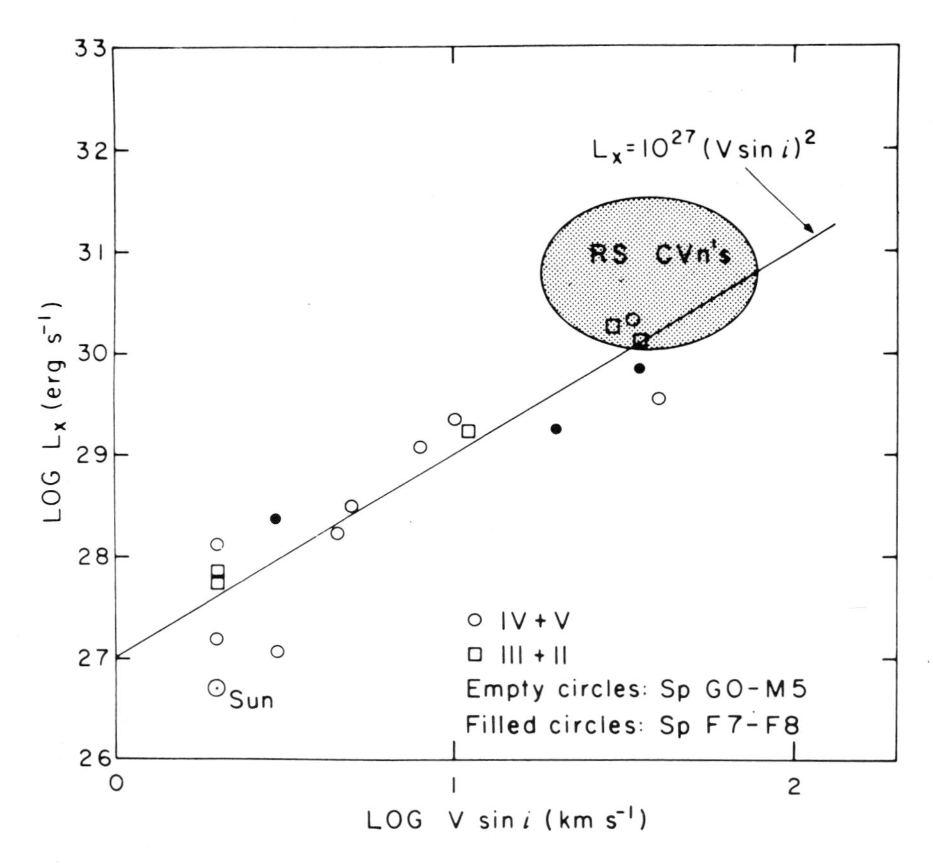

Fig. 2. X-ray luminosities plotted *versus* projected rotational velocities for stars of spectral type F7–F8 (filled symbols) and G0 to M5 (open symbols). The position of RS CVn stars is indicated. (From Pallavicini *et al.*, 1981b.)

orders of magnitude in L_X. We also note that the RS CVn binaries, whose rotation rate is maintained by tidal coupling at a value large compared to single stars of the same age and spectral type, fall close to the curve, at the very top due to their rapid rotation. Also, Walter (1981) has shown that single stars of luminosity class III through V within the same spectral range fall nicely on the same relation as RS CVn's. In other words, rotation rate determines a star's level of coronal activity as measured by L_X/L_{bol}, and not age, which may be very different for single stars and members of binary systems, even if both have the same rotation rate. (This statement is true for widely separate binaries where tidal distortions or mass exchange do not significantly impact the structure or dynamics of their coronas. For close binaries such as Algol-type binaries or WUMa systems, the situation is more complex; see Dupree, 1981, and references therein.)

The more detailed relation between X-ray emission and rotation in lower main sequence stars is by no means clear. As mentioned, Pallavicini *et al.* (1981b) find $L_X \sim (v \sin i)^2$, over a broad range of rotation rates independent of spectral type. However, Walter (1982) finds that the dependence of L_X/L_{bol} on rotation rate is best fit by a power law in rotation period with a break in slope at a 12 day period. He associates this with the Vaughan–Preston 'gap' in emission at this rotation period, to be described velow. At this point, we simply emphasize that much work still needs to be done before the precise dependence can be confidently characterized.

3.3. The Influence of Convection Zone Properties

Rotation is not the only parameter determining magnetic activity levels in convecting stars. Walter (1981) found that the proportionality between L_X/L_{bol} and Ω was different for G8–K5 stars: the earlier stars displayed a mean L_X/L_{bol} about a factor of 10 lower for a given angular velocity, although within each spectral range L_X/L_{bol} was found to vary linearly with Ω. This finding would appear to be consistent with expectations of dynamo theory that the strength of surface fields should increase not only with increasing angular velocity but also with increasing fractional depth of convection zone, i.e. towards later spectral type (e.g. Durney and Robinson, 1982).

Vaiana *et al.* (1981) have reported on results of the extensive Einstein survey of X-ray emission from stars in all parts of the H–R diagram. Some of the implications of those observations have already been discussed. Relevant to the present discussion is the observation is that for late K and M dwarfs the surface flux f_X increases steadily with decreasing mass, so that the mean value in the M stars surveyed is nearly an order of magnitude greater than that for G stars; extreme values are much larger. As pointed out by Vaiana *et al.* (1981; see also Vaiana, 1983, and Rosner and Vaiana, 1980) this disagrees by many orders of magnitude with predictions for acoustic heating of coronae, which is calculated to drop steadily and rapidly for K and M stars. In addition, because angular velocities on the main sequence are generally found to decrease, rather than increase, with advancing spectral type on the main sequence (Slettebak, 1970; Vaughan *et al.*, 1981) the result suggests an increase of surface magnetic activity with increasing fractional depth of the convective zone toward later types (see also Linsky,1981).

Rosner and Vaiana (1980) scale calculations on growth rates of magnetic buoyancy in rotating stars to the run of convection zone depth in late-type stars, assuming that the dynamo is operative at the base of the convection zone. They find the time-scale before flux eruption to increase dramatically towards smaller mass stars, and conclude that there should be a substantial increase of surface magnetic flux in K or M dwarfs over that in G dwarfs of the same rotation rate. Durney and Robinson (1982) have estimated the dependence of the surface area covered by plages, assuming dynamo field amplification time limited by buoyant rise time of fields, and that at the surface the field strength equals $(8\pi P_g)^{\frac{1}{2}}$, where P_g is the photospheric gas pressure. They find that the area increases rapidly towards later spectral types, for any value of Ω, and for a given spectral type the area coverage increases rapidly with Ω.

Thus it appears that, broadly speaking, surface magnetic activity on lower main sequence stars increases with both increasing rotation rate and advancing spectral type (that is, deepening convection zones). The detailed dependence of magnetic activity on rotation rate and convective properties, if it can be determined, should provide important clues to stellar dynamo mechanisms. One problem is that observed activity indicaters, such as Ca II or X-ray emission, depend in a complex way on surface magnetic flux, and this dependence may be expected to vary in a so-far poorly determined way with spectral type. With this caveat in mind, however, Noyes *et al.* (1984; see also Noyes 1983) found that the Ca II chromospheric flux ratio R_{HK} – that is, the ratio of surface flux F_{HK} to the bolometric luminosity per unit area σT_{eff}^4 – depends tightly on the Rossby number $\mathscr{R}$, which is the ratio of rotation period to the convective turnover time, evaluated near the base of the convection zone. (This dependence is tight, however, only for models with $\alpha \sim 2$, where α is the ratio of mixing length to scale height; these models have relatively deep convective zones.) The chromospheric emission ratio increases with decreasing Rossby number, or equivalently, increasing influence of Coriolis forces on the convective motions; this is in qualitative accord with standard dynamo theory (e.g. Durney and Latour, 1978).

3.4. THE VAUGHAN–PRESTON GAP

Vaughan and Preston (1980) have carried out an extensive survey of main sequence stars in the solar neighborhood, which revealed the distribution of Ca II fluxes with spectral type for about 486 field stars. Their results are reproduced as points in Figure 3. The stars with lowest Ca II emission appear to be the oldest as deduced from their kinematical properties (larger velocities perpendicular to the galactic plane, due to 'heating' of old stars by encounters with massive scatterers or gravitational instabilities of the disk). Conversely, the stars with high Ca II emission have small velocities perpendicular to the plane, and are among the youngest of the sample. It is natural to expect stars, during the course of their main sequence lifetimes, to evolve approximately vertically downward in the graph of Figure 3, as their Ca II emission decreases.

Vaughan and Preston noted that there appeared to be a gap in the density of points in Figure 3, dividing 'young' stars with relatively high Ca II emission from 'old' stars with low Ca II emission. Vaughan (1980) identified the same gap in Wilson's (1978) set of stars selected for study of possible cycles (vertical bars in Figure 3). The stars which display clear, Sun-like cycles generally occur below the gap, while those above the gap tend to show more chaotic, irregular long-term behavior (Vaughan, 1980).

If stellar H–K emission really declines smoothly as $(\text{age})^{-1/2}$, and if past stellar birth-rates were uniform or smoothly varying, then the present-day distribution of H–K emission among lower main sequence stars should be smooth, rather than bimodal. The apparent bimodal distribution in Figure 3 could be due simply to some bias in the sample, or to a statistically insignificant fluctuation of the stellar population in the solar neighborhood. Indeed, Hartmann *et al.* (1984) argue from numerical simulations that the statistical significance of the gap is not high.

On the other hand, it is possible that the gap is real, and results from a sudden decreases of dynamo efficiency (and hence Ca II emission) at a critical rotation rate. Under such a conjecture, the critical (perhaps mass-dependent) rotation rate is characterized by Ca II emission levels as a function of $B-V$ that delinate the top of the gap in Figure 3. Upon de-spinning to this rotation rate, the stars experience a rapid decrease in dynamo efficiency, and their H–K emission drops to the level delineating the bottom of the gap. The dichotomy in long-term activity variability, such that regular, Sun-like activity cycles tend to occur only in stars below the gap, supports (but scarcely proves) the conjecture of a change in the operation of the dynamo when stars cross the gap. Following this suggestion Durney *et al.* (1981) have argued that a difference in dynamo behavior above and

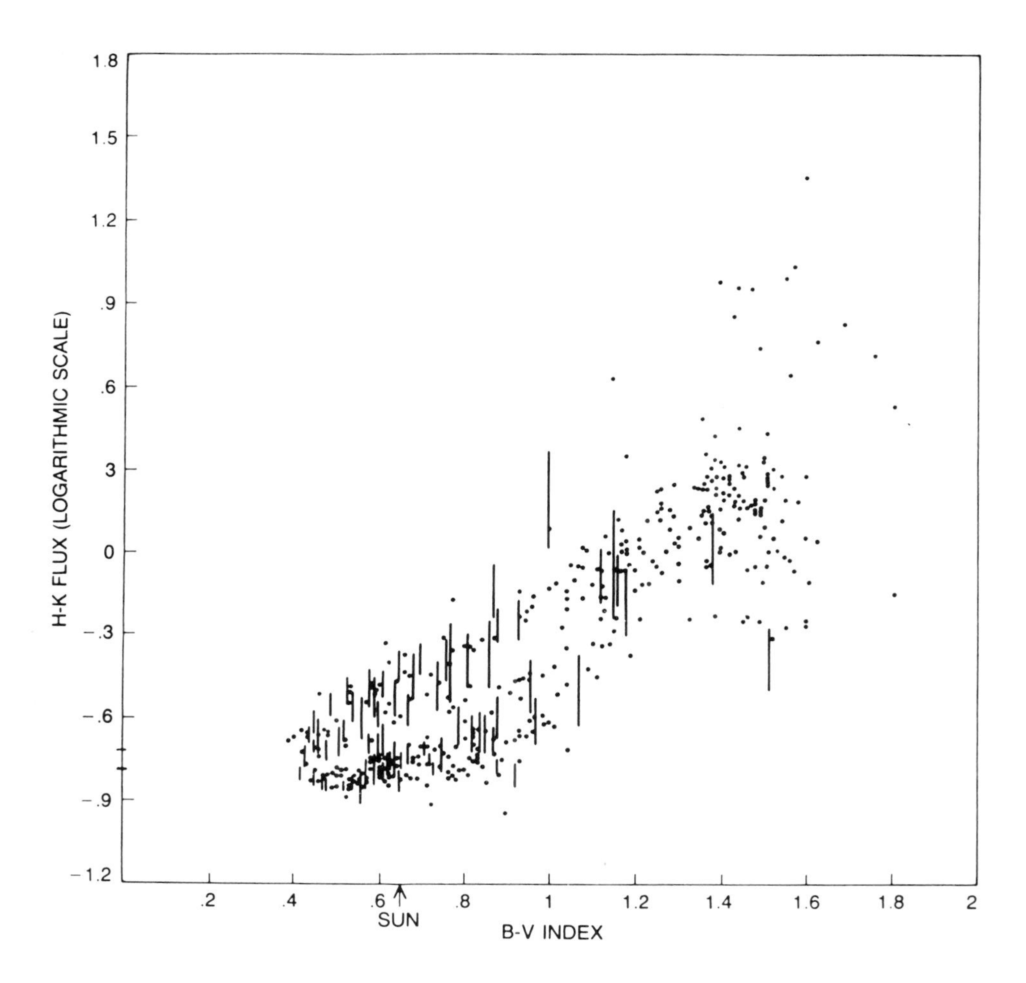

Fig. 3. Points: The Ca II emission index S plotted versus spectral color $B-V$ for 396 solar neighborhood main sequence stars, from Vaughan and Preston (1980). Lines: range of variability of the Ca II emission index for the 91 stars studied by Wilson (1978) over the period 1966–1977. (From Vaughan, 1983.)

below the gap could be produced because stars above the gap are rotating sufficiently rapidly that their dynamo number N_D exceeds a critical value above which higher modes of the dynamo can be excited. Alternatively, Knobloch *et al.* (1981) have suggested that convection in rapidly rotating stars may not be quasi-cellular, as in the Sun, but may rather occur in long convective rolls, with roll axis parallel to the axis of stellar rotation. For such stars, the dynamo would not be reversing (and therefore cyclical variations would not be observed), but at the same time the surface fields could be amplified to a larger value than for cellular convection. As the star de-spins, it reaches a critical rotation rate at which it jumps discontinuously to the cellular convection regime, which produces a solar-type cyclical dynamo and a lower level of overall magnetic activity.

Middelkoop (1982) has presented evidence that stellar rotation rates decrease smoothly rather than discontinuously across the gap (at least for stars in the range $0.52 < B - V \leq 0.63$), thereby supporting the idea of a different dependence of activity on rotation for stars above and below the gap. However, this conclusion is based on measured values of $v \sin i$ near the threshold of sensitivity of these measurements, and it is scarcely conclusive. On the other hand, Noyes *et al.* (1984) find that the dependence of activity ratio R_{HK} on Rossby number is similar for stars above and below the gap; this argues against the idea of a different dynamo mechanism for stars above and below the gap.

In summary, the nature or even the reality of the gap is unclear at this time. If real, its explanation could be of considerable importance for dynamo theory. Even if not real, its discovery has stimulated useful observational and theoretical work.

3.5. THE EVOLUTION OF ROTATION AND MAGNETIC ACTIVITY ON THE SUN

Let us assume that the Sun is a typical G2 dwarf in its magnetic and rotational evolution, just as it appears to be typical in so many of its other properties. Then we may infer some aspects of the earlier (and later) history of the Sun from the study of rotation and magnetic activity of younger and older main sequence stars. We dismiss the pre-main sequence stage of the Sun's activity as being beyond the scope of this review, and take up the history when the Sun was situated on the main sequence at an age comparable to the youngest cluster whose rotation is well studied – namely the Pleiades (Kraft, 1967) with an age of about 7×10^7 yr. At this age the Sun would have had a rotation period of about 3.2 days, judging from extrapolation backward of Skumanich's (1972) $(\mathrm{age})^{-1/2}$ scaling law for rotational velocity. (This also, of course, agrees with Kraft's observations, from which the relation was in part derived.)

In the scenario sketched below, we assume an $(\mathrm{age})^{-1/2}$ scaling law for rotation throughout the lifetime of the Sun. We also assume an $(\mathrm{age})^{-1/2}$ scaling law for Ca II emission (i.e. surface magnetic activity level) at least until the Sun reaches the Vaughan–Preston gap.

At the age of the Hyades (6×10^8 yr; see Soderblom, 1983) the Sun would have slowed to a rotation period of 9 days, so that its rotation velocity and Ca II emission level were both higher than those of the Sun by the ratio 2.8, according to the activity–age scaling relation. It might well be like the well-studied Hyades moving group G2 V star HD 1835, which is observed (Vaughan *et al.*, 1981) to have an angular velocity 3.2 times that of the Sun and a mean Ca II chromospheric emission (corrected for the photospheric contribution) of 2.5 times that of the Sun. The chromospheric emission of both active and quiet regions would be several times larger, as may be seen from Wilson's (1978) data on HD 1835. The Sun would not show a clear 11 years cycle of activity, judging from Wilson's (1978) study of HD 1835, in which the star was found to show large irregular fluctuations of surface activity, with no apparent cycles. Presumably the Sun's surface magnetic field strength, being determined by its photospheric gas pressure, would be similar to present values ($\sim 10^3$ gauss), but the total magnetic flux (without regard to sign) would be several times greater, being brought to the surface in larger active regions, more concentrated flux tubes within active regions, or both. We have already noted that HD 1835 exhibits photometric fluctuations some 30 times as large as those of the Sun, so we may presume the aggregate spot area of the 600-million-year-old Sun was 30-fold larger than at present. Presumably, X-ray emission levels from active regions were greater, as well as the energy released in solar flares.

In principle, records of particle implantation in the surface rocks of the moon and meteorites carry information about past solar magnetic activity, but presently available data only permit, rather than require, higher levels of activity and winds early in the life of the Sun (see Newkirk, 1980, for a review).

At an age of 1.1×10^9 yr, according to the $(\text{age})^{-1/2}$ scaling relation the rotation period of the Sun would have slowed to a value of about 13 days. The chromospheric Ca II relative emission would have reached a value of about twice its present value. This activity level corresponds to the top of the Vaughan–Preston gap for a G2 dwarf.

At this point the Sun may have suffered a rapid drop of chromospheric activity due to a sudden decrease in dynamo efficiency (or, conceivably, due to the onset of greatly enhanced spindown). On the other hand, it may have continued to evolve steadily through ever-decreasing rotation rates and gradually decreasing chromospheric activity. As indicated in the previous section, the evidence is not yet clear on this point. In any case, after the Sun's activity reached the level characteristic of stars below the gap, it would have displayed the characteristic quasi-periodic magnetic and chromospheric activity cycle it shows today.

The Sun does not have the lowest Ca II emission among of the G dwarfs, as noted by Vaughan and Preston (1980); yet some of the even quieter stars still show evidence of chromospheric variability (cf. Vaughan *et al.*, 1981). Thus it appears that the Sun could continue its gradual decline in activity level for a long time.

It has been argued (Smith, 1979; Blanco *et al.*, 1974) that the Sun is anomalous, in having both rotation and Ca II flux well below the mean for old field stars of the same spectral type. In fact, the Sun could always have had a low angular velocity, imposed as an accident of birth. In that case, it would have spent its entire evolutionary history below the Vaughan–Preston gap, and would never have experienced the more active, less periodic mode of field generation characterizing stars above the gap. However, the recent study of Soderblom (1983) comparing rotation, activity and age of cluster and field stars, indicates that while the Sun's angular velocity may lie below the mean for field stars of its spectral type, the difference is not significant (~ 1 standard deviation).

4. Avenues for Future Research

It is abundantly clear that the comparative study of solar and stellar phenomena is reaping rich rewards, both in solar and stellar physics. Nowhere is this more apparent than in the study of magnetic activity. There is little doubt that the new perspectives on magnetic activity gained from studying magnetic fields, spots, active regions, and activity cycles on other stars will profoundly contribute to our understanding of these phenomena on the Sun, even though detailed observations are in their infancy. At the same time, further progress also requires observations with the angular resolution that only the Sun can provide. The recent advances in our knowledge of solar and stellar magnetic activity reviewed above suggest a number of new avenues for observational and theoretical research, which we list below.

4.1. Observational Studies of Solar Magnetic Activity

(1) Observations of spatially resolved magnetic field and velocity structures on the Sun are needed to further tighten constraints on the dynamo mechanism. A few specific examples include:

(a) further study of the torsional oscillation recently reported by Howard and La bonte (1980);

(b) further studies of differential rotation as a function of latitude, depth, and possibly phase of the activity cycle (using the technqiue of solar seismology – see Chapter 7 in Volume I of this work); and

(c) studies of the latitude, longitude, and time-dependence of ephemeral active region flux emergence and the overlying coronal bright points.

(2) Studies are needed of specific solar magnetic phenomena which can also be seen or inferred on other stars, with the aim of developing scaling relations valid for other late-type stars. Some specific examples include:

(a) study of the detailed relation between chromospheric emission from point to point on the Sun as it depends on underlying photospheric magnetic flux; through quantitative comparison of emission not only in H and K but also other chromospheric lines like Lα, Mg II h and k, etc., one may hope to get a better calibration of chromospheric emission as a 'stellar magnetometer';

(b) further study of closed magnetic loops, ranging in size from X-ray bright points through active region loops to large-scale quiet Sun structures, to refine scaling laws for the dependence of coronal emission measure and temperature on geometrical parameters of the loop, and upon magnetic field strength, flux, and footpoint shear;

(c) similar studies of open field regions, with the goal of identifying the source in the photosphere of Alfvén wave flux or other energy and momentum input to the solar wind.

(3) Study of the Sun seen as a star.

(a) The monitoring of disk-integrated Ca II emission carried on by White and Livingston (1981) and others, should be continued, at least through an entire solar cycle. A main purpose is to provide a dependable comparison point for similar stellar observations.

(b) Continued observations should be made of the solar constant, to a level better than 10^{-4} and ideally 10^{-5}, extended over a complete cycle. This is of importance for understanding the general influence of solar magnetism on the structure of the convection zone and, hence, the solar luminosity variation on times comparable to the activity cycle, and also to explore the photometric effects of specific features (sunspots) on short time-scales, so that the amplitude of stellar photometric rotational modulation can be interpreted quantitatively.

(4) Studies of the solar wind aimed at more accurate determination of its braking torque. These studies may require observations made from out-of-the-ecliptic platforms. They should be extended to cover an entire solar cycle, and should be closely related to data on coronal hole sizes and distribution over the activity cycle.

(5) Studies of solar magnetic activity at earlier epochs, using 'proxy' data from terrestrial, lunar, or meteoritic samples. Two important goals are to determine:

(a) more information about the intermittency of magnetic activity in the present Sun, as is revealed by Maunder-type minima and corresponding maxima of activity; and
(b) information about the level of magnetic activity (e.g. as a driver of flares and winds) in the young Sun.

Both types of information will becomes increasingly important as our body of knowledge about similar stellar phenomena steadily increases.

4.2. Observational Studies of Stellar Magnetic Activity

(1) The pioneering work of Wilson (1978) in studying activity cycles in late-type stars still continues at Mt Wilson Observatory. It is very important that this be extended for some time into the future, and extended to include other stars, such as, if possible, important southern hemisphere stars (e.g. α Cen). Primary needs are:
(a) to determine the frequency of occurrence of longer-period cycles than those presently detected;
(b) to get better information on the shape of activity curves; and
(c) to extend the range of spectral types monitored to later-type (M) stars.

(2) The survey of chromospheric activity in neighborhood stars (Vaughan and Preston, 1980) which gave rise to the discovery of the Ca II emission gap (Figure 3) should be extended so that the question of the reality of the gap is clearly resolved, and the population density of stars with respect to the gap in the chromospheric emission-color plane is well defined.

(3) Stellar rotation measurements, either through spectral line-broadening techniques or rotational modulation measurements, should be made for many lower main sequence stars with varying activity levels, masses, and ages. Particular attention should be paid to stars above, below, and within the 'gap' in order to explore the ideas discussed in Section 3.4. Also, rotation rates should be obtained for all stars with well-determined activity cycle periods, and for stars which give evidence for irregular long-term activity behavior.

(4) The observation of rotational modulation, either in chromospheric emission lines such as Ca II or broad-band light in the case of 'starspot' modulation, can in principle be used to infer differential rates and latitude migration of active regions. At the same time, such data may be used to determine the surface distribution of magnetic activity, from detailed analysis of light curves.

(5) Magnetic fields should be measured in a variety of stars – for example, by using the techniques proposed by Robinson (1980) or others. An important goal is to verify the relation between magnetic flux and Ca II emission level, which at present can only be assumed by extrapolation from the solar case. Another goal is to explore whether surface fields do indeed scale as $(8\pi P_g)^{1/2}$, where P_g is the photospheric gas pressure.

It may be possible to determine the magnetic fields of starspot umbrae, by obtaining Zeeman broadening data in low-excitation spectral lines, which are formed preferentially in cooler regions, and also by choosing lines in the infrared, where cool spots are at their brightest relative to the photosphere. Such an observation would not only verify the magnetic nature of starspots, but would also give useful information on magnetic field concentration level in other stellar photospheres.

(6) The study of stellar coronae and transition zones, so well initiated with the aid of

the Einstein and International Ultraviolet Explorer satellites, should be extended. It would be useful to ascertain better the relation between coronal and chromospheric (e.g. Ca II) activity for stars of different ages and rotation rates. For example, if surface patterns of activity change as stars age and spin down, there may well be a correspondingly different ratio of the amounts of coronal and chromospheric heating.

4.3. Theoretical Studies

It is clear that a major impediment to more rapid progress in understanding solar and stellar magnetic activity is our inadequate theoretical knowledge of the generation of solar magnetic fields. As detailed elsewhere in this work, standard α–ω dynamo modeling may be able to reproduce many observed aspects of the solar cycle, but when MHD relations are considered self-consistently, the model fails to predict a number of basic properties (see also Gilman and Miller, 1980). The only velocity flow which has been actually observed on the Sun to be connected with field migration during the activity cycle – namely, the torsional oscillation reported by Howard and La Bonte (1980) – fails to be reproduced by standard dynamo theory, and raises doubts (La Bonte and Howard, 1982) whether standard dynamo theory is applicable at any level. Clearly we cannot expect to make real progress in extending theories of magnetic field generation to other stars until we understand better the clues which our own Sun is giving us.

In addition, it is important to develop theories for the field strength and area coverage of plages on stars of different spectral types (i.e. convection zone properties) and rotation rates (see Durney and Robinson, 1981); such predictions would be a useful point of contact with emerging capabilities to mearure stellar magnetic fields.

Finally, it is important to determine more quantitatively how, for stars with different surface field geometries and convection zone dynamics, magnetic field energy is deposited in the overlying atmosphere. It should ultimately be possible to apply these theoretical treatments (such as, for example, the work of Golub *et al.*, 1980) to observations of chromospheres and coronae of young and old stars, with different degrees of rotation and, perhaps, rather different surface field geometry, reflecting different modes of dynamo operation.

Acknowledgements

The author thanks C. Zwaan, D. Mihalas, S. P. Worden, J. Harvey, A. H. Vaughan, R. Rosner, and O. C. Wilson, for enlightening discussions during the preparation of this review.

References

Ayres, T. and Linsky, J.: 1980, *Astrophys. J.* **241**, 279.
Babcock, H. W.: 1961, *Astrophys. J.* **133**, 572.
Baliunas, S. L., and Dupree, A. K.: 1982, *Astrophys. J.* **252**, 668.
Baliunas, S. L., Vanghan, A. H., Hartmann, L., Middelkoop, F., Mihalas, D., Noyes, R. W., Preston, G. W., Frazer, J., and Lanning, H.: 1983, *Astrophys. J.* **275**, 752.

Belvedere, G., Paterno, L., and Stix, M.: 1980, *Astron. Astrophys.* **91**, 328.
Blanco, C., Catalano, S., Marilli, E., and Rodono, M.: 1974, *Astron. Astrophys.* **33**, 257.
Bopp, B. W. and Espenak, F.: 1977, *Astron. J.* **82**, 916.
Brandt, J.: 1966, *Astrophys. J.* **144**, 1221.
Brown, D. N. and Landstreet, J. D.: 1981, *Astrophys. J.* **246**, 899.
Chugainov, P. F.: 1980, *Isv. Krymskoi Astrofiz. Obs.* **61**, 124.
Dearborn, D. S. P. and Blake, J. B.: 1980, *Astrophys. J.* **237**, 616.
de Campli, W. M. and Baliunas, S. L.: 1979, *Astrophys. J.* **230**, 815.
Delhaye, J.: 1953, *Compt. Rend.* **237**, 294.
Dupree, A. K.: 1981, in R. M. Bonnet and A. K. Dupree (eds), *Solar Phenomena in Stars and Stellar Systems*, D. Reidel, Dordrecht, p. 407.
Durney, B. R.: 1972, in *Solar Wind*, NASA SP-308, Washington, D.C., p. 282.
Durney, B. R. and Latour, J.: 1978, *Geophys. Astrophys. Fluid Dyn.* **9**, 241.
Durney, B. R. and Robinson, R. D.: 1982, *Astrophys. J.* **253**, 290.
Durney, B. R., Mihalas, D., and Robinson, R. D.: 1981, *Publ. Astron. Soc. Pacific* **93**, 537.
Durney, B. and Stenflo, J.: 1972, *Astrophys. Space Sci.* **15**, 307.
Eaton, J. A. and Hall, D. S.: 1979, *Astrophys. J.* **227**, 907.
Foukal, P. V. and Vernazza, J.: 1979, *Astrophys. J.* **234**, 707.
Galloway, D. J. and Weiss, N. O.: 1981, *Astrophys. J.* **243**, 945.
Gershberg, R. E.: 1975, *IAU Symp.* **67**, 47.
Gilman, P. A. and Miller, J.: 1981, *Astrophys. J. Suppl.* **46**, 211.
Golub, L., Maxson, C., Rosner, R., Serio, S., and Vaiana, G. S.: 1980, *Astrophys. J.* **238**, 343.
Hall, D. S.: 1981, in R. M. Bonnet and A. K. Dupree (eds), *Solar Phenomena in Stars and Stellar Systems*, D. Reidel, Dordrecht, p. 431.
Hartmann, L.: 1981, in R. M. Bonnet and A. K. Dupree (eds), *Solar Phenomena in Stars and Stellar Systems*, D. Reidel, Dordrecht, p. 487.
Hartmann, L., Davis, R., Dupree, A., Raymond, J., Schmidtke, P., and Wing, R.: 1979, *Astrophys. J.* **233**, L69.
Hartmann, L., and Rosner, R.: 1979, *Astrophys. J.* **230**, 802.
Hartmann, L., Londono, C., and Phillips, M.: 1979, *Astrophys. J.* **229**, 183.
Hartmann, L. W., Soderblom, D. R., Noyes, R. W., Burnham, N. S., and Vaughan, A. H.: 1984, *Astrophys. J.* **276**, 254.
Holt, S. S., White, N. E., Becker, R. H., Boldt, E. A., Mushotzsky, R. F., Serlemitsos, P. J., and Smith, B. W.: 1979, *Astrophys. J.* **234**, L65.
Howard, R. and La Bonte, B.: 1980, *Astrophys. J.* **239**, L33.
Kelch, W. L.: 1978, *Astrophys. J.* **222**, 931.
Kelch, W. L., Linsky, J., and Worden, S. P.: 1979, *Astrophys. J.* **229**, 700.
Knobloch, E., Rosner, R., and Weiss, N.: 1981, *Monthly Notices Roy. Astron. Soc.* **197**, 45p.
Kraft, R. P.: 1967, *Astrophys. J.* **150**, 551.
La Bonte, B. J. and Howard, R.: 1982, *Solar Phys.* **75**, 161.
Landini, M. and Monsignori-Fossi, B. C.: 1975, *Astron. Astrophys.* **42**, 213.
Leighton, R. B.: 1959, *Astrophys. J.* **130**, 366.
Leighton, R. B.: 1969, *Astrophys. J.* **156**, 1.
Linsky, J.: 1977, in O. R. White (ed.), *The Solar Output and its Variations*, Colorado Assoc. Univ. Press, Boulder, Ch. 7.
Linsky, J.: 1981, in R. M. Bonnet and A. K. Dupree (eds), *Solar Phenomena in Stars and Stellar Systems*, D. Reidel, Dordrecht, p. 39.
Lüst, R. and Schluter, A.: 1958, *Z. Astrophys.* 38, 190.
Marcy, G. W.: 1981, *Astrophys. J.* **245**, 624.
Marcy, G. W.: 1982, *Publ. Astron. Soc. Pacific* 94, 989.
Mestel, L.: 1968, *Monthly Notices Roy. Astron. Soc.* **138**, 359.
Mewe, R., Schrijver, C. J., and Zwaan, C.: 1981, *Space Sci. Rev.* **30**, 191.
Middelkoop, F.: 1981, *Astron. Astrophys.* **101**, 295.
Middelkoop, F.: 1982, *Astron. Astrophys.* **107**, 31.
Mullan, D. J.: 1977, *Solar Phys.* **54**, 183.

Newkirk, G.: 1980, in Pepin, Eddy and Merin (eds), *Proc. Conf. on Ancient Sun*, p. 293.
Noyes, R. W.: 1982, in *Proc. 2nd Camb. Workshop on Cool Stars, Stellar Systems, and the Sun*, SAO Special Rept. No. 392, II, 41.
Noyes, R. W.: 1983, *IAU Symp.* **102**, 133.
Noyes, R. W., Hartmann, L. W., Baliunas, S. L., Duncan, D. K., and Vaughan, A. H.: 1984, *Astrophys. J.* **279**, 763.
Oranje, B. J., Zwaan, C., and Middelkoop, F.: 1982, *Astron. Astrophys.* **110**, 30.
Oskanyan, V. S., Evans, D. S., Lucy, C., and McMillan, P. S.: 1977, *Astrophys. J.* **214**, 430.
Pallavicini, R., Peres, G., Serio, S., Vaiana, G., Golub, L., and Rosner, R.: 1981a, *Astrophys. J.* **247**, 692.
Pallavicini, R., Golub, L., Rosner, R., Vaiana, G., Ayres, T., and Linsky, J.: 1981b, *Astrophys. J.* **248**, 279.
Parker, E. N.: 1955, *Astrophys. J.* **122**, 293.
Parker, E. N.: 1978, *Astrophys. J.* **221**, 368.
Phillips, M. and Hartmann, L.: 1978, *Astrophys. J.* **224**, 182.
Preston, G. W.: 1971, *Astrophys. J.* **164**, 309.
Radick, R. R., Hartmann, L., Mihalas, D., Worden, S. P., Africano, J. L., Klimke, A., and Tyson, E. T.: 1982, *Publ. Astron. Soc. Pacific* **94**, 934.
Ramsey, L. W. and Nations, H. L.: 1980, *Astrophys. J.* **239**, L121.
Raymond, J. C. and Foukal, P. V.: 1982, *Astrophys. J.* **253**, 323.
Robinson, R. D.: 1980, *Astrophys. J.* **239**, 961.
Robinson, R. D., Worden, S. P., and Harvey, J. W.: 1980, *Astrophys. J.* **236**, L155.
Rosner, R., Tucker, W. H., and Vaiana, G. S.: 1978, *Astrophys. J.* **220**, 643.
Rosner, R. and Vaiana, G. S.: 1980, in R. Giacconi and G. Setti (eds), *X-Ray Astronomy*, D. Reidel, Dordrecht, p. 129.
Schatzman, E.: 1959, *IAU Symp.* **10**, 129.
Schatzman, E.: 1962, *Ann. d'Astrophys.* **25**, 18.
Schatzman, E.: 1965, *IAU Symp.* **22**, 153.
Scherer, P. H., Wilcox, J., Svalgaard, L., Duvall, T., Dittmer, P., and Gustafson, E. 1977, *Solar Phys.* **54**, 353.
Severny, A., Wilcox, J. M., Scherer, P. H., and Colburn, D. S.: 1970, *Solar Phys.* **15**, 3.
Sheeley, N. R.: 1967, *Astrophys. J.* **147**, 1108.
Simon, T., Kelch, W. L., and Linsky, J.: 1980, *Astrophys. J.* **237**, 72.
Skumanich, A.: 1972, *Astrophys. J.* **171**, 565.
Skumanich, A. and Eddy, J. 1981, in R. M. Bonnet and A. K. Dupree (eds), *Solar Phenomena in Stars and Stellar Systems*, D. Reidel, Dordrecht, p. 349.
Skumanich, A., Smythe, C., and Frazier, E. N.: 1975, *Astrophys. J.* **200**, 747.
Slettebak, A.: 1970, in A. Slettebak (ed.), *IAU Coll. on Stellar Rotation*, D. Reidel, Dordrecht, p. 3.
Smith, M. A.: 1979, *Publ. Astron. Soc. Pacific* **91**, 737.
Soderblom, D. R.: 1983, *Astrophys. J. Suppl.* **53**, 1.
Spiegel, E. A. and Weiss, N. O.: 1980, *Nature* **287**, 616.
Spruit, H. C.: 1979, *Solar Phys.* **61**, 363.
Spruit, H. C. and Zweibel, E. G.: 1979, *Solar Phys.* **62**, 15.
Swank, J. H., White, N. E., Holt, S. S., and Becker, R. H.: 1981, *Astrophys. J.* **246**, 208.
Torres, C. A. O. and Ferraz-Mello, S.: 1973, *Astron. Astrophys.* **27**, 231.
Unno, W.: 1959, *Astrophys. J.* **129**, 375.
Vaiana, G. S. *et al.*: 1981, *Astrophys. J.* **245**, 163.
Vaiana, G. S.: 1983, *IAU Symp.* **102**, 165.
Vaughan, A. H.: 1980, *Publ. Astron. Soc. Pacific* **92**, 392.
Vaughan, A. H.: 1983, *IAU Symp.* **102**, 113.
Vaughan, A. H. and Preston, G. W.: 1980, *Publ. Astron. Soc. Pacific* **92**, 235.
Vaughan, A. H., Baliunas, S. L., Middelkoop, F., Hartmann, L. W., Mihalas, D., Noyes, R. W., Preston, G. W.: 1981, *Astrophys. J.* **250**, 276.
Vogt, S. S.: 1975, *Astrophys. J.* **199**, 418.
Walter, F. M.: 1981, *Astrophys. J.* **245**, 677.
Walter, F. M.: 1982, *Astrophys. J.* **253**, 745.

Walter, F. M., Cash, W., Charles, P. A., and Bowyer, C. S.: 1980, *Astrophys. J.* **236**, 212.
Weber, E. J. and Davis, L.: 1967, *Astrophys. J.* **148**, 217.
White, O. R. and Livingston, W. C.: 1981, *Astrophys. J.* **249**, 798.
Willson, R., Gulkis, S., Johnson, H., Hudson, H., and Chapman, G.: 1981, *Science* **211**, 700.
Wilson, O. C.: 1968, *Astrophys. J.* **153**, 221.
Wilson, O. C.: 1978, *Astrophys. J.* **226**, 379.
Wilson, O. C. and Wooley, R. V. D. R.: 1979, *Monthly Notices Roy. Astron. Soc.* **148**, 463.
Withbroe, G. L.: 1981, in F. Orrall (ed.), *Solar Active Regions*, Colorado Assoc. Univ. Press, Boulder, p. 199.
Zwaan, C.: 1977, *Mem. Soc. Astron. Ital.* **48**, 525.
Zwaan, C.: 1978, *Solar Phys.* **60**, 213.
Zwaan, C.: 1981a, in R. M. Bonnet and A. K. Dupree (eds), *Solar Phenomena in Stars and Stellar Systems*, D. Reidel, Dordrecht, p. 463.
Zwaan, C.: 1981b, in S. Jordan (ed.), *The Sun as a Star*, NASA SP-450, Washington, D. C., p. 163.

Harvard-Smithsonian Center for Astrophysics,
Cambridge, MA 02138,
U.S.A.

CHAPTER 20

EFFECTS OF SOLAR ELECTROMAGNETIC RADIATION ON THE TERRESTRIAL ENVIRONMENT

ROBERT E. DICKINSON *

1. Introduction

Solar electromagnetic radiation is surely the most important and pervasive resource of our terrestrial environment. It is the primary and ultimate source of energy for a multitude of natural processes whose functioning has led to life on Earth as we know it. The human race is now at a crossroads in its search for self-realization and self-improvement. Our population has more or less reached the carrying capacity of our planet. This capacity, of course, is not a firm number. Better social organization and technological improvements would increase the numbers Earth can support. However, growth beyond a doubling of present population could place impossible strains on our natural ecosystems and resource base (cf., e.g., Ehrlich *et al.*, 1977). Therefore, it seems likely that the global population cannot grow from its current 4.5 billion to much more than 10 billion or so. This demographic transition from a growing to a steady-state human population may be the most profound change in human civilization since the transition from hunter-gatherer to agricultural economies 10 000 years ago (Sassein, 1980).

There is now considerable doubt as to how successfully we can steer this great demographic transition over the next 50 years into a higher form of human civilization. Success will require improved knowledge and application of this knowledge in a wide range of scientific disciplines. Studies of processes which are intimately tied to a supply of solar terrestrial radiation will be of especially great importance. Failure to apply adequately our human intellectual resources to these studies could well increase our momentum towards Malthusian mass starvation and social breakdown as the dominant limitations on population growth.

In this essay I hope to convey to the reader a sense of both the unity and high scientific interest in the wide panoply of disciplines where solar electromagnetic radiation is the ultimate energy source.

The short wavelength end of the spectrum (wavelengths less than 320 nm), although containing little total energy, has enough energy per photon to dissociate molecules and so can readily destroy living cells (e.g. Harm, 1980). Hence, one of our atmosphere's most important beneficial functions is the absorption of these more energetic solar photons. Figure 1 indicates schematically the chemical constituents and altitude levels responsible for this absorption at different wavelengths. About 0.01% of the incident flux

* The National Center for Atmospheric Research is sponsored by the National Science Foundation.

Peter A. Sturrock (ed.), Physics of the Sun, Vol. III, pp. 155–191.

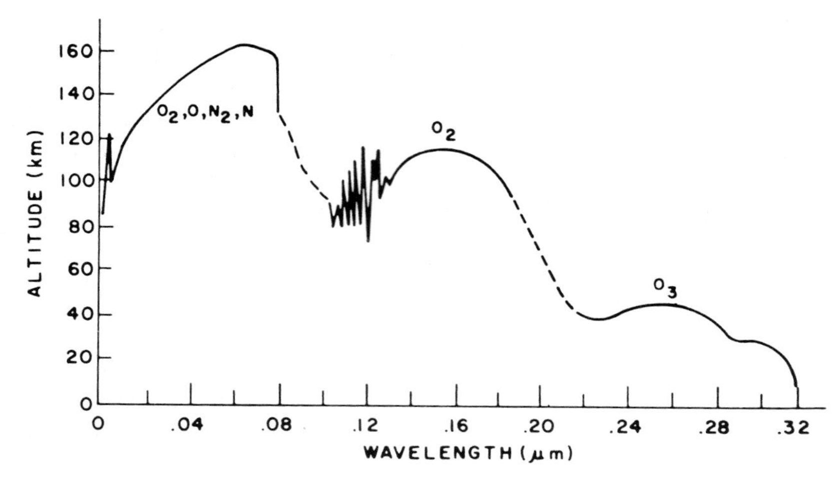

Fig. 1. Molecules principally responsible for solar absorption of various wavelengths and the altitude by which 0.63 of the radiation has been absorbed. (From Coulson, 1975; adapted from an earlier figure of Friedman, 1960.)

is absorbed in the thermosphere above 80 km, and about 0.2% above the ozone peak at 50 km. From the viewpoint of the atmospheric sciences, however, this absorption is not an end in itself but the beginning of a complex of chemical, physical, and dynamical responses to the energy made available by the solar radiation and ultimately degraded to heat, which is removed by thermal infrared radiation. Because the small densities of the atmosphere at high levels, the rates of energy deposition per unit mass are generally much larger at high levels in the atmosphere than they are in the lower atmosphere, in spite of the small fraction of the solar constant absorbed there. The processes driven by radiation less than 320 nm are discussed further in the next section.

Most of the solar radiation at wavelengths longer than 320 nm is channelled directly into heat, and more than half of it is absorbed at the Earth's surface. This energy maintains the Earth's surface at a temperature compatible with life. It also is responsible for other aspects of the climate system essential for the functioning of our human economics. Table I shows the disposition of solar heating within the atmosphere and Earth's surface.

TABLE I

Fraction of intercepted solar flux absorbed in atmosphere and at surface

Absorber	
Ozone	0.03
Water vapor	0.24
Other molecules	0.02
Clouds and aerosols	0.05
Surface	0.36
Total	0.70

The fraction of solar radiation reflected to space is of major significance in the climate system. Clouds are the primary reflectors, but the Earth's surface also reflects a significant amount of radiation, especially regions of ice and snow or deserts. Clouds, water vapor, and other trace gases in the trosposphere compensate in part for the reflection losses by elevating much of the system's infrared emission from the surface to higher and, hence, colder levels of the atmosphere.

The variation of absorbed solar radiation with latitude and season is, of course, a consequence of the orientation of the Earth's axis of rotation relative to its orbit about the Sun and, to a lesser extent, the ellipticity of the Earth orbit. Dynamic meteorology is concerned with the complex of motion systems that are fed by these heating gradients and thereby redistribute thermal energy from regions of relative excess to regions of relative deficit. The oceans, in turn, respond not only to the thermal energy they receive, but also to transfers of momentum from atmospheric winds at their surface. The climate system, as another important consequence of solar radiation, is treated later in more detail.

Only a small fraction of solar energy incident at the Earth's surface supplies energy for chemical processes. But, nevertheless, this supply is crucial, for among the processes sustained are those of photosynthesis. First developed in the early history of earth about 4 billion years ago, and under refinement ever since, photosynthesis is a marvelous sequence of biochemical energy transfer processes that allows plants to reduce the carbon dioxide of air into carbohydrates and other plant components. Most of the net primary production by plants is eventually consumed by other, nonphotosynthetic, life forms such as microorganisms in the soil or ocean sediments. With enough oxygen for respiration, these microorganisms are able to use all forms of reduced carbon for energy until the carbon of the photosynthate is eventually returned to the environment as carbon dioxide. However, some of the organic compounds are much more decay-resistant than others and may persist to be buried deeply in an anoxic decay-free environment. The net effect of this fate has been the conversion of atmospheric CO_2 into sedimentary carbon, e.g. $CO_2 + H_2O \rightarrow CH_2O + O_2$. One of the consequences of this burial of organic carbon has been the formation of commercially valuable deposits of fossil fuels such as coal, gas, and oil.

Photosynthesis, as fueled by solar radiation, is the primary basis for life on Earth. Its action in the geological past has not only supplied fossil fuels but has profoundly modified the composition of the atmosphere and the Earth's surface rocks. It produced the oxygen in our atmosphere, and even earlier, the oxygen evolved by photosynthesis was used to concentrate iron from weathering of rocks into most of today's commercially valuable iron ore deposits. We return to these topics later.

The general intent of this essay is to discuss the effect of solar electromagnetic radiation on the terrestrial environment. A systematic approach to this question would involve our considering all environment processes where solar emission is the primary energy source and all important materials which have been generated by solar driven processes. By the time we had considered as well the important processes linked to solar driven processes or generated materials, we would have examined a large part of the terrestrial environment and much of the atmospheric, geological, oceanic, and biological sciences. It is impractical for me to treat these subjects comprehensively, so I shall try to sketch an impression of the range of the effects of solar radiation on the environment by surveying a number of topics of particular current interest, in varying levels of detail. These include atmospheric chemistry, some aspects of the transfer of radiation within the atmosphere, global energy balance and climate feedbacks, especially those due to clouds, impacts of fossil fuel energy use, evolution of early life processes, photosynthesis and plant productivity as it relates to photosynthesis and the global carbon cycle.

It should be realized that not only are all these topics closely linked to the supply of solar radiation, but they are also highly dependent on each other.

2. Atmospheric Structure and Composition

2.1. Thermosphere

It is convenient here to consider the terrestrial environment downward from the top. Figure 2 shows schematically the different atmospheric regions. The thermosphere is

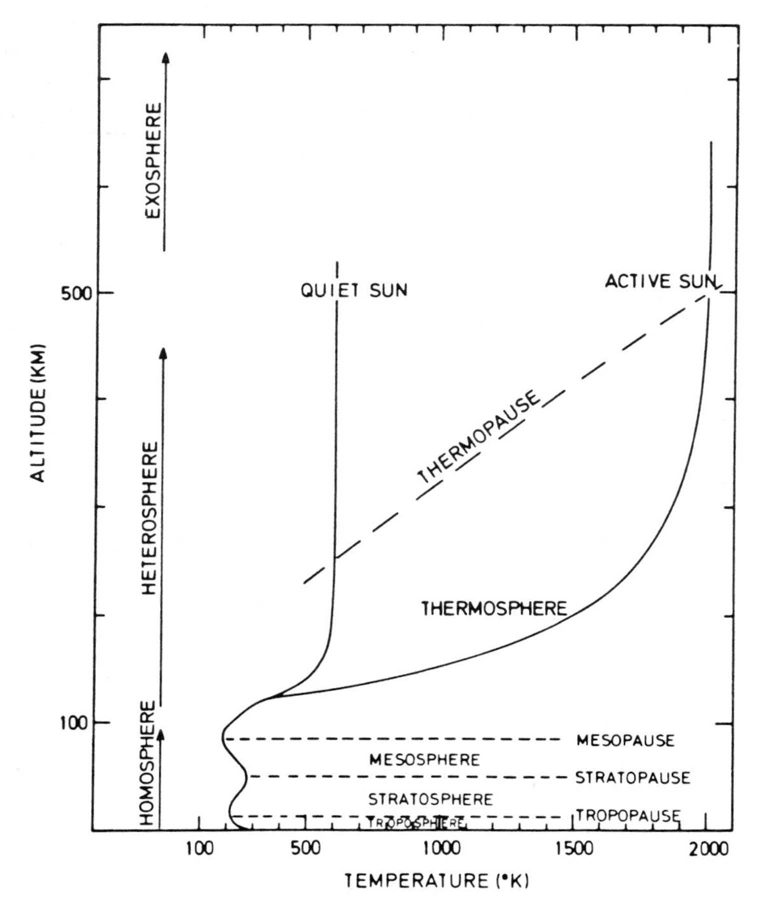

Fig. 2. Regions of the atmosphere and vertical distribution of temperature. (From Banks and Kockarts, 1973.)

the atmosphere at altitudes above 80 km or so and receives its name for its characteristic increase of temperature with altitude. Above the thermosphere is the exosphere where neutral molecules collide so infrequently that bulk temperatures and horizontal velocities are essentially constant with altitude and individual molecules soar on ballistic orbits into space. The lower thermosphere extends from about 80 to 120 km in altitude, or in pressure from 10^{-5} to 10^{-8} of surface pressure. The upper thermosphere, extending from 10^{-8} to about 10^{-11} of surface pressure, and with altitudes of 350–700 km depending on the solar cycle, is the 'near space' environment through which orbit many satellites, space shuttles, etc. Indeed, much of what we know of this region has been learned in the last decade or two through satellite technology. The drag exerted by the thermosphere on satellites was the first means of estimating the densities of this tenuous region.

The thermosphere is the sink of solar extreme ultraviolet radiation (EUV). It is the absorption of this radiation and, to a lesser extent, energy from auroral processes that determines the thermal, compositional, and ionic structure of the upper thermosphere. This structure has a large variation with the solar cycle due to the factor of 2 to 3 or more variation of the incident EUV fluxes and energy released by auroral processes. Temperature in the thermosphere decreases from 800–1500 K at the base of the exosphere to 300–400 K at 120 km and down to 200 K at 80 km. This temperature structure on the global average may be accounted for in the upper thermosphere simply as a balance between the deposition of thermal energy due to EUV and downward molecular conduction of heat. The picture is complicated in the lower thermosphere by infrared radiative losses from CO_2 and NO, and turbulent transfers of thermal energy and chemical constituents.

The absorption of EUV by thermospheric molecules leads to ionization of the major species O_2, N_2, and O and a consequent array of chemical reactions that decompose O_2 and N_2, recombine the ions and electrons and produce such minor species as NO. Peak electron concentrations of 10^{12} m^{-3} are generated near 300 km (the F-layer). Figure 3 summarizes the most important energetic processes in the thermosphere.

The spatial and temporal variations in mixing ratios of thermospheric constituents and the dominance of atomic oxygen at high levels are perhaps the most distinctive physical differences between the thermosphere and lower atmospheric layers. These variations are due first to the dissociation of O_2 upon exposure to EUV above 120 km and to 120–170 nm radiation in the lower thermosphere; and due, second, to the comparable rates of transport by molecular diffusion and by macroscopic atmospheric motions, especially between 100 and 200 km, as contrasted to the lower atmosphere where molecular diffusion is negligible except on very small spatial scales. Hence, as atomic oxygen migrates downward to the lower thermosphere for recombination and as molecular oxygen moves upward to dissociation, diffusion and large-scale mixing compete to determine their concentrations. For a global average, molecular diffusion is dominant above 120 km and each species has its own scale height. However, hydrodynamic transport can impose horizontal compositional variations throughout the thermosphere.

One of the current major thrusts in studies of the thermosphere is the construction of large comprehensive computer models that are used to test our understanding of individual processes and establish better how they interact. More thorough descriptions of the thermosphere may be found in Banks and Kockarts (1973) and Roble (1977).

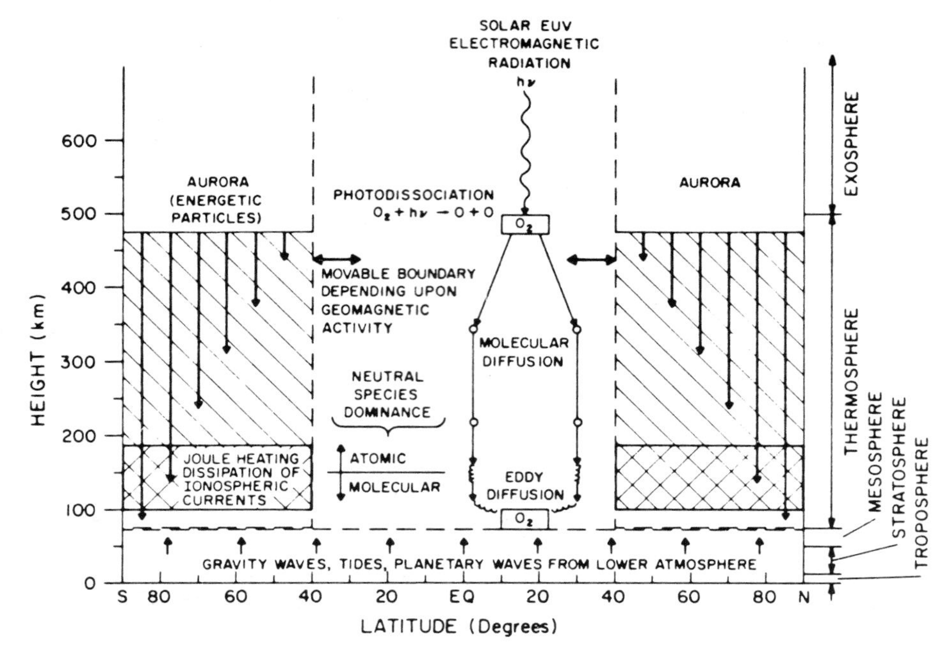

Fig. 3. Energy sources for thermospheric process. (From Roble, 1977.)

2.2. STRATOSPHERE AND MESOSPHERE STRUCTURE

The stratosphere and mesosphere extend from approximately 15 km and a pressure of 0.1 surface pressure to 80 km and 10^{-5} of surface pressure. They have, in a sense, the simplest global average temperature structure of the various atmospheric regions. The global average temperature structure can be described as approximately a consequence of the radiative balance between absorption of ultraviolet radiation (250–320 nm) by ozone and infrared cooling by various gases, but especially the 15 μm CO_2 bands.

Near the top of the mesosphere and above, 15 μm vibrational quanta escape more rapidly to space by radiation transfer than they are removed by collisional deactivation, and so the radiative cooling processes are not in local thermodynamic equilibrium (non-LTE). About half the cooling in the mesosphere is due to transitions whose lower state is an excited level ('hot bands') which are in non-LTE down to 65–70 km because of their weak band strengths.

The latitudinal variation of temperature at different levels is shown in Figure 4. One of the most remarkable features of the mesosphere up to the mesopause is its much warmer winter than summer temperatures. These contrasting temperatures are believed due to the magnitude of upward motions in the summer and downward motions in the winter hemisphere. Upward motions cool by adiabatic expansion and downward motions warm by adiabatic compression. In the winter hemisphere, downward motions provide additional energy by carrying into the mesosphere the chemical energy of atomic oxygen molecules. This energy is transformed to heat upon recombination into molecular oxygen.

The absorbed solar energy per unit mass is much greater at the stratopause around

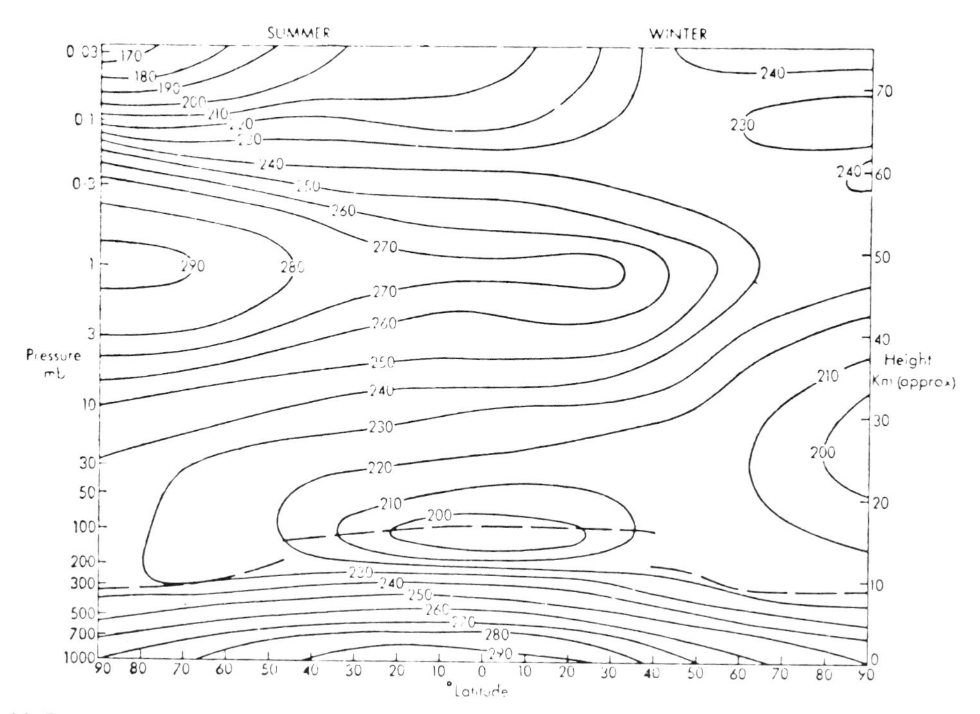

(a) Representative latitudinal mean cross section of temperatures, K, for the solstices. Heavy dashed lines represent the tropopause

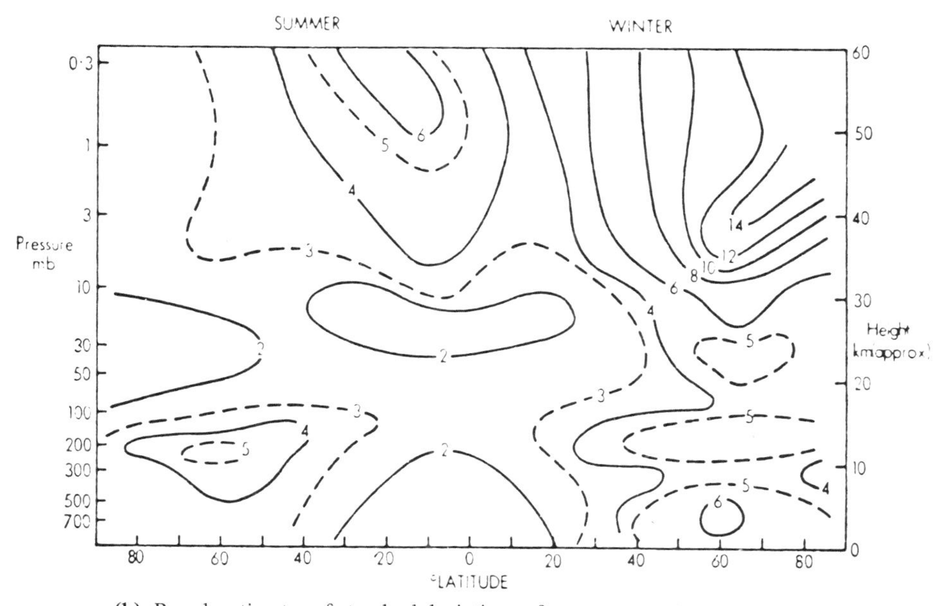

(b) Rough estimates of standard deviations of temperature, K, at the solstices

Fig. 4. Longitudinally averaged temperature field for Northern Hemisphere summer and winter from surface to 75 km. (From Murgatroyd, 1969.)

50 km than it is in the mesosphere. It is common for thermally driven circulations to 'overshoot' into regions of little or no heating and so reverse temperature gradients. The detailed mechanism responsible for the reverse temperature gradients in the mesosphere is now an active research question. It is believed to involve a large frictional drag due to the breaking of small-scale waves propagating upward from the troposphere.

2.3. Stratosphere and Mesosphere Chemistry

The most important minor constituent of the stratosphere and mesosphere is ozone (O_3). Although present in only a few parts per million, it is the dominant absorber of solar radiation and thus, in large part, is responsible for the observed temperature and wind structure just discussed. Furthermore, from the viewpoint of biological processes, it is indeed fortunate that this absorber is present, for otherwise the UV radiation would reach the surface and have a highly deleterious effect on various life forms.

The initial step in the process of forming ozone is the dissociation of molecular oxygen to atomic oxygen

$$O_2 + h\nu(\lambda < 242 \text{ nm}) \longrightarrow O + O \tag{2.1}$$

which then combines with O_2 with the help of a third body M to give ozone,

$$O_2 + O + M \longrightarrow O_3 + M. \tag{2.2}$$

Upon absorption of solar radiation in the 240–320 nm range, the ozone dissociates back to form atomic oxygen

$$O_3 + h\nu \longrightarrow O_2 + O. \tag{2.3}$$

Removal of ozone occurs through a wide array of catalytic pathways which predominantly bring about the reaction

$$O_3 + O \xrightarrow{\text{catalysis}} 2O_2. \tag{2.4}$$

Much of the effort in stratospheric chemistry over the last decade has gone towards improving our understanding of the various important cycles leading to the recombination of ozone. These efforts have been prodded by the realization that it was possible for human activities on a global scale to perturb significantly these cycles and hence the concentrations of ozone in the stratosphere.

The primary catalytic cycles destroying ozone are of the form

$$\begin{aligned} X + O_3 &\longrightarrow XO + O_2 \\ XO + O &\longrightarrow X + O_2 \end{aligned} \tag{2.5}$$

where the most important X's are atomic chlorine Cl, nitric oxide NO, and the hydroxyl radical HO. Besides these primary catalytic loops, it is necessary to consider various important crosslinking reactions, of the form

$$XO + Y \longrightarrow X + YO. \tag{2.6}$$

The most important such reactions have XO as HO_2, or ClO and Y as NO.

Catalytic species X can also be moved into inactive reservoirs by combining with other species Z, i.e.

$$X + Z \longrightarrow XZ, \tag{2.7}$$

where XZ represents one or more stable species. The most important such reaction pairs are X, Z = (ClO, NO_2), (ClO, HO_2), (HO, HO_2), (HO, NO_2), (HO, HNO_3), (HO_2, NO_2), (HO, HNO_4), (Cl, 'H'), where 'H' refers to an H in CH_4. Finally, active species X can be released from reservoirs provided by (2.7) or from tropospheric sources through dissociation, either by photolysis, i.e.

$$XZ + h\nu \longrightarrow X + Z \tag{2.8}$$

or by reaction, e.g.,

$$HCl + HO \longrightarrow Cl + H_2O.$$

Particular tropospheric sources XZ of special interest include N_2O, CH_3Cl, $CFCl_3$, an CF_2Cl_2 (the latter two also known as 'Freons' or CFMs), and H_2O. Currently, about two dozen photolysis reactions are considered in models of stratospheric chemistry and well over 100 individual reactions.

One focal point of research has been the question of the amount of depletion of ozone that would occur if the current release rate of CFMs were continued indefinitely into the future. This assumed release rate is perhaps the minimum release that is likely. If no further controls are applied to production rates, releases could continue to increase in the future. For example, the scenarios for CO_2 release from fossil fuel burning usually assume a factor of 2 to 5 growth in emission in the next 50 years in spite of their severe resource constraints. In addition to their use in spray cans, now banned in the U.S., the CFMs are employed in a large variety of other applications, in particular for refrigeration and air-conditioning devices and for forming the bubbles in the various foams used for funiture, mattresses, fast food containers, etc. Until 1974, CFM usage was growing exponentially with a doubling time of less than a decade, and further increases over the present global usage (about 0.3 megatonnes $CFCl_3$ and 0.4 megatonnes CF_2Cl_2 per year) seem likely. Molina and Rowland (1974) suggested on the basis of crude calculations that present release of CFMs, if continued long enough, would reduce global stratospheric ozone concentrations by about 10%. A reduction of stratospheric ozone by 10% implies a 20–30% increase in biologically damaging UV radiation at the Earth's surface. This issue has now become one of the most thoroughly studied environmental questions and perhaps one of the most interesting examples of possible future increased damage from solar output. The sun is implicated not only as the source of the damaging UV radiation, but also as the primary energy source for chemical reactions in the stratosphere and in particular for the production of the radicals that destroy ozone.

Up to now, much of the scientific progress made has been toward characterizing the complexity of the problem and the major sources of uncertainty. Milestone summaries of progress along these lines have been given in NAS (1982, 1983a). The best estimates of the ozone depletion which would result from continuing current emission rates of CFMs have changed somewhat over the years, largely because of changes in the estimates of production and lifetimes of various inert reservoirs of stratospheric chlorine, such as is characterized schematically by Equations (2.7)–(2.8). In particular, NAS (1976) gave a best estimate of about 7% global ozone reduction, whereas NAS (1979) raised the estimate to about 16%, but current models are giving ozone depletion for current emission rates of at most 5%. However, increases in CFM emissions of much beyond current rates leads to saturation of chlorine reservoirs involving nitrogen and so to much larger ozone reduction (Prather, 1984). It is now realized that realistic projections of future stratospheric ozone concentrations must include not only increases of chlorine compounds but also of other trace gases as well. In particular, increases in carbon dioxide act to lower stratospheric temperatures and so increase ozone. Changes in methane also affect ozone. Other improvements are needed in treating this problem. Meteorologists, in particular, would like to see more realistic treatments of the transport of the various species whose exchange between troposphere and stratosphere is crucial. One of the questions of current interest is the possible magnitude of fluctuations in ozone in the upper stratosphere due to variations in the solar radiation responsible for ozone generation, especially that with wavelengths around 200 nm.

The current decrease in estimates of the ozone change is largely a consequence of lowered estimates of the HO and ClO radical concentration in the lower stratosphere as prompted by several kinds of observational information (e.g. Wine *et al.*, 1981; Turco *et al.*, 1981). The lower estimates of HO concentrations imply a further possible serious reduction of ozone concentration by increases in oxides of nitrogen. A 30% increase in the concentration of stratospheric oxides of nitrogen would also be expected to reduce ozone by about 7%. The issue as to how harmful would be the likely increases in UV at the surface to biological systems has received less attention. The primary evidence for damage has been epidemiological studies showing how skin cancers relate to UV exposure and some experimental studies of the reductions in crop growth effected by UV exposures.

2.4. TROPOSPHERIC CHEMISTRY

In the troposphere, the first photochemical processes to be explored in detail were those in urban areas where heavy releases of hydrocarbons and oxides of nitrogen are acted on by solar radiation to produce the chemical soup known as smog (e.g. Dermerjian *et al.*, 1974; McEwan and Phillips, 1975). It is no accident that some of the sunniest cities are also the smoggiest. One of the actions of urban smog and solar radiation is to produce ozone, which unlike stratospheric ozone is regarded as a hazard rather than a benefit. In other words, the deleterious effects of its toxicity to plants and animals greatly outweigh any benefit that could come from its absorption of UV radiation.

One of the current major thrusts in atmospheric chemistry is to ascertain the role of similar processes in maintaining background ozone concentrations outside urban areas. Ozone acts through Equation (2.3) to generate $O(^1D)$, an electronically excited state of atomic oxygen, which dissociates water into hydroxyl radical

$$H_2O + O(^1D) \longrightarrow 2HO \qquad (2.9)$$

The HO together with NO act to oxidize catalytically various hydrocarbons as, for example, methane which reacts as follows:

$$\begin{array}{ll} CH_4 + HO & \longrightarrow CH_3 + H_2O \\ CH_3 + O_2 + M & \longrightarrow CH_3O_2 + M \\ CH_3O_2 + NO & \longrightarrow CH_3O + NO_2 \\ CH_3O + O_2 & \longrightarrow H_2CO + HO_2 \\ H_2CO + h\nu & \longrightarrow H_2 + CO \\ \hline \text{net: } CH_4 + HO + NO + 2O_2 & \longrightarrow CO + H_2 + H_2O + HO_2 + NO_2 \end{array} \tag{2.10}$$

The HO_2 and NO_2 regenerate the HO and NO and make O, hence O_3 in the processes

$$\begin{array}{ll} HO_2 + NO \longrightarrow HO + NO_2 & \\ NO_2 + h\nu \longrightarrow NO + O & \text{(twice)} \\ O + O_2 + M \longrightarrow O_3 + M & \text{(twice)} \end{array} \tag{2.11}$$

The CO is further oxidized by HO to CO_2 making more HO_2

$$\begin{array}{l} CO + HO \longrightarrow CO_2 + H \\ H + O_2 + M \longrightarrow HO_2 + M \end{array}$$

Hence, if only the above paths were followed, the oxidation of each CH_4 to O_3 would generate three ozone molecules. However, it is also necessary to consider the competing reaction

$$HO_2 + O_3 \longrightarrow HO + 2O_2$$

that not only destroys one ozone directly but another by scavenging the O from HO_2 and so short-circuiting the processes described in (2.11).

One of the major themes of tropospheric chemistry is the radicals generated by solar photons. Another theme is the intimate links between atmospheric gases and biological processes. For example, the key species methane is mostly due to anaerobic fermentation of plant carbohydrates by microorganisms in swamplike environments or stomachs of cattle. Figure 5 shows schematically the major sources and sinks of methane in the lower atmosphere.

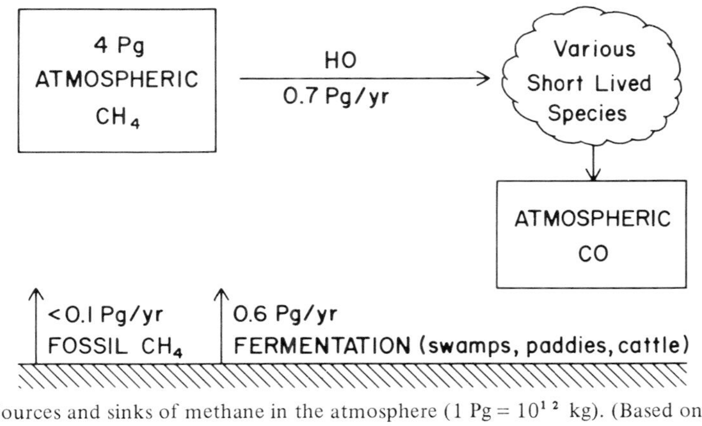

Fig. 5. Sources and sinks of methane in the atmosphere (1 Pg = 10^{12} kg). (Based on Ehhalt and Schmidt, 1978.)

In quantifying tropospheric chemistry, it is crucial to determine global 'clean air' concentrations of hydroxyl radical HO and nitric oxide NO as baseline data. Mapping the global distributions of these species is now a high priority research objective. Since these concentrations are typically 0.1 and 10 parts per trillion, respectively, and the species are highly reactive, it has been necessary to develop highly sophisticated instrumentation for their measurement.

The NO_X oxides of nitrogen – i.e. NO, NO_2, and HNO_3 – are generated in the troposphere not only by natural processes (e.g. lightning and soil respiration) but also by fossil fuel combustion. The total generation rate is estimated to be between 30 and 100 million tonne NO per year. These species may be incorporated into tropospheric aerosols, and either the gas or the aerosol may be removed by incorporation into raindrops or by dry deposition at the surface. Atmospheric chemists are just now beginning to emphasize the development of the theoretical methods and numerical tools necessary in determining the sources of tropospheric NO_X to model adequately these removal processes. An additional complication is the downward transport of NO_X from the stratosphere after generation by oxidation of N_2O. Although this source is small (~1% of the tropospheric source terms), it appears to control the mixing ratio of NO_X in the upper troposphere and so may be of major importance for tropospheric ozone chemistry (Liu *et al.*, 1980).

As I shall discuss later, ozone and methane are of modest but not entirely insignificant importance for the tropospheric heat budget. Since the sources of NO_X and CO due to human activities appear as large as natural sources, and that of CH_4 only a factor of 10 smaller, it is of considerable interest to ask how these gases and hence tropospheric ozone would change as fossil fuel consumption increases in the future. Table II summarizes

TABLE II

Response of ozone and methane concentrations relative to current values for hypothetical future increases in fossil fuel pollutants

Increase relative to current levels of human production	$\Delta O_3/O_3$	$\Delta CH_4/CH_4$
4 × CO	13%	35%
4 × CH_4*	12%	90%
4 × NO	12%	–17%
All of above	44%	91%

* Assumes currently anthropogenic CH_4 is 20% of total production. This is an upper limit obtained from C^{14} measurements (cf. Ehhalt and Schmidt, 1978).

some recent calculations of Hameed *et al.* (1980). These questions are linked, although indirectly, in many ways to solar electromagnetic radiation. Not only are the chemical transformations of tropospheric trace species driven by solar photons, but also the very existence of fossil fuels and methane is testimony to past photosynthetic utilization of solar radiation.

3. The Climate System

3.1. Current Questions

Two practical areas of inquiry now provide the stimulus for much of climate research. First is the challenge of making useful seasonal forecasts of possible anomalous conditions (e.g. will it be cold and dry next winter), and second is to determine the average long-term climate change that would result from changes to the global heat balance due to human activities. There have long been attempts to link the first question to short-term solar variability, but most such attempts have foundered due to lack of plausible physical mechanisms and convincing statistical treatments.

If solar variability modulates climate on a month-to-month basis, it is extremely unlikely that this modulation depends on any direct heating effect. The climate system has too much thermal inertia to respond significantly in a few weeks to solar heating variations much less than order of 10%, which is much larger than the 0.1% level of variability suggested by observations of solar output (Willson *et al.*, 1981). On the other hand, on a time-scale of decades to centuries, the time-scale of the second question, it is likely the climate system can respond significantly to solar heating changes of a percent, and possibly considerably less on even longer time-scales. Much of current climate research is concerned with the identification, modeling and quantification of the various interactive processes of energy feedbacks in the system that determine how large such a response would be. Likewise, the remainder of this section shall emphasize the energy feedbacks believed to be of greatest importance for long-term climate change. Some of the climate processes important for establishing the response to external changes are shown schematically in Figure 6.

3.2. Introduction to Simple Climate Models

The simplest climate models treat only global average conditions and focus on global average temperature as their fundamental dependent variable. The basic concept these models express is that of global average energy balance; neglecting geothermal heat reaching the surface of order 0.04% of the solar heating, we have a balance between net global heating Q and the change of global heat storage, i.e.

$$\frac{\partial H_s}{\partial t} = S - \mathscr{F} \equiv Q, \tag{3.1}$$

where H_s = total heat storage of the climate system (i.e. the net storage of internal energy in all the reservoirs, $H_s = \Sigma H_{si}$, where H_{si} refers to the internal energy of the oceans, snow–ice and surface land layers), S is the absorbed solar radiation, and $\mathscr{F}$ the net flux of thermal infrared radiation out of the top of the atmosphere. To the extent that changes in heat content of a thermal reservoir are linear in some temperature which characterizes the reservoir, a ΔH_s can be expressed as

$$\Delta H_s^i = c_i \, \Delta T_i, \tag{3.2}$$

where ΔT_i denotes the change of a reservoir's mean bulk temperature, and,

$$c_i = \frac{\partial H_s^i}{\partial T_i} \, . \tag{3.3}$$

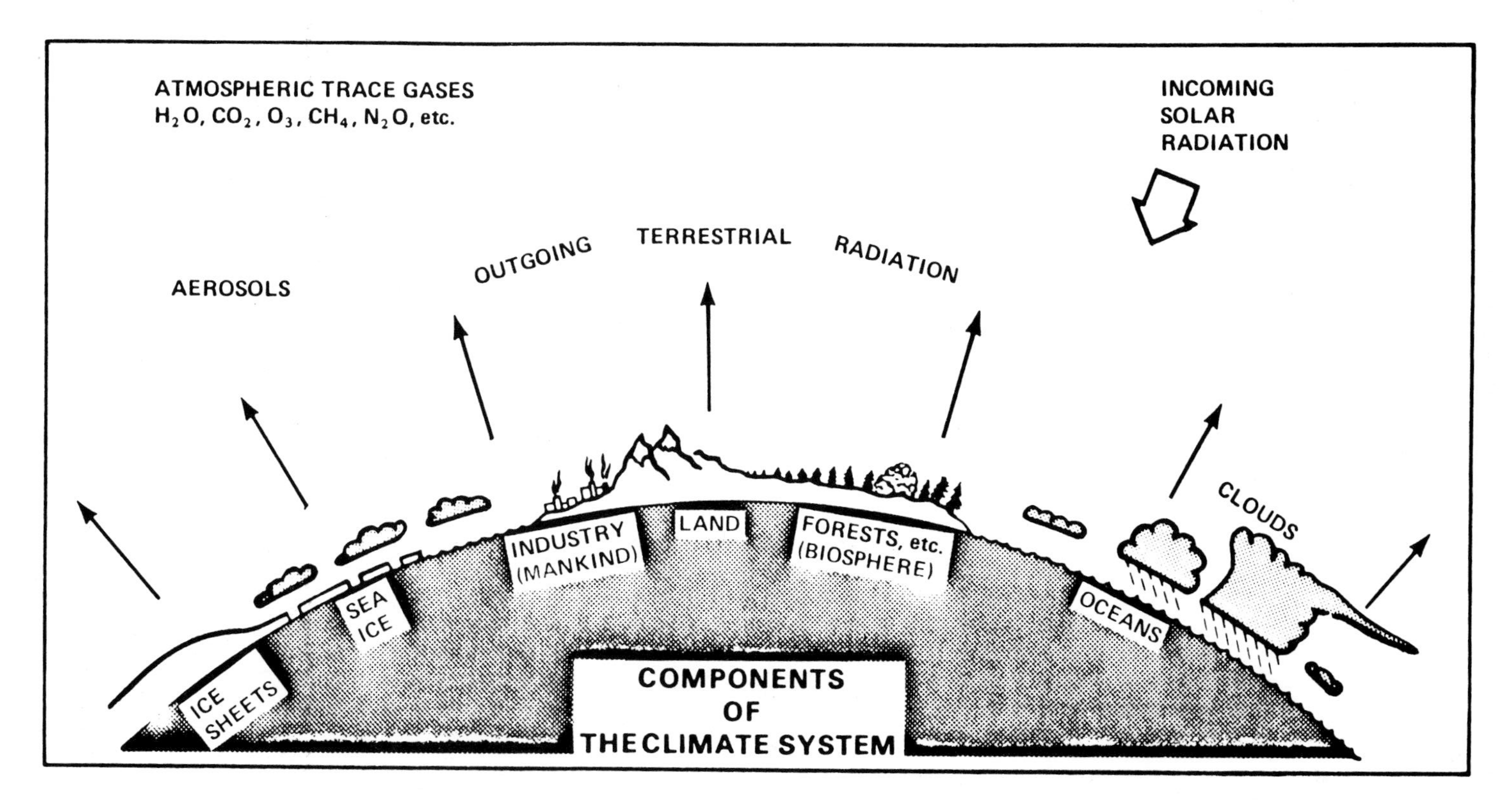

Fig. 6. Schematic of surface and radiative aspects of climate system. (After Kellogg, 1980.)

A frequent, convenient, but not very accurate assumption is that the sum of all internal energy changes can be approximated by

$$\Delta H_s = c\ \Delta T, \tag{3.4}$$

where T is some kind of global average temperature and c is the total system heat capacity, With somewhat greater accuracy, we can also assume that the net global heating Q depends only on T and various externally prescribed parameters such as the solar constant, concentrations of various atmospheric species, etc. These external parameters are denoted e_i, i.e.

$$Q = Q(T, e_i). \tag{3.5}$$

Hence, linearizing Q about some basic equilibrium state where it vanishes, and denoted by bars, we have

$$\delta Q = \frac{\partial Q(\bar{T}, \bar{e}_j)}{\partial T}\, \Delta T + \frac{\partial Q(\bar{T}, \bar{e}_j)}{\partial e_j}\, \Delta e_j\,. \tag{3.6}$$

Combining Equations (3.1), (3.4), and (3.6) we have the commonly considered global climate model,

$$c\ \frac{\partial \Delta T}{\partial t} + \lambda\ \Delta T = \Delta Q, \tag{3.7}$$

$$\lambda = -\left(\frac{\partial Q}{\partial T}\right)_{e_j}, \tag{3.8}$$

$$\Delta Q = \left(\frac{\partial Q}{\partial e}\right)_T \Delta e_j.$$

It is often appropriate to consider a quasisteady state for which Equation (3.7) reduces to

$$\Delta T = \frac{\Delta Q}{\lambda}. \tag{3.9}$$

Hence, we see the remarkable result that in order to estimate global average temperature change, we need two parameters: (1) the change of heating of the climate system for fixed temperature, i.e. present-day conditions; and (2) the feedback of the system λ, calculated independently of the perturbation.

Note that it is not essential that Δe_j be of small amplitude provided we can obtain $\Delta Q(\Delta e_j)$. Also, the formalism can be extended to large ΔT provided we can obtain λ as a function of ΔT. This is done in the more elaborate energy balance models (as reviewed by North *et al.*, 1981). More serious is the neglect of the coupling between global average and spatially varying changes.

It is generally believed that λ should lie in the range 1–2 $W\ m^{-2}\ K^{-1}$, at least for timescales of decades to centuries. For convenience in further discussion, we shall use

$$\lambda = 1.5\ W\ m^{-2}\ K^{-1}.$$

The solar constant is about 1370 W m^{-2}. Due to the Earth's spherical geometry, the average incident flux on earth is one-quarter of this. About 30% of this flux is reflected to space, leaving about 240 W m^{-2} average solar radiation absorbed by the climate system. For example, a 1% change of solar constant would provide

$$\Delta Q \text{ (1\% solar)} = 2.4 \text{ W m}^{-2}$$

from which it may be inferred that

$$\Delta T \text{ (1\% solar)} = 2.4/1.5 = 1.6 \text{ K}.$$

3.3. TAPPING OF THERMAL RADIATION BY ATMOSPHERIC CONSTITUENTS

Within the troposphere, the variation of temperature with altitude (usually a decrease) is controlled primarily by a mix of radiative and dynamic processes. It is therefore convenient in simple modeling approaches to assume that atmospheric temperature is a known function of surface temperature. See, for example, the review of radiative-convective models by Ramanathan and Coakley (1978). Atmospheric constituents absorb thermal infrared radiation emitted from the surface and lower layers of the atmosphere and reradiate it at a generally lower temperature. The primary effect of this trapping process (sometimes referred to as the greenhouse effect) is to reduce the net loss of thermal radiation from the top of the system, or equivalently to require the surface to be warmer than it would be for thermal balance to be achieved without the radiatively active atmospheric constituents. This is the primary reason that the net radiative balance of the system [e.g. as formalized in Equation (3.5)] can change as the composition of atmospheric gases changes.

The transfer of energy through the lower atmosphere by atmospheric gases follows the usual formalism of radiative transfer for a gas in local thermodynamic equilibrium. However, the atmospheres of the Earth and other planets are highly 'non-grey'. The emission (or absorption) of radiation as a function of wavelength occurs in pressure-broadened lines determined by the vibrational–rotational quantized degrees of freedom of the molecules. The emission for a particular vibration transition is referred to as a band. Line centers saturate for paths of radiation as short as a few centimetres for the most strongly emitting bands, which hence transfer radiant energy primarily in the far wings of their rotational lines. There is considerable overlap between various bands but not so much as to reduce significantly the high spectral variability of atmospheric absorption-emission at thermal wavelengths. Furthermore, not only the strongest bands, but dozens of weak bands are of significance, down to bands with band strengths as small as 10^{-5} that of the strongest bands. With all this complexity, it has been necessary for formulations of infrared radiative transfer in the troposphere to make heavy use of laboratory transmission measurements for obtaining band absorptivities over various paths.

For some constitutents, especially water vapor, it is possible to use an emissivity fomulation, i.e. a slab of atmosphere of temperature T emits upward and downward as a grey body with emissivity ϵ. One of the simplest such formulations for water vapor has been derived by Cess (1974) who obtains

$$\epsilon = 0.75[1 - \exp(0.096\sqrt{P_w \tilde{P} H_w})], \qquad (3.10)$$

where P_w is water vapor pressure, $\tilde{P}$ is atmosphere pressure averaged with respect to P_w along a vertical path, and H_w is water vapor scale height. The 0.75 indicates that water vapor can trap at most 75% of the surface emission. The main gap in the water vapor spectrum is between 8 and 20 μm. Carbon dioxide plugs most of this gap at wavelengths longer than 12 μm so that the major 'window region' is over 8–12 μm. In this part of the spectrum, trace gaes in concentrations of the order of 1 part per million or less, contribute significantly to further thermal trapping. The most important of these constituents are ozone (O_3), methane (CH_4), nitrous oxide (N_2O), the chlorofluoromethanes (CFMs = CF_2Cl_2 and $CFCl_3$) and ammonia (NH_3), listed according to their current relative importance. The concentrations of all these gases are increaseing due to human activities (Ramanathan *et al.*, 1985).

The absorption bands of carbon dioxide and these other gases are generally quite narrow compared to the width of the Planck function black-body emission curve appropriate to terrestrial temperatures. Consequently, simple treatments of their radiative transfer which integrate over their individual vibrational–rotational lines regard their absorption and emission to occur at a single wavenumber. For very weak lines (or equivalently, small mixing ratios) such that the absorption and emission are by line centers, the bands become optically thin; e.g. Dickinson *et al.* (1978) show that the infrared heating due to CFMs in concentrations of several parts per billion or less of an atmospheric layer of temperature T can be expressed for cloudless sky as

$$Q_{\mathrm{CFM}}(\mathrm{K\ day^{-1}}) = Sr[0.5\exp(\text{-}1440\ T_g) - \exp(\text{-}1440/T)]\,, \tag{3.11}$$

where T_g is the lower boundary effective emitting temperature, r is mixing ratio in parts per billion, and $S \simeq 3$ is a nondimensional average band strength.

To summarize the net effect of atmospheric absorbers, we note that the atmosphere–Earth system emits radiation equivalent to that of a black body of temperature T_e = 255 K. Since observed surface temperature is nearly 288 K, the surface is on the average warmer by 33 K than it would have been without the atmosphere. Equivalently, since the average lapse rate (decrease of temperature with altitude) in the lower atmosphere is 5–6 K for each kilometre, the system radiates from an average altitude of about 6 km.

The relative contribution of the various atmospheric constituents to the trapping of infrared radiation is $H_2O \simeq 50\%$, clouds $\simeq 26\%$, $CO_2 \simeq 18\%$, $O_3 \simeq 4\%$, all other gases $\simeq 2\%$. These estimates, based on calculations by Ramanathan and Coakley (1978) of the relative increase of outgoing IR if one or another constituent is removed, are oversimplified, in part because we are ignoring the contributions of these constituents to absorption and reflection of solar radiation. Also, the contributions do not add linearly; if we removed clouds but not H_2O vapor, or vice versa, the other would partially compensate for this removal. Hence, removal of both H_2O vapor and clouds would not increase the loss of thermal infrared radiation by the sum of the increased losses due to their individual removal.

Clouds and water vapor, whose radiative effects are treated in the next section, largely adjust to changes of the climate system on a time scale much shorter than a year. By contrast, the less abundant gases generally change in concentration with climate change much more slowly, so these gases are usually treated as externally specified variables. However, their concentrations may change over decades as a consequence of human activities, as addressed later.

3.4. THERMAL FEEDBACK BY CLOUDS AND WATER VAPOR

As evident from the discussion of the previous subsection, we can crudely model the global temperature by considering the Earth–atmosphere system to be a black-body radiator in which the greenhouse effect of the atmospheric constituents effectively raises the radiating surface 6 km above the ground. The thermal feedback of such a system would be

$$\frac{d\mathscr{F}}{dT} = 4\sigma T_e^3 \simeq 4 \text{ W m}^{-2} \text{ K}^{-1}.$$

To improve upon this estimate, atmospheric scientists utilize 'radiative–convective' models. In such a model, the temperature of the Earth and that of the atmosphere are calculated together. Within the troposphere, the lapse rate is fixed according to some semi-empirical convective-adjustment criterion, e.g. 6.5° km^{-1}. The stratospheric temperature is assumed to adjust to radiative equilibrium; that is, within each layer the absorbed solar radiation is balanced by the divergence of infrared thermal radiation.

Such models with *fixed composition* have been found to give essentially the same black-body sensitivity as found in the simpler models described above, namely

$$\left(\frac{d\mathscr{F}}{dT}\right)_{\substack{\text{radiative–convective}\\ \text{fixed composition}}} \simeq 4 \text{ W m}^{-2} \text{ K}^{-1}. \quad (3.12)$$

In reality, many atmospheric and surface parameters are likely to change with temperature and so drastically modify this result. It is likely that the three most important such quantities are: (a) atmospheric water vapor, (b) clouds, and (c) tropospheric lapse rates.

In order to circumvent the theoretical difficulty of identifying how these processes would change with temperature change, Cess (1976) took the following empirical approach. He assumed that the flux of thermal radiation out of the atmosphere is given by

$$\mathscr{F} = A_1 + B_1 T_s - A_2 C, \quad (3.13)$$

where A_1, A_2, B_1 are constants, to be obtained by regression on data, T_s is surface temperature, and C is the fractional cloud cover. Ideally, one would obtain these constants from analysis of data on the variation of global average $\mathscr{F}$ with global average T_s and C. Unfortunately, such variations are too small to be obtained from available data sets. Hence, Cess made the crucial assumption that *latitudinal* variations of $\mathscr{F}$ occur because of the same feedback processes that determine *global* variations of $\mathscr{F}$. In this way he found: $A_1 = 260 \text{ W m}^{-2}$, $B_1 = 1.6 \text{ W m}^{-2} \text{ K}^{-1}$, 90 W m^{-2}.

Physically, it seems unlikely that the assumed equivalence between global and latitudinal variations in flux would be correct. For example, there are large latitudinal variations of the atmospheric radiative constituents that result primarily for other reasons than temperature variation, such as, for example, the differences between the tropical rain belt and the subtropical dry zone. Furthermore, later studies have shown that a wide range of infrared feedbacks could be obtained with the approach by Cess depending on what detailed assumptions were made. Nevertheless, the inferences of Cess have been very valuable in providing a benchmark for further studies of thermal infrared feedback and have stimulated considerable further research.

The value of the temperature feedback parameter determined by Cess, $B_1 = 1.6$ W m^{-2} K, is much smaller than the black-body value of 4 W m^{-2} K mentioned earlier, implying temperature responses 250% larger than that of a black body. In other words, it implies that the climate system has important positive feedbacks in its thermal infrared emission which act in response to changing external conditions. One such feedback was long ago identified by Manabe and Wetherald (1967) who pointed out that as global temperature varies, the atmospheric water vapor content will also change. Their suggestion for treating this effect in climate models was to assume that relative humidity remained constant. Both observations and general circulation modeling (GCM) experiments with a hydrological cycle have shown that relative humidity changes much less with temperature than does absolute humidity. Thus, the humidity content of the atmosphere follows temperature roughly according to the Clausius–Clapyron relationship, namely, a 1% increase of temperature is accompanied by approximately a 20% increase in water vapor. This added water vapor increases the amount of radiation absorbed both by enhancing the trapping of thermal infrared and by increasing the solar absorption, with infrared trapping contributing nearly 90% of the total increase of absorbed radiation.

When Manabe and Wetherald assumed constant relative humidity in their radiative–convective models, the temperature change for a given external change essentially doubled, giving

$$\lambda = -\frac{\partial Q}{\partial T} = 2.2 \text{ W m}^{-2} \text{ K}^{-1},$$

where $Q = S - \mathscr{F}$ is again the globally averaged absorbed radiation. In comparison, Cess's estimate (1.6 W m^{-2} K^{-1}) is less than 80% of that derived by radiative–convective models. To explain this discrepancy, Cess and Ramanathan (1978) have pointed out that the conventional radiative–convective models assumed a fixed cloud-top altitude. They studied alternative models where, instead, cloud-top temperature was held fixed, and found $\partial \mathscr{F}/\partial T \simeq 1.3$ W m^{-2} K^{-1}.

It seems plausible that actual values of $\partial \mathscr{F}/\partial T$ would lie between the value appropriate for fixed cloud-top temperature and that for fixed cloud-top altitude. Perhaps 1.6 is as likely as any, although many scientists lean toward larger values around 2 W m^{-2} K. The albedo feedback term $\partial S/\partial T$ probably is about 0.5 W m^{-2} K, giving the value $\lambda =$ 1.5 W m^{-2} K^{-1} that was earlier suggested.

Another very striking result of Cess's analysis is the strong dependence of $\mathscr{F}$ on cloudiness ($A_2 = 90$ W m^{-2}). Several other investigators have by different approaches found values of A_2 about half as large. For example, Ellis (1978) compared statistically the outgoing IR radiation of clear and cloudy sky regions and found

$$A_2 = 40 \text{ W m}^{-2}.$$

A possible reconciliation of this discrepancy is that Cess's result applies only to small changes in cloudiness, whereas that of other investigators applies to the effect of adding clouds where none existed. That is, Equation (3.13) should be written

$$= A_1 + B_1 T_s - A_2 C_0 - \left(\frac{\partial \mathscr{F}}{\partial C}\right) C',$$

where C_0 is an average reference cloudiness and C' is a cloudiness perturbation, such that

$$A_2 \simeq 40 \text{ W m}^{-2}$$

$$\frac{\partial \mathscr{F}}{\partial C} \simeq 90 \text{ W m}^{-2}.$$

If clouds were to change uniformly at all altitudes, $\partial \mathscr{F}/\partial C = A_2$. If A_2 were as large as 90 W m^{-2}, the average cloud-top altitude would be nearer 8 km than the 5 km that is observed. If, as Cess finds, $\partial \mathscr{F}/\partial C \simeq 2 \times A_2$, the net cloud-infrared feedback would be double that inferred from cloud cover changes alone. These conditions could apply were cloud *altitude* to change with cloud cover such that the average cloud height was greater when cloudiness was greater. The effective cloud altitude can change either because the location of the tops of clouds at various levels has changed or because the fraction of clouds at different levels has changed by differing amounts. Changes in the coverage of cirrus clouds may be especially significant since their tops generally lie above 8 km or so.

The following analysis further clarifies the role of cloudiness in a global average climate model. Since fractional cloud cover C is part of the internal dynamics of the climate system, its impact on the climate feedback parameter λ should be considered in a complete model; that is, we might evaluate the feedback parameter λ as

$$-\lambda = \left(\frac{\partial Q}{\partial T}\right)_C + \frac{\partial Q}{\partial C}\frac{\partial C}{\partial T}. \tag{3.14}$$

Attempts have been made to analyze this expression but they have foundered on the difficulty of evaluating $\partial C/\partial T$. Cloudiness generally occurs on small scales of space and time and depends on other physical processes occurring on these scales. Thus, it seems unlikely that global average cloudiness could be simply related to global temperature change independent of smaller scale processes. In other words, neglecting coupling between global averages and spatial variations of clouds is extremely dubious. Hence, when a global model is used to evaluate the possible effects of clouds on climate, it is more straightforward to regard cloudiness as an external parameter and simply ask how the global radiative balance would change with changing cloud cover, with the climate otherwise remaining the same; that is, we consider

$$\frac{\partial Q}{\partial C} = \frac{\partial S}{\partial C} - \frac{\partial \mathscr{F}}{\partial C}. \tag{3.15}$$

In such a scheme, the change in solar absorption ΔS due to clouds is simply proportional to the change in planetary albedo when clouds are added. This dependence is established in global average models by the following relationships: $\Delta S = -\frac{1}{4}\mathscr{S}\Delta\alpha_p$, where $\mathscr{S}$ is the solar constant and α_p, the planetary albedo, is obtained from $\alpha_p = (1 - C)\,\alpha_{nc} + C\alpha_c$; hence, $-\Delta\alpha_p = (\alpha_{nc} - \alpha_c)\,\Delta C$, where α_{nc} is the planetary albedo in the absence of clouds and α_c is the planetary albedo for a cloud-covered planet. Cess found empirically that $\alpha_{nc} = 0.18$, and that $\alpha_c = 0.43$. Similar values are found by theoretical calculation.

Taking Cess's values, we find $\partial S/\partial C = \mathscr{S}/4\ (\alpha_{nc} - \alpha_c) = -86$ W m^{-2} so that with $\partial \mathscr{F}/\partial C = -90$ W m^{-2},

$$\frac{\partial Q}{\partial C} = 4 \text{ W m}^{-2} \simeq 0.$$

In other words, if $\partial \mathscr{F}/\partial C$ for global climate change were as large as found by Cess, the global energy budget would be quite insensitive to cloud changes. Alternatively, if $\partial \mathscr{F}/\partial C \simeq A_2$, cloudiness change could play a major role in climate feedback and sensitivity. It appears unlikely that simple empirical approaches will give any definitive answers to this question. Rather, their main value is in helping to state clearly and quantitatively the problem. The 'final' answer will likely await the development of credible models of global cloudiness and their effects on radiative fluxes. Such developments have been and are currently of high priority in the climate modeling community.

3.5. ANTHROPOGENIC MODULATION OF TRACE GASES IMPORTANT FOR CLIMATE

Over the last decade, considerable concern has arisen about modification of the global energy balance by changes in the concentrations of atmospheric absorbers. Some suggestions invoke possible changes in atmospheric cloudiness, for example, by inadvertent cloud seeding through pollutant aerosols or cirrus cloud initiation by high-flying aircraft. What cloud changes might occur and what their effect would be on the global heat balance are still poorly known. It is much simpler to estimate the heat balance effects of changes in atmospheric gases. Changes in most minor atmospheric gases are closely linked to the usage of fossil fuels; that is, solar energy captured by photosynthesis and stored as preserved organisms in the geological past. Hence, before we consider directly their effects, we briefly discuss the question of global energy use.

Between 1860 and 1975, the rate of global energy use has grown at about 5% per year from 0.1 to 8 terawatts (1 TW = 10^{12} W). If this exponential growth were to continue for another 90 years, the use would be nearly 900 TW (about 2 W m^{-2}) and would represent a major perturbation of the global energy cycle. Such continuing growth is no longer possible because of the finite availability of energy sources and the probable tapering off of population growth. However, even with no growth, current fossil fuel use will still release enough CO_2 to give incremental heating of more than 1 W m^{-2}.

Barring global catastrophe, the likely amount of global energy use 50 years from now can be estimated to lie between 10 (Lovins, 1980) and 30 TW (Sassein, 1980; Rotty and Marland, 1980); the use in the subsequent 50 years and later is, of course, much more uncertain. In the U.S., energy growth is now expected to be small, barely enough to keep up with increasing population, so that per capita use will remain around 10 kW. Much of the global increase in energy may occur in those developing nations with indigenous supplies or the ability to generate enough foreign exchange to pay for imported energy. Since the current energy usage of developing nations is only a few percent of that of the U.S., from the standpoint of improving human welfare, some growth in their energy usage would appear highly desirable to raise their standards of living. Nevertheless, this growth is bound to have major environmental consequences that must be quantified in order to minimize them where possible and adapt to them if necessary. For example, the current debate as to the relative amounts of coal versus nuclear energy to be used in the near future will have to be resolved as much on environmental as on narrow economic grounds. For nations with high present rates of energy usage, conservation may be the most attractive path for providing more 'end-use' energy (NAS, 1983b).

Carbon dioxide has been monitored in the atmosphere at Mauna Loa Observatory and elsewhere since 1958. During that time, it has increased from 312 ppm (by volume) to the current level of about 345 ppm. Superimposed on this long-term trend is an annual cycle whose amplitude at Mauna Loa is about 6 ppm. It is estimated that the concentrations in preindustrial times (1860 or before) were about 290 ppm. Hence, about half the increase of CO_2 in the atmosphere has occurred since 1958. In addition to fossil fuel burning, other human activities have a major effect on the global carbon cycle, and hence alter the CO_2 content of the atmosphere. To put these and possible fossil fuel effects in perspective, it is necessary to consider the global carbon cycle (e.g. Bolin *et al.*, 1979). The amount of carbon in various reservoirs is usually measured in gigatonnes (1 Gt = 10^{12} kg) of carbon, and transfer rates between reservoirs are estimated in gigatonnes per year. The atmosphere and ocean surface water each contain 700 Gt, living land plants and dead plants – e.g. humus in soil – together contain two to four times this amount. Compared to the amount of carbon in the atmosphere as CO_2, about 10 times as much is held in recoverable fossil fuels and about 50 times as much is found in the deep ocean waters. Much larger amounts of carbon are locked up in the Earth's rocks in small concentrations, but this reservoir exchanges with the other reservoirs only slowly on time-scales of thousands of years and hence can be ignored in considering the effects of fossil fuel burning in the next thousand years.

Currently, vegetation on land exchanges about 50 Gt yr^{-1} of carbon with the atmosphere as live plants take it up, die, and decay, Likewise, in the oceans warm supersaturated surface waters desorb about 50 Gt yr^{-1} of CO_2 which is transported by the atmosphere to high latitudes where it is absorbed again by cold unsaturated surface waters.

The fossil fuel input is now 6 Gt yr^{-1}. The impact of human activities on the biosphere may release as much as 3 Gt yr^{-1} of carbon as estimated by inventorying the various biospheric carbon reservoirs (Bolin *et al.*, 1979; Hampicke, 1980), but the contribution could be much smaller (Seiler and Crutzen, 1980). The total release of excess carbon to the atmosphere is hence between 6 and 9 Gt yr^{-1}. Of this, about 3 Gt remains in the atmosphere. Much of the rest presumably goes into the oceans. However, it is difficult for oceanographers to account for more than 2–3 Gt yr^{-1}; hence there is now considerable interest in identifying other sinks and in studying in more detail the biospheric contribution.

It appears to me that the current difficulties in reconciling the global carbon cycle with possible anthropogenic additions probably do not introduce drastic errors in estimating the atmospheric buildup over the next 50 years, provided the models are tuned to reproduce the past CO_2 data. The future atmospheric CO_2 concentrations over this time will still be determined largely by the rate at which fossil fuel will be burned. Since the use of natural gas and oil is expected to start to decline over the next several decades, any significant increases in fossil fuel use 50 years from now compared to today will necessarily involve increased burning of coal. Unfortunately, the burning of coal releases significantly more CO_2 per unit of energy than do the other fossil fuels and has other serious environmental consequences.

The increases of other atmospheric trace species are more speculative, and their impacts on global temperature are likely to be smaller than that of CO_2. However, they are all expected to add to the warming due to CO_2. If current releases of the CFMs continue for the next century, $CFCl_3$ and CF_2Cl_2 are expected to increase to 1 and

2 ppb, respectively. The measured year-to-year increase of these compounds in the atmosphere is striking. Nitrous oxide is increasing at a rate of 0.2% per year (Weiss, 1981) and with projected increases in fertilizers and fossil fuel burning, this trend could increase. Furthermore, methane has increased from 0.7 ppm in preindustrial times to current levels of 1.7 ppm and is currently increasing at a rate of nearly 1.5% per year (e.g. Rasmussen and Kahil, 1984). Burning of fossil fuels may contribute to increases in tropospheric ozone and methane and thus further warming of global climate (e.g. Hameed *et al.*, 1980; Ramanathan, 1980).

As stated earlier, the question of climate change due to increases of various atmospheric constituents, reduces to how much more radiative heating there will be for a given increase of a constituent, and what the response of the climate system is to a given heating. In such a formulation, the effects of these gases can be directly compared with the effect of variations of solar heating. Table III compares the heating due to increases of various constituents with a 1% (2.4 W m^{-2}) increase in solar heating. With

TABLE III

Increased trapping of thermal infrared radiation by hypothetical increases in various atmospheric trace gases (as inferred from Ramanathan and Coakley, 1978, and references therein). The numbers in parentheses indicate the percentage increase from current concentrations for the assumed perturbation.

Trace gas	Perturbation	ΔQ relative to a 1% solar constant increase
CO_2	30 ppm (+10%)	0.25
CO_2	300 ppm (+100%)	1.75
F-12	2 ppb (×10)	0.3
F-11	0.75 ppb (×5)	0.1
CH_4	1 ppm (+60%)	0.1
N_2O	200 ppb (+60%)	0.2
NH_3	1 ppb (+150%)	0.01
Troposphere O_3	(+25%)	0.2

the value for λ inferred earlier, the right-hand column can be multiplied by 1.6 to obtain an estimate of the implied steady-state global ΔT. For relatively rapid increases in globally averaged absorbed radiation such as that due to CO_2, there is a time lag of several decades or more between the applied heating and the steady-state response. In particular, for the present and near-future rate of CO_2 increase, the transient ΔT is about 0.5 the steady-state one. For example, atmospheric CO_2 has increased by about 10% in the last half century, which would imply a ΔT of about 0.4° in steady state, but only 0.2° if oceanic heat uptake is properly considered.

3.6. ATMOSPHERIC AND OCEANIC CIRCULATION AND THE SEASONS

The presence of ice and snow in winter in northern latitudes and the seasonal cycle of climate are manifestations of the role of variations of absorbed solar radiation in fabricating our climate system. In discussing climate, I have considered up to now only the global energy balance aspects. Solar radiation also drives the redistribution of energy within the climate system. Differential solar heating and the continental configurations set up gradients of temperature and moisture in the atmosphere which, in turn, drive our weather systems and their statistical manifestation as regional climate systems, and produce the poleward transport of heat. Without the atmospheric and oceanic redistribution of energy, the seasonal and latitudinal climate contrasts would be much more severe than they are. Variation of solar heating with latitude is, furthermore, one of the major sources of energy for oceanic circulations, which also move this energy from equatorial to polar latitudes. Variations in these energy transfers can lead to important fluctuations in climate.

These processes have been extensively studied over the last quarter-century by meteorologists and oceanographers, and analyzed with comprehensive three-dimensional numerical models or simpler energy balance models (as, for example, recently reviewed by North *et al.*, 1981). However, there is still considerable uncertainty as to the detailed mechanisms whereby the oceans transport internal energy, and this question is now of high priority for further research (Anderson, 1983).

3.7. Primitive Climate, the Carbon Cycle and the Faint-Early-Sun

It has always been very interesting to attempt to understand processes that occurred on Earth during its very early history. The overriding themes in the early history of the Earth are the close link between solar radiation, climate, atmospheric chemistry and geological and biological processes with the carbon cycle one of the key elements (e.g. Walker, 1977). Models of early climate have been strongly influenced by 'the faint-early-Sun paradox'. The solar emission (e.g. as reviewed by Newkirk, 1980) is thought to have been 20–30% smaller than today. The early solar emission with today's atmosphere would imply an ice-covered Earth. The inevitable conclusion is that the atmosphere must have had a much larger opacity in the thermal infrared than today. It was first suggested (Sagan and Mullen, 1972) that copious amounts of ammonia and methane were the likely source of this excess opacity, but now carbon dioxide is more favored. On geological time-scales, CO_2 is maintained in the atmosphere in large part by balances between volcanic sources and net loss due to chemical weathering of calcium aluminum silicates (e.g. Garrels *et al.*, 1976; Holland, 1978). The rock weathering, in turn, depends on the supply of water by rainfall. Walker *et al.* (1982) have shown that the temperature dependence of the weathering implies large increases of CO_2 in the atmosphere as a response to lowered global temperatures, on long geological time-scales. They infer that atmospheric concentrations of CO_2 in the early history of the Earth must have been many tens of times larger than they are now. The composition and climate of the early atmosphere must have strongly influenced the origin of life on Earth, the topic we turn to next.

4. Solar Radiation Drives the Biosphere

4.1. Origins of Photosynthesis

The earliest forms of life on Earth must have arisen over 4 billion years ago (or within 0.5 billion years of the origin of Earth). All the details of the early evolution of life are highly speculative but many general principles have been established by studies of current cellular processes, fossil records, and the composition of sedimentary rocks. Reduced carbon compounds are believed to have been supplied to the oceans of the early Earth by the actions of ultraviolet radiation and perhaps geological processes. These compounds apparently clumped into large molecules that were capable of self-replication (e.g. Eigen *et al.*, 1981). The oceanic organic 'soup' not only provided building blocks but also some chemical energy through rearrangements of the constituent molecules. Such energy conversions would have been fermentation-like processes where some pieces of organic molecules were oxidized by further reduction of other pieces. The productivity of a biota based on fermentation would have been limited both by the supply of carbon and of energy and so would be very low by current standards. All organisms that developed before the last billion years or so were prokaryotes; that is, were single-celled and without nucleus, as, for example, are bacteria. Any of these creatures that found a way to increase its energy or carbon intake would have been capable of more rapid growth and reproduction and so would gain an important evolutionary advantage. Perhaps the first autotrophs (self-feeders) were cells that used highly reduced nonorganic compounds such as H_2 and H_2S to generate energy that would allow CO_2 to be used for carbon supply.

At some yet very early date over 4 billion years ago, some cells developed pigments that had metastable excited electronic levels which could be populated as a result of absorbed solar radiation. The first use of this excited electronic energy may have been to remove hydrogen from various organic or inorganic compounds to set up gradients of proton energy across cellular membranes. Such gradients could facilitate the movement of nutrients across the membranes. However, at this stage it was not possible for these bits of life to obtain protons from water because the consequent byproduct, hydrogen peroxide (H_2O_2), was highly toxic. The prokaryotic cells must soon have discovered the enzymatic apparatus needed to use the downgradient proton flow for generation of the energetic phosphate bonds of ATP (adenosine triphosphate), or some predecessor of this ubiquitous energy 'currency of life' now used for the energy supply of all living cells, and whose generation by proton gradients is reviewed by Hinkle and McCarty (1978). The creatures that developed this fantastic application of solar electromagnetic radiation must have indeed flourished until the molecules supplying protons to them became scarce.

At some point a progenitor of the blue-green algae developed an enzyme to facilitate the transformation of H_2O_2 into water and oxygen, thus overcoming the shortage of proton (or equivalently, electron) suppliers. The consequent discarding of oxygen to the oceans presumably encouraged the ancestors of the purple nonsulfur bacteria to learn how to run their photosynthetic apparatus in reverse, as an alternate metabolic pathway. by doing so, they were able to supply protons from high-energy carbon compounds to the oxygen, and so set up the proton gradients to make ATP, as they already did in photosynthesis (Dickerson, 1980). This 'respiratory' apparatus greatly enhanced the energy available from organic foods.

Meanwhile, evolutionary forces led to the development of optimum light catchers, the chlorophyll pigments. Hundreds of individual light-collecting pigment molecules linked together into the 'antennae' such as characterize modern plants, and somehow in the blue-green algae two sets of energy-converting apparatus, referred to as photosystem I and photosystem II, became coupled together.

The earliest sediments testify to a remarkably high productivity achieved by primitive organisms within a billion years of the origin of the Earth. One of the distinctive indices of early life is the amount of organic carbon buried in sedimentary rocks. Direct quantitative measurement of this buried carbon is made difficult because of variability from sediment to sediment and lack of preservation, but it can be inferred indirectly from carbon isotope ratios. Living cells incorporate less C^{13} than C^{12} into their structure and hence leave inorganic carbon sediments with a surplus of C^{13}. The largest crustal carbon reservoir is carbonate, but about one-quarter of the crustal carbon is organic. This organic carbon represents about 10 atmospheres (10^5 kg m^{-2}) of CO_2 that have been removed from the atmosphere by photosynthesis over the eons. At current rates of photosynthesis, such an amount of carbon is incorporated into plants in about 200 000 years. However, at present, most photosynthetic carbons are reoxidized through the respiration of the animals that consume them, so that with current rates of carbon burial of less than 0.5% of net primary productivity, renewal of the crustal reservoirs would require about 100 million years – that is, a time comparable to the age of oceanic crust imposed by plate tectonics, and of land crust imposed by weathering-erosion cycles.

One of the *most amazing findings* of carbon isotope chemists is that the C^{13}/C^{12} ratio of carbonate rocks has on the average remained nearly constant as far back as 3.8 billion years, the age of the oldest rocks accessible to geologists (Schidlowski *et al.*, 1979). Most of the original crustal CO_2 gas would have been incorporated into sediments by then, so that it appears there must have been an extensive ecosystem of primitive prokaryotic organisms before that time, with photosynthetic fixation of carbon proceeding at least at 1% the current rate. At least some of the organic carbon produced by these synthesizers would not have been buried but would have been used as food by anaerobic fermenters. The net result of such fermentation could have been $2CH_2O \rightarrow CO_2 + CH_4$. Without oxygen in the atmosphere, the relative if not the absolute biological generation of methane should have been much greater in the early atmosphere than it is now. About 1% of the carbon fixed by plants is now so converted to methane, largely in swamps and the digestive systems of cattle (e.g. cf. Figure 5).

In spite of the early proliferation of photosynthetic activity, until about 2.2 billion years ago, the copious amounts of oxygen evolved by the primitive photosynthetic cells were apparently all retained within the oceans. Geological evidence indicates anoxic conditions on land until that time.

The missing oxygen is known to have gone into the oxidation of ferrous ion, FeO, to ferric iron, Fe_2O_3, and of reduced crustal sulfur compounds to sulfate (SO_4^{--}) compounds. It is quite likely that this deposition of oxidized compounds onto the ocean sediments was facilitated by bacteria which could have gained energy and electrons in the process.

The reason for the initial appearance of significant amounts of oxygen in the atmosphere about 2 billion years ago is not known, but is more likely due to oceanic mechanisms for loss of O_2 becoming less efficient rather than a rapid growth in the rate of photosynthesis. Perhaps the continental crust was by that time sufficiently oxidized that further deposition of oxygen into oceanic sediments could not keep up with photosynthetic production. It seems to me, however, more likely that the growth of atmospheric oxygen was a response to an evolutionary shift in the homeostasis of the biosphere of that time.

One likely possibility is the ability to fix nitrogen in an oxic environment. The enzyme nitrogenase that is generally used for nitrogen fixation is destroyed by oxygen. Many very primitive prokaryotes including the blue-green algae have this enzyme, suggesting that biological processes were nitrogen-limited in very early times. Present-day blue-green algae have their nitrogen-fixing machinery in 'heterocysts' to protect them from oxygen. The earliest blue-green fossils with this feature are in the 2-billion-year-old Gun Flint chert. Broda (1975) and Fenchel and Blackburn (1979) give other useful reviews of the likely evolution of prokaryotic cellular machinery.

After the development of eukaryotic cells (cells in the nuclei as in all higher organisms) about a billion years ago, further evolutionary advances in plant structure included the development of multicelled specialized structures, such as leaves, roots and reproductive systems. Of special note is the development of flowering plants in the Cretaceous about 100 million years ago. These evolutionary changes have increased the productivity of plants in response to solar radiation in otherwise hostile land environments.

4.2. Photosynthesis in Action

The photosynthetic machinery of modern plants was first developed with the prokaryotic blue-green algae at least several billion years ago. The operation of that machinery on a cellular basis is now briefly reviewed.

Photosynthesis in plants occurs within the disk-shaped subcellular bodies called chloroplasts. There are about 50 chloroplasts in a typical plant cell, and these are roughly 5–8 μm in diameter and 1 μm in width. Each chloroplast contains roughly 10^9 molecules chlorophyll that absorb light by raising an electron to an excited electronic level. These molecules are situated on an elaborate internal membrane system and are hooked up into groups of several hundred. These 'antennae' funnel by resonant transfer the energy of the individual photons into the chlorophyll of reaction centers. The reaction centers have the structure of either photosystem I or photosystem II. Photosystem I traps 0.70 μm photons whereas photosystem II traps 0.68 μm photons. Shorter wavelength photons lose their excess energy primarily to heat, as the other chlorophylls move their excitation energy to the reaction centers (for further discussion cf., e.g., Nobel, 1974).

In photosystem II, hydrogens are detached from water; that is, absorption of 4 photons promotes $4H_2O \rightarrow 4(HO + H^+ + e)$, thence $4HO \rightarrow 2H_2O_2 \rightarrow 2H_2O + O_2$. The electrons fill in the photosystem II chlorophyll ground state as the excited electrons are transferred from the inner side of the membrane to other excited electron carriers on the outer side of the membrane. These molecules pick up for each electron an H^+ and pull it through the inner chloroplast membrane. The electrons then can supply the photosystem I complex which uses its absorbed photons to move the electrons back to the outer side of the membrane, where their reducing power is stored in a molecule known as NADPH. Figure 7 (Hinkle and McCarty, 1979) illustrates these processes. Finally, all the H^+ migrate through the 'CF_1–F_0 complex' and in doing so transform 1 ADP to ATP for every 3 protons. It is sufficient here for the reader to remember that ATP and NADPH are molecules used by all living creatures to store chemical energy within their cells in a readily available form.

The primary use of the NADPH and ATP in plants and generated as described above is to fuel the conversion of CO_2 molecules through the Calvin cycle, a simplifed version of which is shown in Figure. 8. The plant uses the 5-carbon phosphate sugar (ribulose

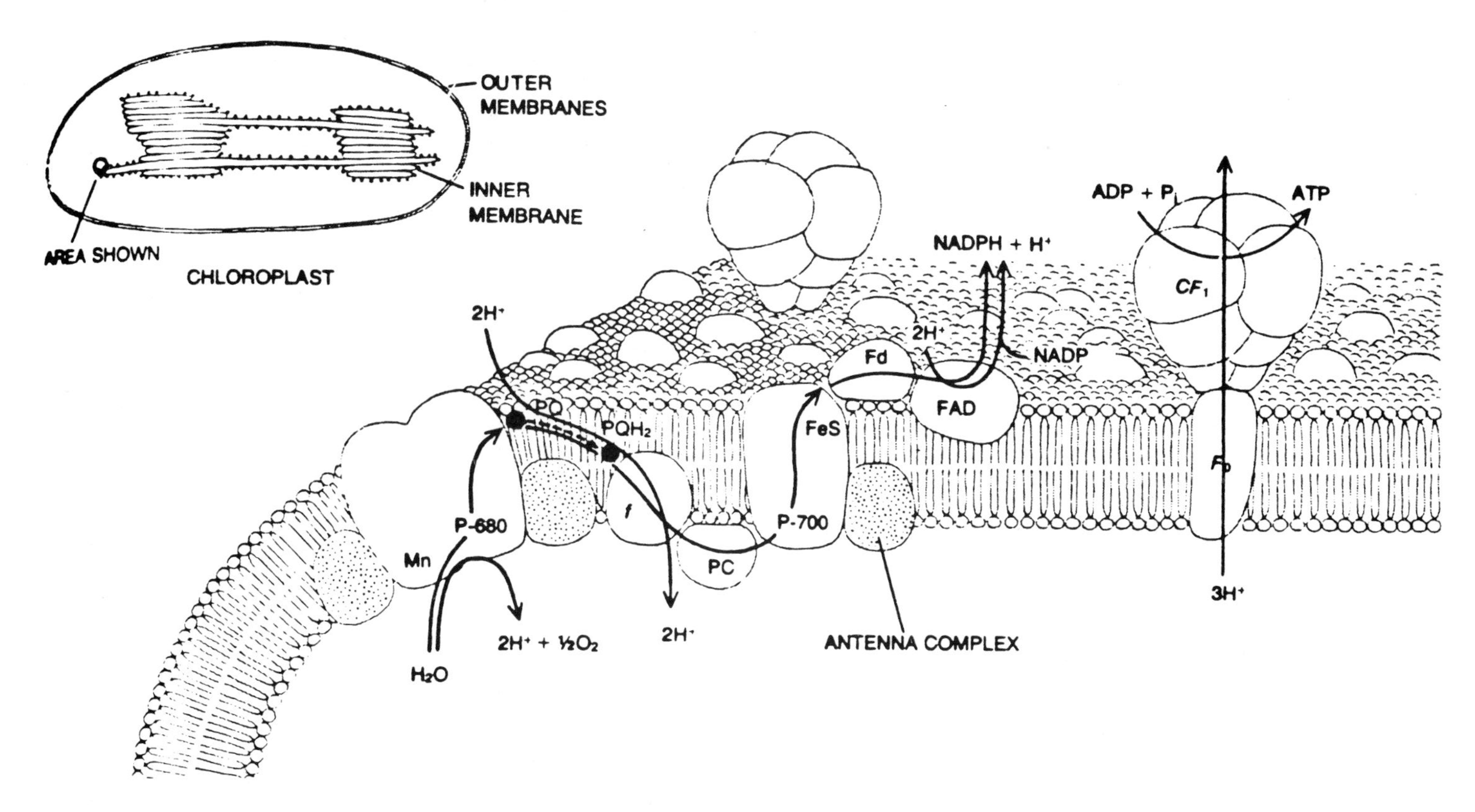

Fig. 7. Chloroplast membrane, showing light driven movement of protons and electrons to generate proton gradient; movement of protons through CF_1–F_0 generates the ATP which together with the NADPH are used to reduce CO_2 to carbohydrates. (From Hinkle and McCarty, 1978.)

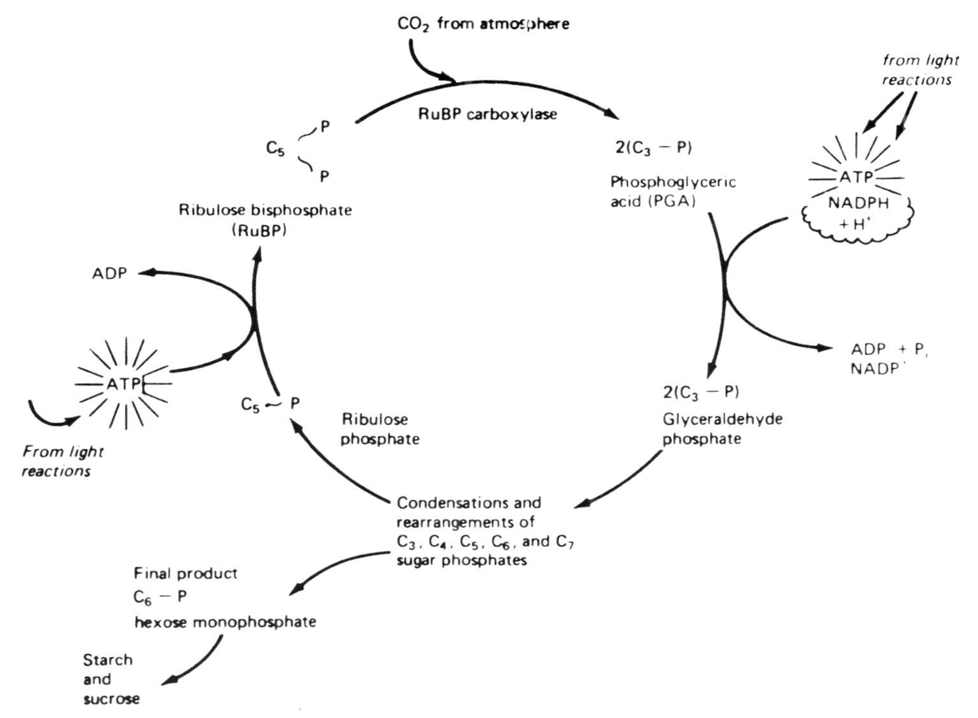

Fig. 8. Outline of Calvin cycle for reduction of CO_2 to carbohydrate using energy of ATP and NADPH. (From Galston *et al.*, 1980.)

biphosphate) to capture the CO_2, and the whole immediately splits into two 3-carbon phosphoglyceric acids. By using the energy and reducing power of the 3 ATP and 2 NADPH molecules per CO_2 taken up, the cycle is able eventually to regenerate the ribulose biphosphate and incorporate the new carbon into starch and sugars. The starch is the stored form of carbohydrate, whereas the sugars are the mobile forms transported to other parts of the plant.

Several complications need be mentioned. First, the ATP and NADPH molecules are energy carriers not only for the fixation of carbon but also for all the other plant metabolic processes requiring energy. The energy for these other metabolic processes would, in the absence of supplies of photons, be provided by the plant's cellular respiratory cycles. The plant's respiration is fueled by converting its carbohydrate supplies back into water and carbon dioxide. It is known, for example, that the reduction of nitrate ions to ammonia proceeds much more rapidly in sunlight than darkness and so presumably utilizes available solar energy. See Penning de Vries (1975) for further discussion of respiratory requirements.

Second, the enzyme ribulose biphosphate carboxylase that first incorporates CO_2 into the plant molecule ribulose biphosphate also can attach O_2 to ribulose biphosphate, initiating the so-called photorespiratory cycle that eventually oxidizes plant carbon back to CO_2. Photorespiration reduces the plant's productivity not only because of the above-mentioned carbon loss but also because it competitively inhibits the pickup of CO_2. This latter effect becomes especially pronounced at low levels of CO_2.

Third, some plants have developed additional mechanisms to minimize deleterious environment effects accompanying their pickup of CO_2. The C-4 plants first pick up CO_2 with a 3-carbon compound which then is delivered to the site of the Calvin cycle. The term C-4 refers to the fact that a labelled CO_2 first found after assimilation is in a 4-carbon compound as contrasted with the more common 'C-3' plants where the first observed carbon compound after pickup is the 3-carbon phosphoglyceric acid. The C-4 plants manage to reduce sufficiently O_2 concentrations and increase CO_2 concentrations at their Calvin cycle reaction sites as to virtually eliminate photorespiration. However, this is done at an additional cost of photosynthetic energy. Ehleringer and Björkman (1977) have measured the number of photons needed per CO_2 molecule for a number of C-3 and comparable C-4 species. The measurement refers to leaves attached to plants grown in the laboratory under comparable conditions and at sufficiently low light levels that the light supply limits the rate of photsynthesis.

They found that under normal O_2 and CO_2 concentrations both C-3 and C-4 plants assimilate one CO_2 molecule for about 19 photons of light absorbed. But with 2% O_2 concentration, they found that only about 14 photons were needed for C-3 plants, whereas 19 photons were still needed for the C-4 plants. Their results thus suggest that C-3 and C-4 plants are both of comparable photosynthetic efficiency under normal conditions.

The C-4 plants, nevertheless, do generally seem to have a competitive advantage over C-3 plants under hot, dry, and high light conditions, as indicated by the prevalence of the C-4 mechanism in tropical and subtropical grasses. We shall return to the question of photosynthetic efficiencies.

The CAM (crassulacean acid mechanism) plants operate similarly to the C-4 plants except they open their stomates to pick up CO_2 at night but use it the next day in the Calvin cycle while their stomates are closed to conserve water. Many desert succulents have developed this mechanism to help conserve water. The cost is a considerable retardation of their rate of CO_2 uptake.

4.3. Harvesting the Sunlight, Net Primary Productivity

Another key issue in contemplating the impact of solar radiation on the biosphere is the net rate at which carbon is incorporated into plants. Two questions may be asked: 'What are the maximum achievable rates?' and 'What are the average rates in natural and cultivated ecosystems?'

The theoretical maximum efficiency is one CO_2 molecule per eight or nine photons for a C-3 plant, and for a C-4 plant, that plus an additional 2 ATP (about two more photons). The resonant transfers between the different chlorophyll molecules in the antennae system are quite efficient, losing at most about one photon per CO_2 (Nobel, 1974).

If we take the measurements of Ehleringer and Björkman (1977) as typical, it would appear that 5–10 additional photons are lost in the plant prior to carbon uptake. Some of the light is absorbed by other pigments that are less efficient than the chlorophyll molecules in resonantly transferring the energy into the reaction centers. Furthermore, the ATP and NADPH or carbohydrate generated from about two photons may be used for other plant functions as, for example, the reduction of nitrates to amino groups (Penning de Vries, 1975).

Under natural conditions, C-3 plants need at least several more photons for producing the carbon lost to photorespiration and C-4 plants need at least two more photons to provide the energy required to drive their additional CO_2 assimilation machinery. It would appear, therefore, that under natural conditions at least 15–20 photons should be needed per CO_2 assimilated in approximate agreement with Ehleringer and Björkman (1977). Rapidly growing plants need to further expend the energy of about 30% of their assimilated carbohydrates in manufacturing more complex plant products at night and in other parts of the plant (Penning de Vries, 1975), but the final energy content of the plant dry material is less than the initial energy assimilated by perhaps half of that (15%). If one CO_2 molecule were converted into plant dry matter per 20 solar photons (0.4–0.7 μm), this would represent conversion of about 10% of the solar energy captured into plant chemical energy. However, since only about 45% of the incident solar radiation is photosynthetically active radiation (PAR), and at least 10% is reflected or transmitted to the ground (e.g. Good and Bell, 1980), the effective efficiency is then about 4%. Efficiencies for rapidly growing crop plants of about 4% of incident total solar radiation have been reported but the more usual maximum rates obtained for field crops are around 2% (Cooper, 1975). Global maps of average gross and net productivity have been developed (e.g. Box, 1978).

A 1% energy conversion of global average solar radiation incident at the Earth's surface would generate annually 4 kg m^{-2} dry matter. Reported maximum yields of grain crops (wheat, corn, rice, etc.) are in the range 2–3 kg m^{-2} and about 0.4 of this is incorporated in the grain (Loomis and Gerakis, 1975). The maximum grain yields of 1 kg m^{-2} (10 t ha^{-1}) can be compared with the human nutritional need of the per person energy equivalent of about 250 kg of carbohydrates per year (4.5×10^3 kJ yr^{-1}).

The total global annual net production of plant dry matter on land is about 1.3×10^{14} kg (Bolin *et al.*, 1977) or, on the average, about 1 kg m^{-2} of land surface. This is also about the average yield of cultivated lands. Cultivated lands use about 10% of the total global land area, but typically 30–40% of cultivated land must lie fallow unless irrigation and chemical fertilizers are used.

The global biomass accumulation of cultivated fields is about 10^{13} kg yr^{-1}. About half the potential productivity is lost to insects, rodents, diseases, and weeds and the net harvested yield is about 2×10^{12} kg yr^{-1} in grain equivalent food. About 60% of this production is in cereal crops. Harvests of the major grains are shown in Figure 9. About 20% of the harvested food is lost in storage and a comparable amount in the use of grains for animal food, with a consequent factor of 10 loss in its energy content (Pimentel *et al.*, 1975).

Revelle (1976) has estimated that the grain equivalent yield of global food production could be increased to about 10^{13} kg yr^{-1} by extensive agricultural development in the underdeveloped countries using application of fertilization, irrigation, and multiple cropping in order to more than triple the land effectively planted to crops. Presumably, the energy and environmental costs for this increased productivity would be severe (Ehrlich *et al.*, 1977).

Summarizing global plant productivity in terms of efficiencies, on the average about 0.25% of the solar energy incident cover land is incorporated into biomass energy. Over cultivated lands, which cover 3% of the Earth's surface, about 0.05% of the incident solar energy finds its way into food usable to humans. In other words, less than 0.002% of the

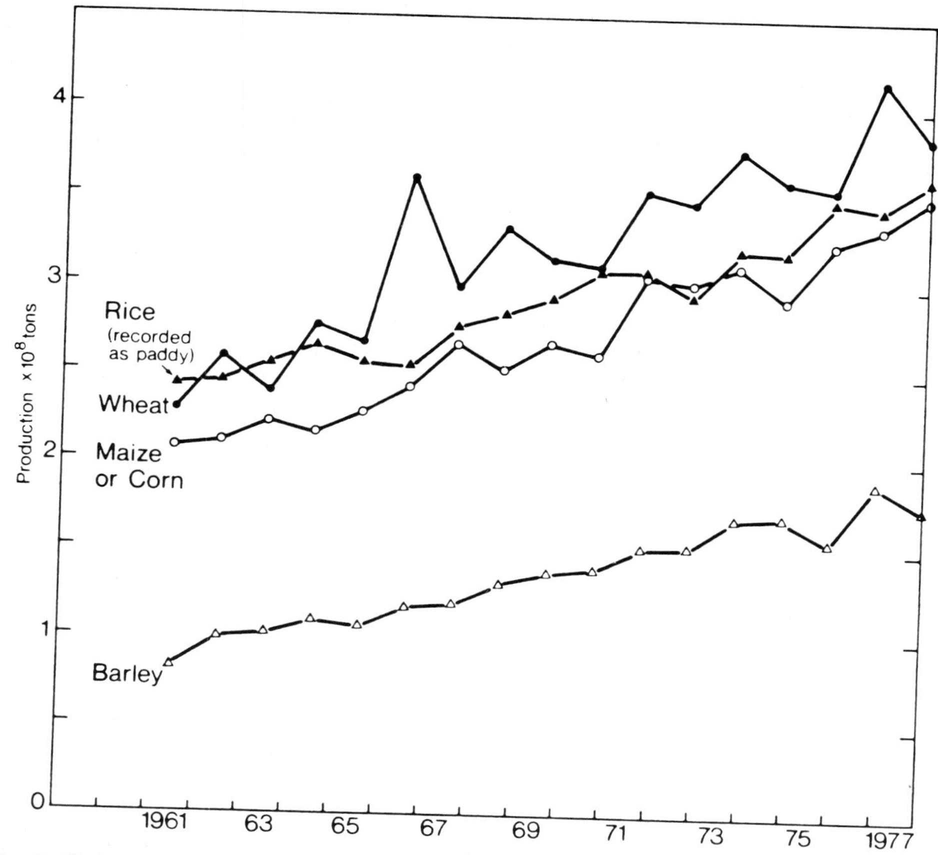

Fig. 9. Global production of major cereals according to FAO Production Yearbooks. (Taken from Tarrant, 1980.)

solar radiation incident at the Earth's surface (i.e. about 4×10^{16} kJ yr^{-1}) supplies the needs of human nutrition. The total biomass produced in the oceans is comparable to that produced over land, but only a small fraction (less than 10^{11} kg yr^{-1}) is available for human food because of harvesting high on the predator food chain (e.g. Ehrlich *et al.*, 1977).

Besides production of crops, man harvests biomass solar energy for animal grazing, fuel, and forest products. Livestock grazing on grasslands (about 20% of the land surface) harvests about 3×10^{12} kg yr^{-1}, which in turn generates about 5% of human food supplies. Estimated global biomass fuel and forest product harvests amount to another 10^{13} kg yr^{-1}, representing about as much carbon as burned in fossil fuels (e.g. Bolin *et al.*, 1979).

If we average over oceans as well as land, the average solar conversion into plant energy is 0.12% of the incident energy, only about 0.1 of which is harvested for all human applications. Most of this energy incorporated into plants is consumed by microdecomposers. In returning the solar energy captured by plants to the environment, the decomposers recycle mineral nutrients that would otherwise rapidly become depleted (Swift *et al.*, 1979; Fenchel and Blackburn, 1979).

Where plant products are removed, it is necessary to fertilize or allow the slow natural processes to accumulate again the previous fertility from nitrogen fixation, rock weathering, and dissolved substances in rainfall (Bormann and Likens, 1978).

In brief summary, solar radiation was the major evolutionary force determining the development of biological processes over geological times and now is the primary source of energy for the production of all plant products.

5. Concluding Remarks

This essay has been intended to convey an impression of the range of the effects of solar electromagnetic radiation on the terrestrial environment. I have considered both the present environment and that of the early Earth and have particularly emphasized atmospheric chemistry, climate, and photosynthetic processes. Studies of the terrestrial environment in the early geological past help us to understand better the close connections between climate, atmospheric chemistry, the biosphere, and geological processes.

The evolution of early life must, at first, have been quite dependent on the supply of reduced carbon and nitrogen compounds to the oceans from atmospheric processes. But then after establishment of a global biosphere, the living creatures induced major changes in atmospheric composition which, in turn, greatly modified geological processes and perhaps climate (e.g. Margulis and Lovelock, 1978). As this was done, biological processes coevolved to benefit from the changes they wrought. Of particular significance was the buildup of oxygen within the atmosphere which improved the efficiency of respiratory cellular energy conversion processes and introduced the ozone layer which shields the Earth's surface from most harmful UV solar radiation.

The theme of the interrelatedness of solar radiation, climate, geological and biological processes continues into more recent geological history. However, it is somewhat muted on the million year time-scale where over the last billion years the ramifications of continental drift, sea floor spreading, and related tectonic processes are highlighted (e.g. Eicher and McAlester, 1980). We can point to the accumulation of biological carbon into fossil fuel deposits (e.g. Hunt, 1979) and sedimentary rock cycles (e.g. Garrels *et al.*, 1976) as examples of major solar-driven processes. The accumulation of sedimentary layers, in particular, is linked to solar radiation through the hydrological aspects of climate. Geological evidence for climate variation over the last billion years has been reviewed by Frakes (1979).

Geologists studying the last million years have recently uncovered a strong direct solar signal – that is, they see evidence for major climatic fluctuations following variations in the Earth's orbital parameters – hence the intensity and spatial distribution of incident solar radiation (Imbrie and Imbrie, 1979). The orbital parameter solar variations are thus believed to have been a major contributor to establishing the last ice age that peaked 20 000 years ago, as well as earlier ice ages over the last million years. Shorter time-scale climatic variations may also be solar related, but the evidence is more tenuous and the influences may be more indirect; for example, Harvey (1980) reviews the possibilities of solar variability impacts on the thousand year time-scales.

In discussing present climate, I have considered primarily the global energy balance aspects. Also of importance is the redistribution of energy within the climate system by atmospheric and oceanic motions.

We can look at the consequences of solar radiation not only on the global and regional environments but on local environments as well. For example, a major area of animal physiology research is the heat balance of animals, and in particular, their temperature response to solar heating with atmospheric ventilation and infrared thermal cooling (as discussed, for example, by Gates, 1980). Such processes determine the range of environmental conditions over which particular animals from dinosaurs to mice can survive.

Another major area of research is the interactions of plant canopies with solar radiation, the atmosphere, and the soil (as sketched in Figure 10). The absorption, reflection,

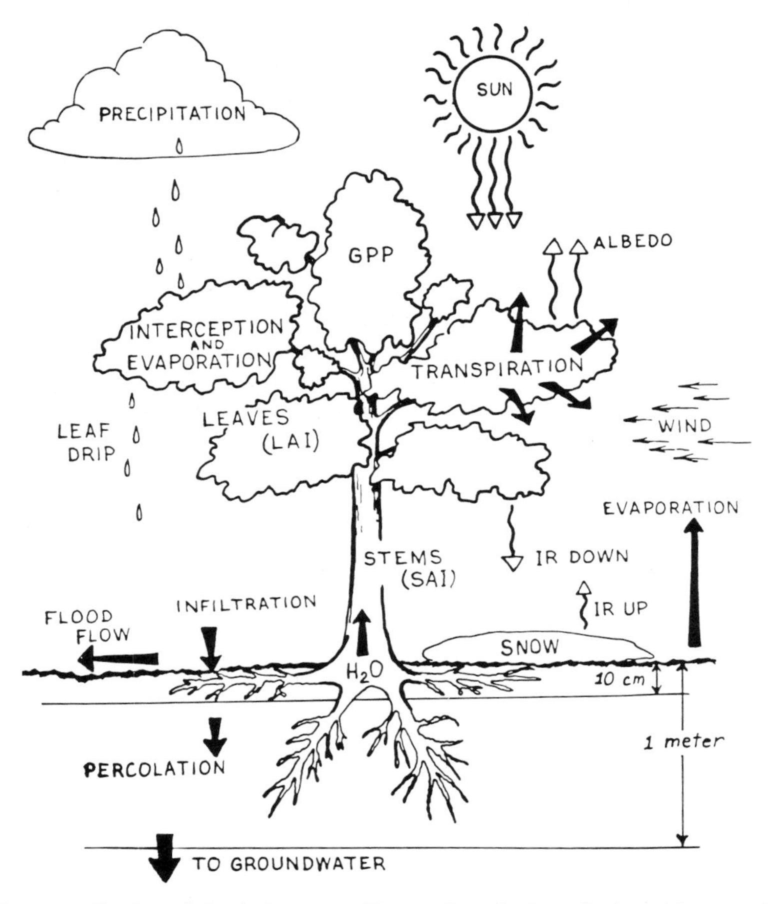

Fig. 10. Conceptualization of physical processes' interaction of solar radiation with vegetation and the surface hydrological cycle.

and transmission of solar radiation through a plant canopy depend not only on the optical properties of the vegetation but also on the detailed leaf architecture. These dependencies, in turn, determine the solar radiation supplied to particular leaves for photosynthesis (cf., e.g., Norman, 1980). The daytime temperature of the leaves is determined by the balance between absorbed solar radiation, net thermal radiation, sensible and latent heat losses. The latent heat lost by transpiration is frequently the largest heat sink for a plant canopy. Thus transpiration helps to retain low enough leaf temperatures to avoid excessive thermal destruction of plant enzymes.

However, transpiration also poses a threat to the plant's continued well-being since it desiccates the plants at a rapid rate. Plants have evolved roots to pick up water from the soil to maintain their water balance. If water is not resupplied to the soil, the soil water potential eventually drops below that maintained osmotically by leaf cells and permanent wilting ensues. Plants have evolved many further strategies to minimize their water loss. In doing so, they must balance their water requirements with the needs of photosynthesis, since water loss from leaf cells to the atmosphere follows the same path as carbon dioxide uptake from the atmosphere to the leaf cell. The assimilated CO_2 encounters further mechanical and biochemical resistances before incorporation into carbohydrate within a chloroplast. (Tenhunen *et al.*, 1980, review models of whole leaf photosynthesis.) There is now considerable interest in combining models of plant absorption of solar radiation, carbon uptake, thermal stress, and water stress into comprehensive time-dependent models of crop growth (various aspects of which are discussed, for example, in the articles in Hesketh and Jones 1980). In contemplating the connections between plant growth, solar radiation and climate, I am impressed by the crucial role of climate in supplying water to the soil. It has long been known to farmers that climate variations cause large variations in their crop productivity, but only recently have agricultural economists realized that climate variations can generate year-to-year variations not only in national but in global harvests and food supply (cf., e.g., Hare, 1979).

The hydrological cycle besides profoundly structuring the biosphere also sculpts the surface of the land through rock weathering and erosion cycles and the formation and removal of soils (e.g. Birkeland, 1979). Water in excess of that returned to the atmosphere by evapotranspiration flows into rivers which, in turn, pass to the seas with their load of dissolved minerals and detritus.

In conclusion, without solar electromagnetic radiation, Earth would be a cold, dead rock.

Acknowledgements

Helpful reviews of the manuscript were given by R. Chervin, J. Eddy, J. Geisler, J. Kasting, M. Mitchell, R. Roble, and S. Schneider.

References

Anderson, D. L. T.: 1983, in B. J. Hoskins and R. P. Pearce (eds.), *Large-Scale Dynamical Processes in the Atmosphere*, Academic Press, New York, p. 305.

Banks, P. M. and Kockarts, G.: 1973, *Aeronomy Part A*, Academic Press, New York.

Birkeland, P. W.: 1974, *Pedology, Weather, and Geomorphological Research*, Oxford Univ. Press, New York.

Bolin, B., Degens, E. T., Kempe, S., and Ketner, P.: 1979, *The Global Carbon Cycle*, Wiley, New York.

Bormann, H. F. and Likens, G. E.: 1978, *Patterns and Process in a Forested Ecosystem*, Springer-Verlage, New York.

Box E.: 1978, *Rad. Environ. Biophys.* **15**, 305.

Broda, E.: 1975, *The Evolution of Bioenergetic Processes*, Pergamon Press, New York.

Cess, R. D.: 1974, *J. Quant. Spectrosc. Radiant. Transfer* **14**, 861.

Cess, R. D.: 1976, *J. Atmos. Sci.* **33**, 1831.
Cess, R. D. and Ramanathan, V.: 1978, *J. Atmos. Sci.* **35**, 675.
Cooper, J. P. (ed.): 1975, *Photosynthesis and Productivity in Different Environments*, Cambridge Univ. Press, p. 593.
Coulson, K. L.: 1975, *Solar and Terrestrial Radiation*, Academic Press, New York.
Dermerjian, K. L., Kerr, J. A., and Calvert, J. G.: 1974, in J. N. Pitts, R. L. Metcalf, and A. C. Lloyd (eds), *Advances in Environmental Science and Technology*, Vol. 4, Wiley, New York.
Dickerson, R. E.: 1980, *Scient. Amer.* **242** (3), 136.
Dickinson, R. E., Liu, S. C., and Donahue, T. M.: 1978, *J. Atmos. Sci.* **31**, 2142.
Ehhalt, D. H. and Schmidt, U.: 1978, *Pageoph* **116**, 452.
Ehleringer, J. and Björkman, O.: 1977, *Plant Physiol.* **59**, 86.
Ehrlich, P. R., Ehrlich, A. H., and Holdren, J. P.: 1977, *Ecoscience, Population, Resources, and Environment*, Freeman, San Francisco.
Eicher, D. L. and McAlester, A. L.: 1980, *History of the Earth*, Prentice-Hall, New Jersey.
Eigen, M., Gardiner, W., Schuster, P., and Winkler-Oswatitsch, R.: 1981, *Scient. Amer.* **244** (4), 88.
Ellis, J.: 1978, Ph.D. Thesis, Colorado State Univ. Fort Collins, Colorado.
Fenchel, T. and Blackburn, T. H.: 1979: *Bacteria and Mineral Cycling*, Academic Press, New York.
Frakes, L. A.: 1979, *Climate Throughout Geological Time*, Elsevier, New York.
Friedman, H.: 1960, in J. A. Ratcliffe (ed.), *Physics of the Upper Atmosphere*, Academic Press, New York, p. 133.
Galston, A. W., Davis, P. S., and Satter, R. L.: 1980, *The Life of the Green Plant*, Prentice-Hall, New Jersey.
Garrels, R. M., Lerman, A., and MacKenzie, F. T.: 1976, *Amer. Scient.* **64**, 306–316.
Gates, D. M.: 1980, *Biophysical Ecology*, Springer-Verlag, New York.
Good, N. E. and Bell, D. H.: 1980, in P. S. Calson (ed.), *The Biology of Crop Productivity*, Academic Press, New York.
Hameed, S., Cess, R. D., and Hogan, J. S.: 1980, *J. Geophys. Res.* **85**, 7537.
Hampicke, U.: 1980, In W. Bach, J. Pankrath, and J. Williams (eds), *Interactions of Energy and Climate*, D. Reidel, Dordrecht, p. 149.
Hare, K.: 1979, in M. Biswas and A. Biswas (eds), *Food, Climate, and Man*, Wiley, New York.
Harm, W.: 1980, *Biological Effects of Ultraviolet Radiation*, Cambridge Univ. Press.
Harvey, L. D.: 1980, *Prog. Phys. Geog.* **4**, 487.
Hesketh, J. D. and Jones, J. W. (eds): 1980, *Predicting Photosynthesis for Ecosystem Models*, Vols. I and II, CRC Press.
Hinkle, P. C. and McCarty, R. E.: 1978, *Scient. Amer.* **238** (3), 104.
Holland, H. D.: 1978, *The Chemistry of the Atmosphere and Oceans*, Wiley-Interscience, New York.
Hoyt, D. V.: 1979, *Climate Change* **2**, 79.
Hunt, J. M.: 1979, *Petroleum Geochemistry and Geology*, Freeman, San Francisco.
Imbrie, J. and Imbrie, K. P.: 1979, *Ice Ages: Solving the Mystery*, Enslow, Short Hills, N.J..
Kellogg, W. W.: 1980, *Ambio* **9**, 216.
Liu, S. C., Kley, D., and McFarland, M.: 1980, *J. Geophys. Res.* **85**, 7546.
Loomis, R. S. and Gerakis, P. A.: 1975, in J. P. Cooper, (ed.), *Photosynthesis and Productivity in Different Environments*, p. 145.
Lovins, A. B.: 1980, in W. Bach, J. Pankrath, and J. Williams (eds), *Interactions of Energy and Climate*, D. Reidel, Dordrecht, p. 1.
Manabe, S. and Wetherald, R. T.: 1967, *J. Atmos. Sci.* **24**, 241.
Margulis, L. and Lovelock, J. E.: 1978, *Pageoph* **116**, 239.
McEwan, M. J. and Phillips, L. F.: 1975, *Chemistry of the Atmosphere*, Wiley, New York.
Molina, M. J. and Rowland, F. S.: 1974, *Nature* **249**, 810.
Murgatroyd, R. J.: 1969, in G. Corby (ed.), *The Global Circulation of the Atmosphere*, Royal Meteorological Society, p. 159.
NAS: 1976, *Effects on Stratospheric Ozone*, National Academy of Sciences, Washington, D.C.
NAS: 1979, *Stratospheric Ozone. Depletion by Halocarbons: Chemistry and Transport*, National Academy of Sciences, Washington, D.C.
NAS: 1982, *Causes and Effects of Stratospheric Ozone Reduction: An Update*, National Academy of Sciences, Washington, D.C.

NAS: 1983a, *Causes and Effects of Stratospheric Ozone Reduction: Update 1983,* National Academy of Sciences, Washington, D.C.
NAS: 1983b, *Changing Climate, Report of the Carbon Dioxide Assessment Committee*, National Academy of Sciences, Washington, D.C.
Newkirk, G.: 1980, in *Proc. Conf. on Ancient Sun*, p. 293.
Nobel, P. S.: 1974, *Biophysical Plant Physiology*, Freeman, San Francisco.
Norman, J.: 1980, in J. D. Hesketh and J. W. Jones (eds), *Predicting Photosynthesis for Ecosystem Models*, Vol. II, CRC Press, p. 49.
North, G. R., Cahalan, R. F., and Coakley, J. A.: 1981, *Rev. Geophys. Space Phys.* **19**, 91.
Penning De Vries, F. W. T.: 1975, in J. P. Cooper, (ed.), *Photosynthesis and Productivity in Different Environments*, p. 459.
Pimentel, D., Dritschils, W., Krummel, J., and Kutzman, J.: 1975, *Science* **190**, 754.
Prather, M. J., McElroy, M. B., and Wofsy, S. C.: 1984, *Nature* **312**, 227.
Ramanathan, V., Cicerone, R. J., Singh, H. B., and Kiehl, J. T.: 1985, *J. Geophys. Res.* **90**, 5547.
Ramanathan, V.: 1980, in W. Bach. J. Pankrath, and J. Williams (eds), *Interactions of Energy and Climate,* D. Reidel, Dordrecht, p. 269.
Ramanathan, V. and Coakley, J. A.: 1978, *Rev. Geophys. Space Phys.* **16**, 465.
Rasmussen, R. A. and Khalil, M. A. K.: 1984, *J. Geophys. Res.* 89, 11599..
Revelle, R.: 1976, *Scient. Amer.* **235** (3), 164.
Roble, R. G.: 1977, in *The Upper Atmosphere and Magnetosphere*, National Academy of Sciences, Washington, D.C., p. 1.
Rotty, R. M. and Marland, G.: 1980, in W. Bach, J. Pankrath, and J. Williams (eds), *Interactions of Energy and Climate*, D. Reidel, Dordrecht, p. 191.
Sagan, C. and Mullen, G.: 1972, *Science* **177**, 52.
Sassein, W.: 1980, *Scient. Amer.* **243** (3), 118.
Schidlowski, M., Appel, P., Eichmann, R., and Junge, C.: 1979, *Geochim. Cosmochim. Acta* **43**, 189.
Seiler, W. and Crutzen, P. S.: 1980, *Climatic Change* **2**, 207.
Swift, M. J., Heal, O. W., and Anderson, J. M.: 1979, *Decomposition in Terrestrial Ecosystems*, Univ. of California Press.
Tarrant, J. R.: 1980, *Food Policy*, Wiley, New York.
Tenhunen, J. D., Hesketh, J. D., and Gates, D. M.: 1980, In J. D. Hesketh and J. W. Jones (eds), *Predicting Photosynthesis for Ecosystem Models*, Vol. 1, CRC Press, p. 123.
Turco, R. P., Toon, O. B., Hamill, P., and Whitten, R. C.: 1981, *J. Geophys. Res.* **86**, 1113.
Walker, J. C. G.: 1977, *Evolution of the Atmosphere*, Macmillan, New York.
Walker, J. C. G., Hays, P. B., and Kasting, J. F.: 1981, *J. Geophys.* **86**, 9776.
Weiss, R. F.: 1981, *J. Geophys. Res.* **89**, 9475.
Willson, R. C., Gulkis, S., Janssen, M., Hudson, H. S., and Chapman, G. A.: 1981, *Science* **211** (13), 200.
Wine, P. H., Ravishankara, A. R., Kreutter, N. M., Shah, R. C., Nicovich, J. M., Thompson, R. L., and Wuebbles, D. J.: 1981, *J. Geophys. Res.* **86**, 1105.

National Center for Atmospheric Research,
Boulder, CO 80307,
U.S.A.

CHAPTER 21

THE EFFECT OF THE SOLAR WIND ON THE TERRESTRIAL ENVIRONMENT

N. U. CROOKER AND G. L. SISCOE

1. Introduction

When the interplanetary magnetic field (IMF) carried by the expanding solar corona has a southward component for about an hour or longer, a number of remarkable phenomena begin to be perceived at Earth. Invariably there are displays of aurora, usually at high latitudes but sometimes also at low latitudes. Instruments measuring magnetic fields show deviations from normal levels. Disruptions may occur in power distribution systems and in radio and cable communication systems. The deeper the terrestrial environment is probed with both ground-based and spacecraft instruments, the more associated phenoma are discovered. It is perhaps surprising that the impact of the expanding corona can be so pronounced, since at Earth its energy flux is six to seven orders of magnitude smaller than the energy flux radiated by the Sun.

The purpose of this chapter is to report on the present state of understanding of how the expanding solar corona, or solar wind, interacts with the terrestrial environment. Energy and mass from the solar wind follow intricate paths to the atmosphere. The morphology of this process is reasonably well known, whereas its physics is understood to a considerably lesser degree. The organization of the chapter roughly follows the path of energy and mass transfer. Section 2 gives a brief review of the magnetic configurations and plasma structures that result from the interaction between the solar wind and the terrestrial environment, since these are central to the subsequent discussion. The structure and variability of the solar wind as modulators of geomagnetic activity are considered in Section 3. Energy and mass transfer mechanisms at the boundary between the solar wind and the domain of Earth's magnetic field, the magnetosphere, are discussed in Section 4, and the internal magnetospheric response to these transfers in Section 5. Sections 6 and 7 consider the coupling between the magnetosphere, ionosphere, and thermosphere and effects on the middle and lower atmosphere, respectively.

The chapter was written with the intention of providing a straightforward overview for solar physicists and other interested nonspecialists. The approach is qualitative, but representative references throughout the text direct the reader to more detailed studies. Although the referencing is not exhaustive, the chapter may serve as a review for solar–terrestrial physicists.

Peter A. Sturrock (ed.), Physics of the Sun, Vol. III, pp. 193–249.

2. General Morphology

In the absence of ionized matter in space, Earth's dipolar magnetic field would extend to infinity. However, the continual flow of the magnetized plasma of the solar wind acts to confine Earth's magnetic field to a volume which Thomas Gold in 1959 named the magnetosphere. A noon–midnight meridian cross-section of the magnetosphere is illustrated in Figure 1, adapted from Rosenbauer *et al.* (1975). Since the solar wind flows at supersonic

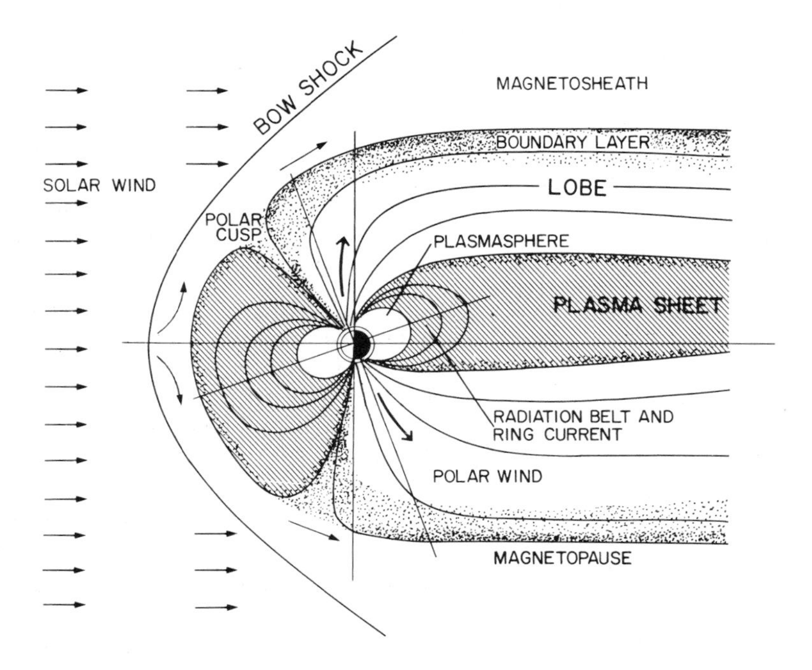

Fig. 1. Noon–midnight meridian cross-section of the magnetosphere. (After Rosenbauer *et al.*, 1975.)

speeds, a shock wave, the bow shock, forms in front of the magnetosphere. The compressed, heated, and slower flowing plasma behind the shock forms the magnetosheath. The boundary between the magnetosheath and the magnetosphere is named the magnetopause. Its position is determined for the most part by the requirement that the solar wind pressure outside balance the pressure of the terrestrial magnetic field inside. But the existence of the tail of the magnetosphere, which extends well beyond lunar orbit, is usually attributed to an additional force acting tangentially along the magnetopause (e.g. Piddington, 1960; Coleman, 1970).

The magnetopause is an effective shield against penetration of solar wind plasma. Only about 0.1% of the solar wind mass flux incident on the boundary is responsible for auroral displays and geomagnetic activity (e.g. Hill, 1979), although more may cross into the magnetosphere and then escape down the tail. Figure 1 illustrates the observed locations of interior plasma structures. The polar cusp, the boundary layers, and the plasma sheet contain plasma of predominantly solar wind origin. The characteristics of the plasma in the polar cusp and boundary layers are similar to those of the adjacent magnetosheath plasma, whereas the plasma in the plasma sheet is hotter and considerably less dense, having undergone some energization process in the magnetosphere. The earthward edge of the plasma sheet, which extends around to the dayside, marks the beginning of the domain of the ring current, where magnetic field gradient and curvature drifts carry the hot ions westward and the electrons eastward in closed paths around Earth. The radiation belts, which consist of trapped energetic particles, also occupy this region of space, but reach their maximum strength earthward of the ring current.

The ionosphere is a region of Earth's atmosphere above about 90 km where free electrons and ions are present in substantial numbers. It is an additional source of magnetospheric plasma. At high latitudes it is the source of the polar wind. Analogous to the solar wind, the polar wind is the continual expansion of ionospheric plasma from the polar cap into the relative vacuum of the lobes of the extended magnetotail. The magnetic field lines in the lobes usually are assumed to be open: that is, they are connected to Earth only at one end. The polar wind plasma is cold, and its density is low.

At lower latitudes, ionospheric particles expand upward along closed magnetic field lines, which are connected to Earth at both ends, to form the plasmasphere. Inside the plasmapause, which is the outer boundary of the plasmasphere, the electric field associated with the rotation of Earth's magnetic field dominates over the general magnetospheric electric field. The corotating magnetic flux tubes fill to a high density approaching the equilibrium value which would be achieved in a static situation. In the closed field line region beyond the plasmapause, cold plasma from the ionosphere is removed by the drift associated with the magnetospheric electric field. It should be noted here that the magnetospheric electric field is itself derived from the motional electric field of the solar wind. Thus, the solar wind also controls the structure of this deep magnetospheric feature.

In addition to cold plasma, the ionosphere also is a source of hot ions. Satellite observations have revealed the existence of beams of heavy ions (primary O^+) leaving the ionosphere at latitudes mapping to the ring current and beyond. These ions have energies comparable to plasma sheet and ring current ions and could represent a sizable fraction of the total energetic particle population of the magnetosphere.

This brief overview of the anatomy of Earth's magnetosphere, as shown in Figure 1, should serve as a glossary of terms which are used throughout this chapter.

3. Solar Wind and Geomagnetic Activity

3.1. SOLAR WIND STREAMS

The corona is not uniform, and as it expands outward it tends to retain its large-scale inhomogeneities. The bulk of solar wind plasma is thought to flow out along open magnetic field lines rooted in coronal holes. (See Hundhausen, 1977, and references therein.) These comparatively cool features occupy primarily the polar regions of the Sun, but they often extend to lower latitudes as well.

Figure 2, adapted from Smith *et al.* (1978), shows a schematic view of the way the magnetic field in coronal holes maps into the interplanetary medium. The dashed lines are hypothetical boundaries between active regions at lower latitudes, over which the field lines are arched and closed, and the coronal holes at high latitudes, over which the field lines are open and extend outward. The simplicity of the figure emphasizes a dipole-like structure of the solar magnetic field. The solar wind flowing from the coronal holes draws the open field lines away from the Sun in a configuration topologically similar to that of Earth's magnetotail (Figure 1). A current sheet, analogous to Earth's plasma sheet, separates the fields of opposite polarity. It is tilted with respect to the solar equator. Thus in the course of one solar rotation an observer at Earth would spend half of the time above the current sheet and half below, sensing two sectors of opposite magnetic polarity. The current sheet forms the sector boundaries. If the current sheet is warped or temporarily disturbed by changing coronal structure, more sectors may be observed.

The pattern of solar wind flow through the magnetic configuration in Figure 2 has

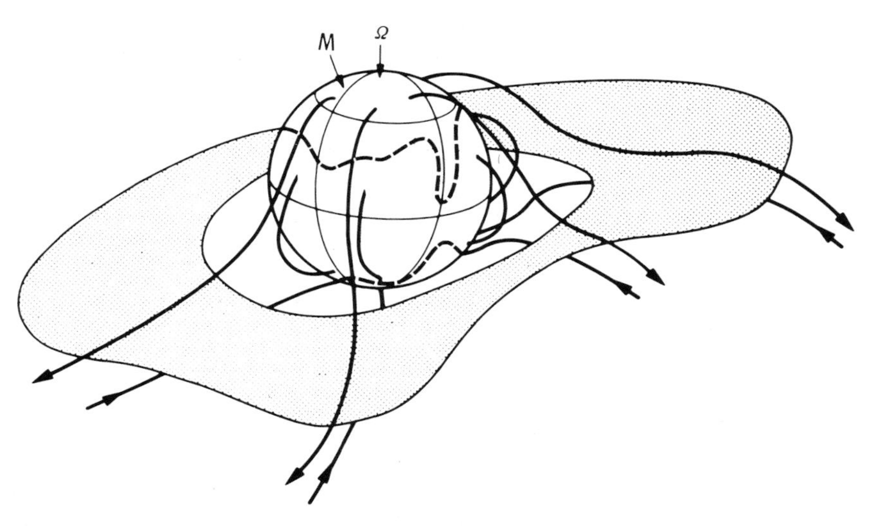

Fig. 2. Model of the heliospheric current sheet responsible for sector structure. The warped sheet is slightly inclined to the solar equator, as indicated by the separation between the rotation axis Ω and the dipole axis *M* (normal to the sheet). The sheet separates magnetic field lines originating in opposite polar coronal holes, outlined by dashed lines. At lower latitudes, in active regions, the field lines are closed. (After Smith *et al.*, 1978.)

substantial north–south components near the Sun but is directed predominantly radially outward beyond a few solar radii. Flow from near the edges of the coronal holes is slower than flow from the central regions. Thus away from the Sun, flow is slowest near the current sheet. Because the current sheet is tilted with respect to the equatorial plane, the pattern of flow which results in that plane has a two-stream structure, one stream in each sector, with highest speeds flowing from the regions where the neutral sheet is furthest from the equatorial plane.

The solar wind stream is the basic structural form of inhomogeneity about which observations relevant to solar–terrestrial coupling are organized. A stream, as observed near Earth, has most of the characteristics illustrated schematically in Figure 3 (from Hundhausen, 1972). The view is a cross-section in the ecliptic plane. Plasma flows radially

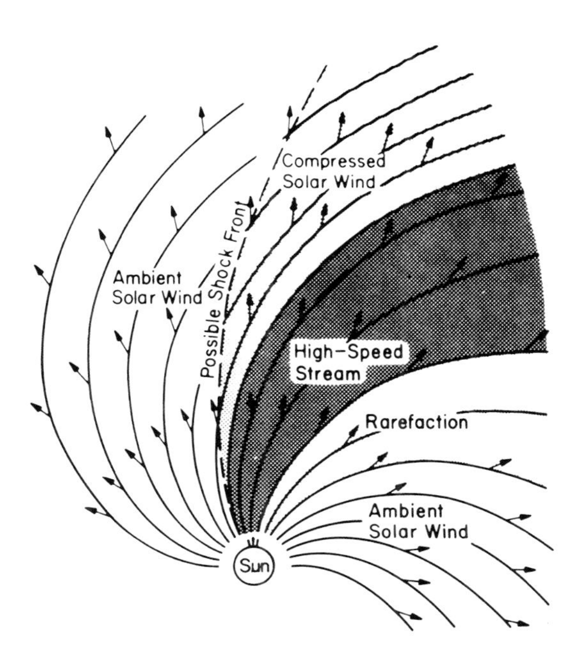

Fig. 3. The interaction of a high-speed stream with the ambient solar wind in the ecliptic plane. In a stationary frame of reference, the solar wind is directed radially outward from the Sun, as indicated by the arrows. The curved magnetic field lines are attached to the rotating sun and form spirals as they are carried outward by the solar wind (Hundhausen, 1972).

outward from the Sun, as indicated by the arrows, and carries with it the solar magnetic field. Since the field lines are connected to a rotating Sun, they become drawn into a spiral pattern (Parker, 1958). The region of high-speed flow also follows a spiral pattern, as illustrated by the darkly shaded region, but its spiral is coiled less tightly than the spiral formed by the slower flowing plasma. The high-speed stream thus flows outward into the slower plasma at its leading edge, resulting in compression there and possible shock formation.

The probability of shock formation increases with increasing distance from the Sun. The shock front illustrated in Figure 3 forms quite close to the Sun, as would be the case for a very large difference in relative speed. A more general diagram of the geometry of shock formation is shown in Figure 4. The view is of the equatorial plane in a frame of reference rotating with the Sun. Unlike the flow in the inertial frame in Figure 3, which is radially outward, the flow in the corotating frame follows the spiral pattern of the magnetic field lines. The central line in Figure 4 is the streamline interface separating the more tightly spiralled streamlines of slow flow from the more loosely spiralled streamlines of fast flow. Streamlines leaving the Sun at first diverge from the interface and then begin to converge towards it. Those streamlines closest to the interface begin to converge sooner than those further away, and they intersect the interface more obliquely than do those further away. Thus along the interface there is an increasing component of flow towards the interface with increasing distance from the Sun. A rough criterion for shock formation is that it occurs where the speed of flow towards the interface is greater than

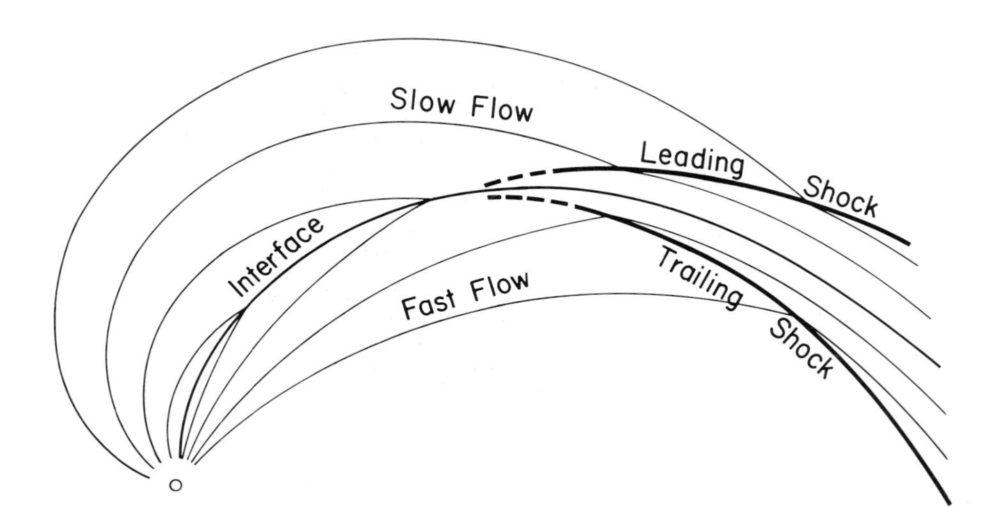

Fig. 4. Geometry of shock formation in the solar equatorial plane in a frame of reference rotating with the Sun. Spiralling streamlines of fast flow approach more tightly spiralled streamlines of slow flow at the interface. Shocks form when the component of flow towards the interface is greater than the local speed of sound.

the speed of sound away from the interface. Since the speed of sound decreases with increasing distance from the Sun, shock formation becomes increasingly likely with increasing distance. In general, two shocks will form, one on each side of the interface. Flow crossing the leading shock increases in speed, and flow crossing the trailing shock decreases in speed, such that in both cases the post-shock flow tends to follow spirals with the same pitch as that of the interface. The shocks are weak near their point of formation and are detached from the interface because of flow alterations at the pressure ridge there. The shocks increase in strength with increasing distance from the Sun.

Figure 5 (from Pizzo, 1980) shows the structure of the interaction front that results when fast coronal hole plasma is forced to push into slower plasma because of a trough-like equatorward protrusion of a polar coronal hole. The profile at 1 AU was obtained by a numerical integration of the applicable nonlinear hydrodynamic equations. The calculation has been carried out for the magnetohydrodynamic case (Pizzo, 1982). The figure shows a symmetrical velocity inhomogeneity outlining the protrusion. It is imposed in the model at a distance of 0.16 AU (approximately 35 solar radii). At 1 AU the inhomogeneity has evolved into an asymmetric velocity profile because the fast plasma encroaches on the slower plasma on the converging edge of the trough and withdraws from the slow plasma on the diverging edge. The convergence creates the pressure ridge with increased density and higher temperature. The high-pressure ridge precedes the trough. Similarly, the divergence results in a broad low-pressure basin that follows the trough. This structure is time-stationary in the frame of reference corotating with the Sun. Thus on every solar rotation Earth is subjected to a short interval of high-density, intermediate-speed flow followed by a longer interval of low-density, high-speed flow.

Historically, solar wind streams reaching Earth were thought to flow from longitudinally confined regions near the solar equator, outward into an ambient solar wind

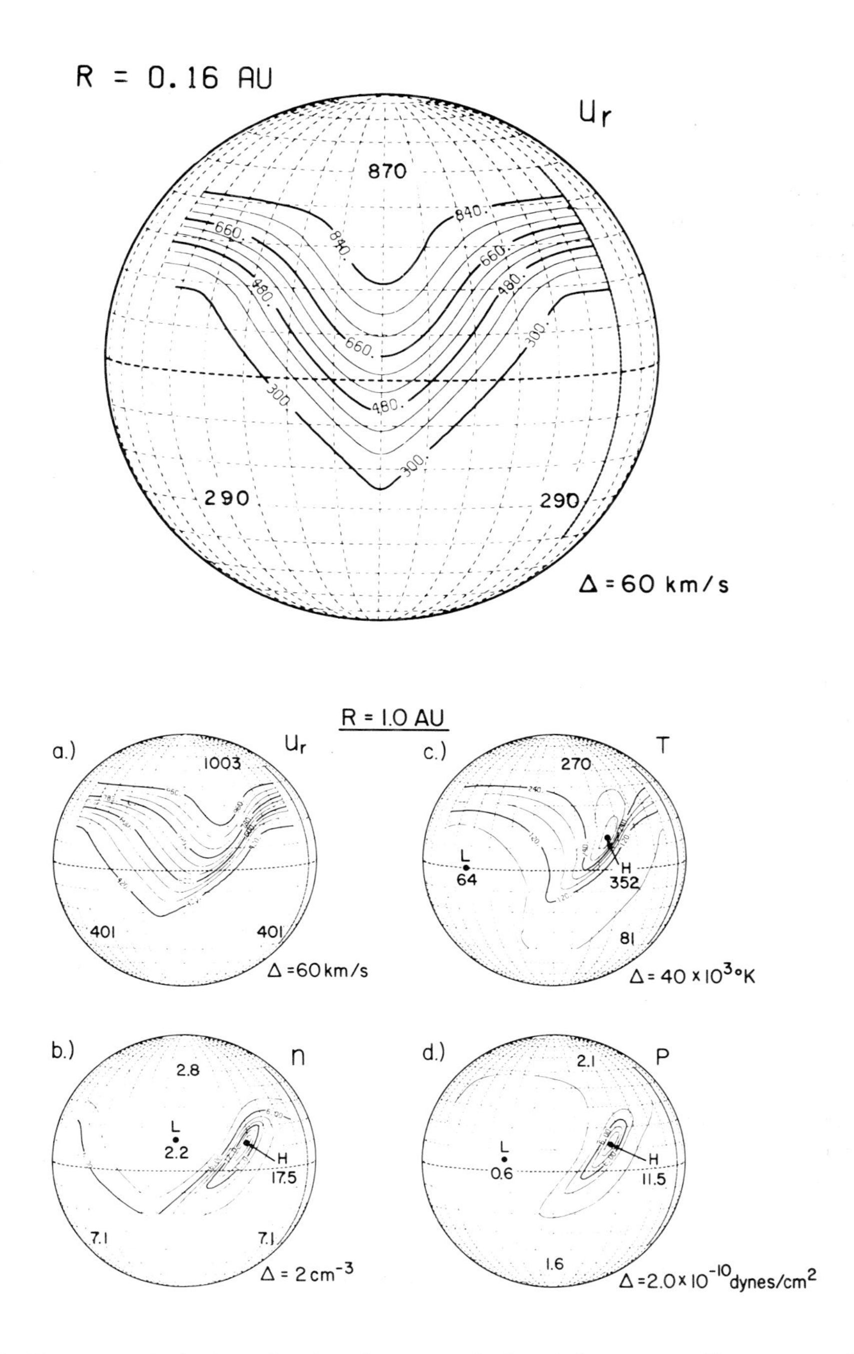

Fig. 5. Views towards the Sun of projected contours of solar wind parameters. The contours of radial velocity u_r at radial distance R = 0.16 AU represent flow from an equatorward protrusion of a polar coronal hole. The contours of (a) u_r, (b) number density n, (c) temperature T, and (d) pressure P show how the flow has evolved into an interaction front at R = 1.0 AU.

plasma, as suggested in Figure 3. More recent research supports the more elaborate, three-dimensional interpretation of solar wind flow shown in Figure 2 and modelled in Figure 5, where a stream in the equatorial plane is simply a characteristic of a two-dimensional cross-section of inhomogeneous flow from a coronal hole at higher latitudes. However, the pattern of coronal holes is often observed to be more complex than the simple dipolar arrangement in Figure 2. Holes extending across the equator are not uncommon. But whether high-speed solar wind plasma reaching Earth has its origin at high solar latitudes or at the equator, the structure of its interaction with slower flowing plasma at 1 AU is well represented by Figure 3.

The structure described up to this point pertains mostly to streams in the recurrent category. The global coronal hole pattern on the Sun evolves slowly compared to the solar rotation period of 27 days. Consequently the pattern of streams intercepted by Earth in the ecliptic plane tends to recur with a 27 day periodicity over intervals of many months. The shocks associated with a recurrent stream corotate with the stream structure, as shown in Figure 4. They are not usually observed with recurrent streams at 1 AU but tend to form at somewhat greater distances, being quite strong and numerous at the orbit of Jupiter at $\sim$ 5 AU (Gosling *et al.*, 1976a; Smith and Wolfe, 1976).

Some streams are observed not to recur. In contrast with recurrent streams, nonrecurrent streams tend to be preceded by strong shock waves at 1 AU and tend to have less pronounced speed profiles. Nonrecurrent streams necessarily are associated with short-lived solar features, but the nature of these features is not well understood. Some clearly are associated with flares and are pictured as blast waves from active regions (Hundhausen, 1972; also, review by Burlaga, 1975). But most nonrecurrent streams are not associated with flares. Gosling *et al.* (1974) have suggested that their source may be the mass ejections from the Sun observed in coronal white light. These coronal transients are associated both with flares and eruptive prominences in active regions and with the eruptive prominences known as disappearing filaments (*disparitions brusques*) away from active regions (Munro *et al.*, 1979). That both flares and disappearing filaments are precursors to geomagnetic storms (e.g. Joselyn and McIntosh, 1981) provides further support for the suggestion that coronal transients are the primary sources of nonrecurrent streams, since storms and streams are well correlated, as will be discussed in Section 3.2. Possible examples of nonrecurrent streams characteristic of coronal transients are presented by Schwenn *et al.* (1980) and Gosling *et al.* (1980).

Coronal holes may be associated with nonrecurrent streams as well as recurrent streams. Hundhausen (1977) and Sheeley and Harvey (1981) have noted periods when individual coronal holes are short-lived but form apparent 28.5 day recurrence patterns by continuously reforming at progressively more eastward longitudes through coalescence in regions of differential rotation. Sheeley and Harvey show the same apparent 28.5 day recurrence pattern for a series of weak solar wind streams at that time (see 1978–1979 period near bottom of Figure 8). Thus it is likely that nonrecurrent streams flow from short-lived coronal holes during periods of apparent 28.5 day recurrence. Consistent with this supposition are observations of the appearance of coronal holes following the disappearance of filaments (Sheeley and Harvey and references therein), since disappearing filaments are associated with coronal transients and nonrecurrent streams. It can be noted that a coronal transient appears to involve an explosive opening of a closed field line region, as in the formation of a coronal hole.

In summary, solar wind streams are associated with coronal holes, coronal transients, disappearing filaments, and solar flares, but these categories are not mutually exclusive. Although there will be exceptions, the following generalization seems appropriate: the coronal hole may be the most basic requirement for the presence of a stream; the coronal transient and disappearing filament may mark the formation of a short-lived hole; and a flare may accompany this formation when it occurs in an active region.

3.2. Geomagnetic Response to Streams

The relationship between two solar wind streams and geomagnetic activity is illustrated in Figure 6 (from Burlaga and Lepping, 1977). The period of time covered is one solar

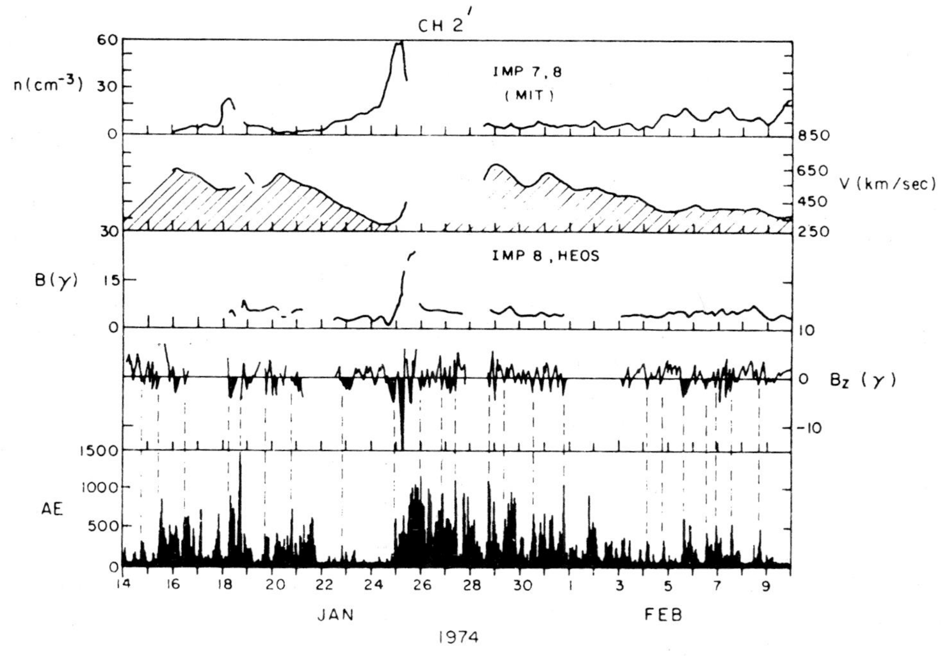

Fig. 6. Time variations of solar wind density n, speed V, magnetic field magnitude B and north–south component B_Z, and the auroral electrojet index AE of geomagnetic activity during passage of two streams. The envelope of geomagnetic activity is correlated with V, whereas individual peaks are correlated with B_Z, as indicated by the dashed lines (Burlaga and Lepping, 1977).

rotation. The plot of speed V outlines the profiles of the two streams. Although a data gap obscures the signatures of the leading edge of the first stream on January 15, the characteristic rise in particle density n and magnetic field strength B beginning on January 24 mark the compression preceding the leading edge of the second stream on January 25. The north–south component B_Z of the IMF (interplanetary magnetic field) also is plotted in the figure. Its variations in time are of much higher frequency than variations of the other solar wind parameters. The passage of the second stream is marked by an increase in magnitude of the highfrequency fluctuations at the leading edge, which is a direct result of the compression and increased field strength there.

Geomagnetic activity is represented in Figure 6 by the AE index. This index is a measure of the magnetic disturbance from the intense east–west ionospheric currents that flow in auroral regions in narrow ribbons called electrojets. Intensification of the auroral electrojets occurs on a time-scale of about an hour and marks the occurrence of the magnetospheric substorm.

Figure 6 clearly illustrates that the AE index is well correlated with the passage of solar wind streams. Qualitatively its response is quite simple. The envelope of activity generally follows the pattern of solar wind speed, and the envelope is modulated by a high-frequency component which closely follows the pattern of southward peaks in B_z, as indicated by the dashed lines. Testing for the quantitative functional form of geomagnetic activity dependence on V and B_z has been the subject of many correlative studies (e.g. Russell *et al.*, 1974; Garrett *et al.*, 1974; Svalgaard, 1977; Akasofu, 1979; Murayama *et al.*, 1980; Baker *et al.*, 1981a; Clauer *et al.*, 1981), but no consensus has been reached between the various authors. In part, disagreement exists because the degree of correlation depends to some extent on the time-scale over which averages are taken (Crooker *et al.*, 1977) and on the index chosen to represent geomagnetic activity.

Although geomagnetic activity correlates best with the southward component of the IMF and with solar wind speed, lesser correlations, independent of these parameters, have been found with solar wind density (Svalgaard, 1977; Maezawa, 1979) and with IMF variance (Ballif *et al.*, 1967, 1969; Garrett, 1974; Maezawa, 1979). The density correlations show an increase in the level of worldwide magnetic disturbance indices with increasing density levels, although the same effect does not appear to be present for the more localized auroral electrojet index. The effect may be interpreted as the result of magnetospheric compression during passage of the leading edge of a stream. Shielding currents on the magnetopause increase in strength as the boundary moves inward and contribute to the level of magnetic disturbance. This interpretation was first applied by Chapman and Ferraro (1931) to the sudden commencement, which is a sudden worldwide increase in the horizontal (~ northward) component of the geomagnetic field that often precedes a geomagnetic storm. It is now known that a sudden commencement marks the passage of the shock front preceding some streams. Streams without shock fronts at 1 AU produce more gradual geomagnetic disturbances. Since disturbances from changes in magnetopause current strength are of the order of tens of gammas ($1\ \gamma = 10^{-9}$ T), it is not surprising that the density effect is negligible in auroral electrojet indices, which measure disturbances of many hundreds of gammas.

Worldwide indices of geomagnetic activity in a sense measure the variance in the geomagnetic field at periods of hours or a day. For example, the K_p index is based on the range of the most active component of the field in a given 3 h interval. Conversely, the degree of variance in the geomagnetic field should be an indication of the level of geomagnetic activity. This is consistent with the data in Figure 7. The variance at short periods has been obtained from spectral analysis of measurements taken at synchronous orbit. These spectra were prepared for the purpose of studying magnetic pulsations (Arthur *et al.*, 1977; see Section 3.3.1 for a discussion of pulsations), but they also show a remarkable pattern in the level of power in response to passage of solar wind streams. Figure 7 shows nine spectra of the eastward component of the geomagnetic field. Each spectrum is constructed from a 43 min data segment centered near 0800 LT. Each row displays spectra from days before, during, and immediately after the passage of the leading edge of a solar wind stream. In all three cases, the overall level of power or variance

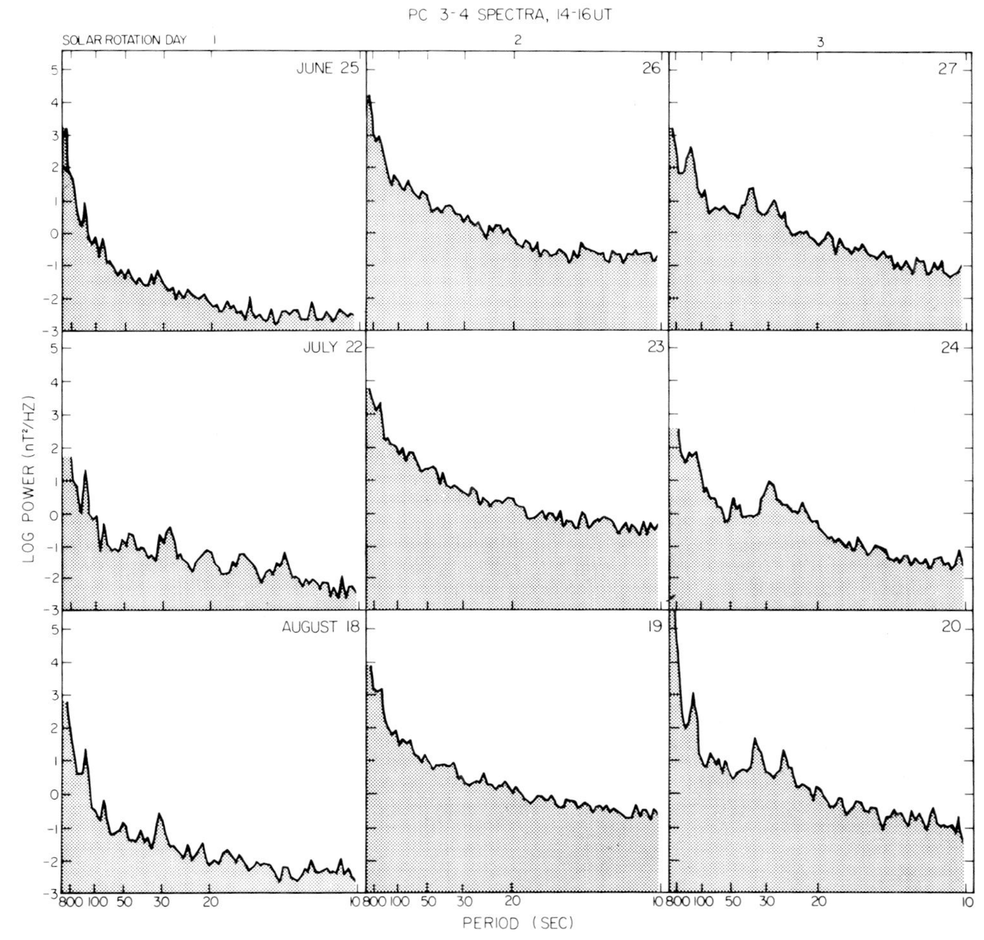

Fig. 7. Spectra of the eastward component of the geomagnetic field at synchronous orbit (ATS 6), constructed by C. W. Arthur from 43 min data segments centered near 0900 LT. Data are shown from the first, second, and third days of three successive solar rotations. In each case the second day marks the passage of the leading edge of a recurrent stream and shows an increased level of power at all periods.

is low preceding the stream, then it increases by nearly two orders of magnitude at the leading edge, and subsequently it decreases slightly in the high-speed region. Since IMF variance also is highest at the leading edge of streams, as illustrated in Figure 6, the correlation between short period geomagnetic and IMF variances appears to be quite good. This correlation and the correlations between geomagnetic activity indices and IMF variance (over minutes) cited above most likely are part of a broad spectrum of correlated geomagnetic and IMF fluctuations at the leading edges of streams. However, over periods longer than about a day, geomagnetic fluctuations correlate better with fluctuations in solar wind speed (Figure 6; Crooker *et al.*, 1977).

Recurrent solar wind streams cause recurrent periods of geomagnetic activity (Neupert and Pizzo, 1974; Hansen *et al.*, 1976), an excellent illustration of which is shown in Figure 8 from Sheeley and Harvey (1981). Each unit of color represents the strength or sign of a particular parameter on a given day. The four columns are records of IMF

Fig. 8. A 27 day recurrence display of IMF polarity, coronal hole occurrence data (plus three days to allow for Sun–Earth transit time), solar wind speed at Earth, and geomagnetic disturbance index C9. A 27 day average sunspot number R is indicated in the narrow column on the right. The color coding is given in Table I. (From Sheeley and Harvey, 1981.)

TABLE I
Color Coding for Figure 8

Code	IMF Polarity	Holes
Red*	Away	Away polarity
Green*	Toward	Toward polarity
Blue	Indeterminate	None
Black	–	No observation

Code	Wind Speed (km s^{-1})	C9	R
Black	No observation	–	–
Dark blue	300–399	0	0
Blue	–	1	1–15
Light blue	400–499	2	16–30
Purple	–	3	31–45
Orange	500–599	4	46–60
Yellow	600–699	5	61–80
Light yellow	700–799	6	81–100
Very light yellow	–	7	101–130
White	800-	8–9	131-

* Weak holes have been assigned darker shades.

polarity, coronal holes, solar wind speed, and the worldwide C9 index of geomagnetic activity. Each column has a width of 27 days, which is the recurrence period for features at fixed longitudes near the solar equator. From top to bottom, the columns show data from 1973 through 1979. In general, the correspondence between all four parameters is good. The most pronounced pattern of 27 day recurrence occurred from late 1973 until mid-1975, when two long-lived coronal holes produced two well-defined streams with very high speeds. The streams in Figure 6 are examples from this time period. The well-correlated response of the C9 index to the two streams is quite clear in Figure 8. The patterns in the speed and C9 columns are nearly identical.

A small, systematic difference in relative intensities between C9 and stream speed within the recurrence patterns in Figure 8 does exist. Sheeley *et al.* (1977) ascribed this difference to an annual variation related to IMF polarity (Russell and McPherron, 1973; Burch, 1973). When the IMF is directed towards the Sun along its average spiral path, geomagnetic activity is most intense near March equinox; when it is directed away, activity maximizes near September equinox. Russell and McPherron interpret this correlation as the result of larger average IMF southward component for 'toward' sectors in March and for 'away' sectors in September owing to systematic differences between the geocentric solar equatorial coordinate system in which the IMF is ordered and the geocentric solar magnetospheric system in which the interaction between the IMF and Earth's dipole is ordered. (See Figure 10 and Section 3.3.2.) During the 1973–1975 period of two high-speed streams per solar rotation, Figure 8 shows that the polarities of the two streams were 'away' and 'toward', respectively. The C9 index shows that intense activity occurs in the first half of the year for the second stream and in the second half of the year for the first stream, which is consistent with the polarity effect.

For the C9 index the polarity effect is clearly of secondary importance in the overall recurrence pattern of Figure 8. A much clearer demonstration of the polarity effect is shown in the 1974–1975 recurrence pattern in Figure 9 for the Dst index of geomagnetic

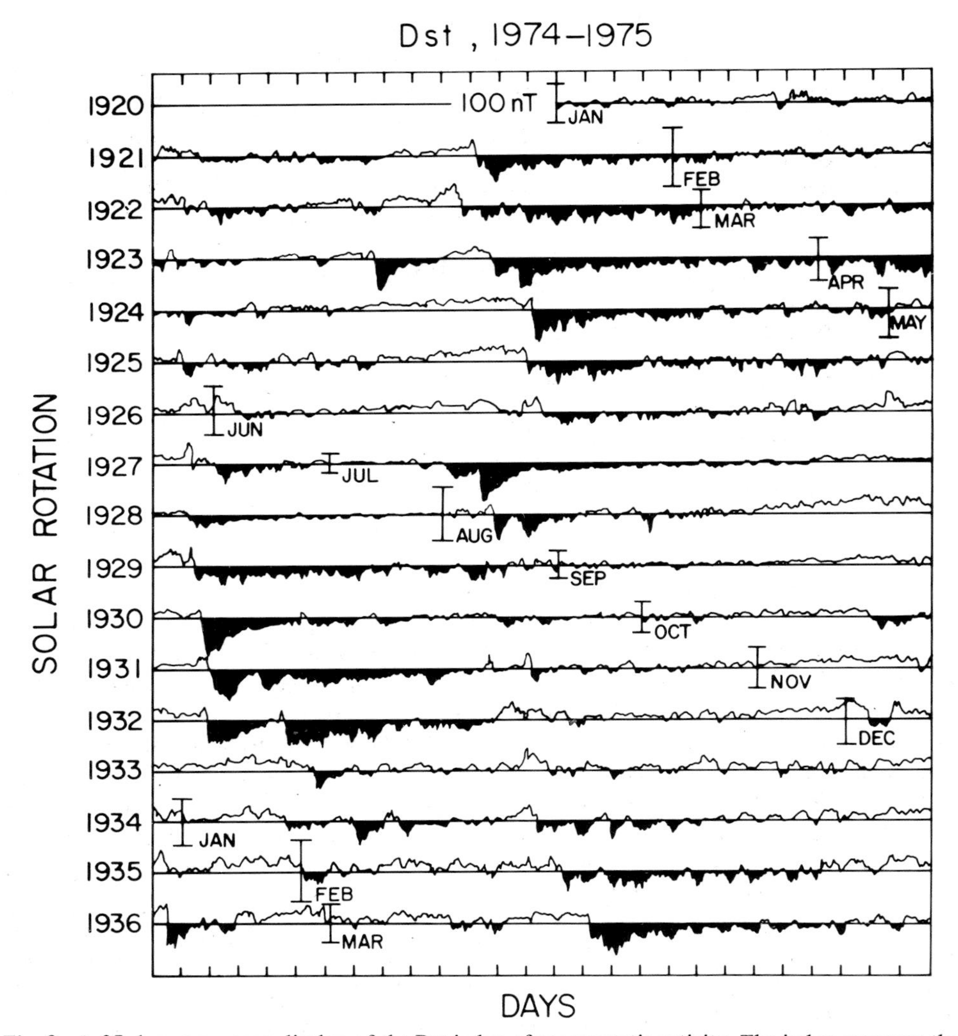

Fig. 9. A 27 day recurrence display of the Dst index of geomagnetic activity. The index measures the average disturbance in the geomagnetic field at the equator. Decreases in Dst mark stormy periods and are shaded for emphasis. The tendency for high activity in the fall on the left half and in the spring on the right half reflects the 'away' and 'toward' polarities, respectively, of two recurrent streams. (The Dst index is prepared by M. Sugiura and R. S. Kennon and is distributed by the National Space Science Data Center.)

activity. Whereas the AE index is a measure of magnetospheric substorms, which as we have noted earlier occur on the time-scale of hours (see beginning of section), Dst is a measure of storms, which occur on the time-scale of days. Specifically, Dst is a measure of disturbance in the horizontal component of the geomagnetic field near the equator. As a solar wind stream sweeps past Earth, Dst first increases in response to compression at the leading edge, as discussed above. This is the initial phase of the storm. Dst then decreases sharply for several hours. The decrease is the main phase of the storm. It is caused by earthward movement of the plasma sheet in the magnetosphere (see Figure 1) and consequent intensification of the ring current. During the main phase, substorm activity is intense. The recovery phase is the slow return of Dst to its original value, over a period of several days. This pattern is apparent in many of the storms present in Figure 9, where negative excursions have been shaded for emphasis. Most of the storms begin with the passage of the leading edge of either the first stream, on the left, or the second stream, in the middle. The most outstanding feature in the figure is the seasonal effect in the Dst response to IMF polarity. Recurrent activity is almost absent during spring for the first stream with 'away' polarity, and during fall for the second stream with 'toward' polarity.

The strong response of Dst to the polarity effect is consistent with a stronger dependence of geomagnetic storm activity on the southward component of the IMF than on solar wind speed. This is in contrast to the dependence on these parameters, of general, worldwide activity, as reflected in C9. The control of the solar wind speed on the C9 index seems to be stronger. Feynman (1980) found that the responses of Dst and the aa index of worldwide activity to solar cycle variations differed in the same sense, as is discussed in Section 3.3.3. These differences in response of two kinds of magnetic activity to two different solar wind parameters may reflect two modes of energy transfer to the magnetosphere from the solar wind. Further evidence of this possibility is discussed in Section 4.

In summary, the geomagnetic response to the passage of a solar wind stream is a geomagnetic storm. The compression preceding the leading edge produces a sudden or gradual commencement disturbance. As the solar wind speed rises at the leading edge, substorm and shorter period geomagnetic fluctuations become intense, and the main phase decrease of the geomagnetic field occurs. The intensity of the substorm activity and the main phase decrease depends jointly upon the polarity of the IMF in the stream and the season of the year. As the region of peak speed passes, worldwide activity is high. Then as the speed decreases, all forms of activity subside.

3.3. Periodic Geomagnetic Activity

Spectral analysis of geomagnetic activity reveals peaks at preferred periods over the entire available range of measurement from seconds to years (e.g. Arthur *et al.*, 1977; Fraser-Smith, 1972). Many of these periodicities are directly related to the solar wind. Some are imposed by periodicities in the solar wind itself, such as the 27 day recurrence of streams, and others, like the polarity effect, result from periodic changes in the orientation of Earth's dipole with respect to solar wind parameters. Prominent among solar wind-related periodicities are geomagnetic pulsations, the durnal and semiannual variations, and solar cycle variations.

3.3.1. *Geomagnetic Pulsations*

Pulsations are regular, small-amplitude fluctuations in the geomagnetic field at periods ranging from tenths of seconds to minutes. Daytime, midperiod pulsations are observed at mid- to subauroral latitudes and at geosynchronous orbit. Examples of some weak midperiod spectral peaks are shown in the third column of Figure 7. Features of these pulsations are correlated with solar wind parameters in a way that is consistent with their being caused by field line resonance with waves generated at the magnetopause and at the bow shock (Gul'elmi, 1974; Greenstadt *et al.*, 1980; references therein). Amplitudes are correlated with solar wind speed and IMF orientation (e.g. Greenstadt *et al.*, 1979).

The speed correlation is consistent with the formation of Kelvin–Helmholz waves on the magnetopause, and the IMF orientation correlation is consistent with the preferred generation of bow shock waves in the subsolar region for small angles between the IMF and the Earth-Sun line. Pulsation periods appear to be correlated with IMF strength (Gul'yel'mi and Bol'shakova, 1973; Russell and Fleming, 1976), which is consistent with the bow shock generated wave mechanism (Greenstadt *et al.*, 1980).

3.3.2. *Diurnal and Annual Variations*

The annual variation of geomagnetic activity is a semiannual waveform with maxima near the equinoxes and minima near the solstices. Parameters which contribute in interrelated ways to this variation are the polarity of the IMF, the orientation of Earth's magnetic axis with respect to the solar equator and the Earth–Sun line, and the heliographic latitude of Earth's position.

The effect of IMF polarity on the annual variation of geomagnetic activity has been described briefly near the end of Section 3.2. Here a more complete description is given. The relevant geometrical parameters are illustrated in Figure 10. The intersection of the solar equatorial and ecliptic planes and Earth's positions in September and December are shown. The arrows indicate the orientations of three geocentric solar coordinate systems: they are the ecliptic (GSE), equatorial (GSEQ), and magnetospheric (GSM) systems. All three share the same x-axis, which is directed towards the Sun, so that transformations between the systems are simply rotations about this axis. The GSE y-axis lies in the

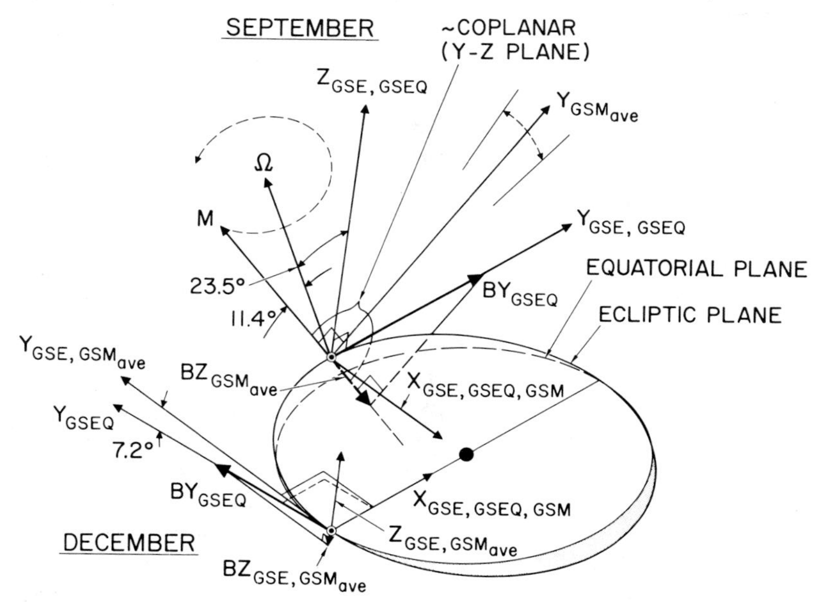

Fig. 10. Relationships between the geocentric solar ecliptic (GSE), equatorial (GSEQ) and magnetospheric (GSM) coordinate systems at September equinox and December solstice. All three systems share the same x-axis which is directed toward the sun. At equinox Earth's rotation axis Ω lies in the $y-z$ plane of all systems. Earth's dipole axis M rotates daily about Ω. The y_{GSM}-axis is defined to be perpendicular to the plane formed by x and M and thus oscillates daily about the $y_{\mathrm{GSM_{ave}}}$ position. The average y-component of the IMF lies in the equatorial plane and is labelled By_{GSEQ}. The figure illustrates how a positive By_{GSEQ} projects a negative z-component in the GSM system ($Bz_{\mathrm{GSM_{ave}}}$).

ecliptic plane, the GSEQ y-axis lies in the equatorial plane, and the GSM y-axis is perpendicular to the plane formed by Earth's dipole axis M and the common x-axis. The illustration for September shows that at equinox the GSE and GSEQ systems are identical, while the GSM system has its maximum displacement from the other systems. As M rotates daily about Earth's spin axis Ω, the GSM y-axis oscillates with an 11.4° amplitude about an average position which is displaced from the GSE y-axis by the full 23.5° angle between Ω and the GSE z-axis, since Ω lies in the y–z plane at equinox. As Earth moves through its orbit from September to December, the GSEQ and GSE systems separate, reaching the full 7.2° angle between the ecliptic and equatorial planes at solstice, as illustrated. At the same time the average separation between the GSM and GSE systems decreases to zero as Ω moves from the y–z to the x–z plane. (The Ω and M axes have been omitted from the December diagram in Figure 10 for the sake of clarity.) The only separation between the GSM and GSE systems at solstice is the 11.4° oscillation resulting from the daily rotation of M about Ω.

It is the negative z-component of the IMF in the GSM coordinate which correlates well with geomagnetic activity (Hirshberg and Colburn, 1969), whereas the IMF reaching Earth along the average spiral direction lies in the x–y plane of the GSEQ system. Russell and McPherron (1973) noted that the GSEQ y-component of an average spiral field with polarity pointing away from the Sun projects a negative z-component in the GSM system in September and December, as illustrated in Figure 10. In March and June an IMF with 'away' polarity projects a positive GSM z-component. The annual variation of this projected z-component is sinusoidal, with maximum negative amplitude occurring shortly after the September equinox. An IMF with polarity pointing towards the Sun produces an annual variation of opposite sign, with maximum negative amplitude occurring shortly after the March equinox. In the case of equally probable polarity throughout the course of a year, the annual variation of the projected negative z-component is a semiannual wave with maxima near the equinoxes. Since Earth's magnetosphere seems to respond to the z-component of the IMF in the manner of a half-wave rectifier, such that geomagnetic activity is proportional to its strength when it is negative but does not occur when it is positive (Burton *et al.*, 1975), Russell and McPherron proposed that the semiannual wave of the negative z-component is the cause of the observed semiannual wave of geomagnetic activity.

An alternative hypothesis for the semiannual variation was suggested by McIntosh (1959). He proposed that the governing parameter is the angle between Earth's magnetic axis and the direction of solar wind flow along the Earth–Sun line. This angle reaches a maximum of 90° at equinox. In support of his hypothesis, McIntosh and, later, Mayaud (1970), showed that the predicted diurnal variation, resulting from the daily rotation of the magnetic axis about the spin axis, agrees with observations.

The relative contributions to the semiannual variation from the McIntosh and Russell–McPherron hypotheses have been considered by several authors (e.g. Svalgaard, 1975, 1977; Berthelier, 1976; Schreiber, 1981). They are in general agreement that although the polarity effect itself is an outstanding feature in data sets separated according to polarity, the net effect of mixed polarities makes only a small contribution to the semiannual variation. When a realistic distribution of the north–south component of the IMF is used in a model of the polarity effect, the annual variation of geomagnetic activity for a given polarity is not at nearly zero level for half of the year, as it would be for an idealized spiral IMF, but instead varies gradually in a sinusoidal-like way. Consequently, the net effect of these two annual variations of opposite phase is a semiannual wave of amplitude considerably smaller than that predicted on the basis of an idealized spiral IMF. The results of Berthelier and Schreiber show that only about one-fourth of the observed amplitude can be attributed to the polarity effect. (Murayama, 1974, estimated a contribution of about one-half of the observed amplitude, but his estimate is based on an assumed ratio of geomagnetic disturbance to southward IMF which is about a factor of 2 larger than that found by Berthelier.) Consistent with this conclusion is the agreement between the observed diurnal variation and the McIntosh hypothesis. The Russell–McPherron hypothesis predicts a different diurnal variation, which is clearly present in data separated according to polarity, but not apparent in data with mixed polarities.

Thus it appears that the semiannual variation results primarily from a mechanism which produces maximum activity when the magnetic axis is perpendicular to the Earth–Sun line, as hypothesized by McIntosh (1959). Boller and Stolov (1970) have proposed that the mechanism is the Kelvin–Helmholtz instability. The flanks of the magnetopause are most unstable to Kelvin–Helmholtz boundary waves when the solar wind is directed perpendicular to the magnetic axis.

Since the Kelvin–Helmholtz instability is favored by high solar wind speed, indirectly the semiannual variation reflects the correlation between solar wind speed and geomagnetic activity. On the other hand, the polarity effect reflects the correlation between the southward component of the IMF and geomagnetic activity. This duality again suggests the existence of two modes of energy transfer to the magnetosphere (Schreiber, 1981), as discussed at the end of Section 3.2.

Murayama (1974) has shown that the amplitude and phase of the semiannual variation shift somewhat systematically in the course of a solar cycle. When sunspot numbers are low, amplitudes tend to be larger and phases earlier than when sunspot numbers are high. The observed phases of peak activity range approximately from the times of maximum heliographic latitude excursion of Earth's position (March 10, September 12), through equinox (March 21, September 23), to the phase predicted by the Russell–McPherron mechanism (April 5, October 5). These observations suggest that contributions to the semiannual variation vary in relative intensity throughout the solar cycle and have some dependence upon the heliographic latitude of Earth's position.

Since geomagnetic activity correlates with solar wind speed, a heliographic latitudinal gradient in speed increasing away from the equator should produce peak activity at maximum latitudinal excursion. Such gradients have been observed (Hundhausen *et al.*, 1971), and apparently their magnitudes have been large enough (~ 11 km s^{-1} deg^{-1}) to make a significant contribution to the semiannual variation (Murayama, 1974). However, measurements over longer time intervals show that large gradients occur only during short periods of 1–3 years and probably are related to the structure of individual streams rather than an overall equator-to-pole gradient (Bame *et al.*, 1977). Most of the time the gradients are too small to be detected above the noise level of 1–2 km s^{-1} deg^{-1} and, thus, could not contribute significantly to the semiannual variation.

Both the amplitude and phase of the contribution to the semiannual variation from the Russell–McPherron mechanism are affected by a heliographic latitude dependence of IMF polarity. Rosenberg and Coleman (1969) found that IMF polarity tends to reflect the dipole-like structure of the solar magnetic field. This structure has been described in Section 3.1 and is pictured in Figure 2. As Earth moves towards and then away from the solar equator in the course of a year, over many solar rotations there will be a statistical tendency to observe the polarity of the southern solar hemisphere during the first half of the year and the polarity of the northern solar hemisphere during the second half. Since geomagnetic activity is favored when the IMF polarity is towards the sun during the first half of the year and away during the second half, the amplitude of the Russell–McPherron mechanism contribution to the semiannual variation should be significantly larger when the solar polarity is in the sense shown in Figure 2 than when it is reversed. Further, when the solar polarity is favorable for activity, the phase of the Russell–McPherron contribution should shift backward in time towards the dates of maximum latitudinal excursion; and when the polarity is unfavorable, the phase should shift to a later time (Russell and McPherron, 1973).

In summary, the semiannual peaks of geomagnetic activity can be attributed to a combination of two primary mechanisms. They are the Kelvin–Helmholtz instability and the Russell–McPherron polarity effect. These mechanisms reflect the well-documented correlations between solar wind speed and the southward component of the IMF. Heliographic latitude dependences of these parameters modulate the phase and amplitude of the semiannual variation over the course of several years.

3.3.3. *Solar Cycle and Longer Period Variations*

The 11 year cycle in solar activity and accompanying changes in solar wind stream structure produce a corresponding 11 year cycle of geomagnetic activity. In general the geomagnetic activity cycle has two peaks. The first peak during sunspot maximum tends to consist of nonrecurrent storms and traditionally has been associated with solar flares. From the discussion in Section 3.1 it now seems appropriate to associate nonrecurrent activity with nonrecurrent streams from short-lived coalescing coronal holes associated either with flares or with coronal transients or with both. Consistent with these associations is the finding that 28.5 day recurrent patterns in IMF polarity tend to occur near sunspot maximum (Svalgaard and Wilcox, 1975), since coalescing coronal holes appear to form such patterns. Further, the solar cycle variation of the occurrence frequency of sudden commencements (Section 3.2), which are caused by interplanetary shock waves, peaks in phase with the sunspot cycle (e.g. Mayaud, 1975), consistent with the presence of nonrecurrent streams with their associated shock fronts during sunspot maximum.

The second peak in the 11 year cycle of geomagnetic activity occurs during the declining phase of the sunspot cycle. It is predominantly the result of storms with a 27 day recurrence pattern. This phenomenon is associated with recurrent solar wind streams from long-lived coronal holes (Section 3.2). During solar cycle 20, the second peak in activity was unusually strong compared to the first peak (Gosling *et al.*, 1977). It corresponded to the presence of the very pronounced and steady double stream pattern during 1973–1975 (Figure 8).

Beginning with solar cycle 20, nearly continuous spacecraft measurements of the solar wind became available, and searches for solar cycle variations of various parameters were made. The IMF magnitude was found to have a broad maximum throughout the cycle and brief minima at sunspot minima (King, 1979). But the parameters which show the most pronounced dependence on solar cycle are those that are best correlated with geomagnetic activity – the north-south component B_z of the IMF and solar wind speed V (Siscoe *et al.*, 1978; Gosling *et al.*, 1976b). The solar cycle variation of B_z is similar to the sunspot cycle in phase but much weaker in amplitude. Variance in the IMF follows a similar variation (Hedgecock, 1975). In contrast, the solar cycle variation of V follows the double-peaked cycle of geomagnetic activity (Crooker *et al.*, 1977). The large peak in V during the declining phase of solar cycle 20 is reflected clearly in the worldwide indices Ap and aa but not in Dst, which shows a weaker solar cycle variation (Feynman, 1980).

Solar cycle variations of geomagnetic activity indices have been reproduced by combinations of the variations of B_Z and V. The variation of Dst follows the product $B_Z V$ (Feynman, 1980). The variation of worldwide indices is well represented by V^2 alone, but the product $B_Z V^2$, which is a more meaningful expression when the geomagnetic response to individual streams is considered (Figure 3), correlates as well, since the solar cycle variation of B_Z is weak (Crooker *et al.*, 1977).

Regression analyses of B_Z and V combinations with Dst and worldwide indices over solar cycle 20 provide a calibration for the activity indices which allows "predictions" of average solar wind conditions in the past when spacecraft measurements were not available (Feynman and Crooker, 1978; Feynman, 1980). For example, the level of geomagnetic activity at solar minimum has increased steadily since 1901. Backward extrapolation to 1901 implies that either the average V was as much as a factor of 2 lower, or the average $|B_Z|$ was as much a factor of 3 smaller, or both parameters were significantly less than at present (see, also, Russell, 1975; Svalgaard, 1977; Suess, 1979).

The polarity effect (Section 3.3.2) has been proposed as the cause of an observed 22 year cycle in geomagnetic activity (Chernosky, 1966; Russell and McPherron, 1973; Russell, 1974). The solar magnetic field changes its polarity every solar cycle about two years after sunspot maximum. Rosenberg and Coleman (1969) predicted that the heliographic latitude dependence of IMF polarity would change in sense at the same time, and this prediction has been confirmed by Wilcox and Scherrer (1972). Thus for alternate 11 year periods beginning just after sunspot maximum, there is a predominance of 'toward' polarity during the first half of the year and 'away' polarity during the second half. These fields project southward IMF components in GSM coordinates and result in higher levels of geomagnetic activity. For the remaining alternate 11 year periods the polarity is reversed and geomagnetic activity is at a lower level.

Although continuous measurements of geomagnetic activity extend backward in time only through the nineteenth century, longer periodicities may be inferred from auroral records (e.g. Feynman and Silverman, 1980), since auroral displays and geomagnetic activity occur together. Auroral records are numerous enough back to about the first century to show a periodicity with an average of 87 years and possibly an additional periodicity of ~ 400 years; the cause of these periodicities most probably is variability in solar activity and the solar wind (Siscoe, 1980, and references therein).

4. Transfer Mechanisms at the Magnetopause

One of the major problems of magnetospheric physics is the determination of the processes by which energy and mass cross the magnetopause from the solar wind to the magnetosphere. The approach of ideal magnetohydrodynamics (MHD) should give at least a good approximate description of the interaction between the solar wind and the magnetosphere, because the magnetospheric scale size and the time-scale for flow past the magnetosphere greatly exceed the ion gyroradius and the ion gyroperiod in the solar wind. Indeed the existence of the bow-shock and post-shock subsonic flow as well as details of magnetosheath plasma properties are adequately accounted for by MHD theory.

A well-known corollary of ideal MHD states that if at any time two plasma elements are not connected by a magnetic field line, they can never be so connected. In this ideal limit, therefore, solar wind plasma could never become connected to field lines within the magnetosphere. The predicted exclusion of solar wind plasma is in fact obeyed to a considerably high accuracy. Only about 1% of the energy flux and 0.1% of the mass flux incident on the magnetopause gain access to the inner magnetosphere (e.g. Hill, 1979). These small percentages, however, are responsible for most magnetospheric and associated ionospheric processes and are therefore very important.

Magnetopause transfer mechanisms must violate the condition of ideal MHD in some way. Strong dissipation occurring in sub-MHD scale size regions has been invoked as well as microscopic diffusion operating over more general portions of the boundary. Magnetic merging and viscous interaction are examples of processes that draw upon localized and dispersed dissipation, respectively. Magnetic merging, the most extensively discussed magnetopause transfer mechanism, is considered here first. (For more general and extensive treatments of the merging problem, see Vasyliunas, 1975, and Sonnerup, 1979.)

4.1. Magnetic Merging

The concept of magnetic merging was first applied in the context of the magnetosphere by Dungey in 1961 to explain the pattern of electric fields in the polar ionosphere which were inferred from ground-based magnetometer data. Basic ideas about the way merging is thought to operate at the dayside magnetopause are illustrated in Figure 11. The inset (Sonnerup *et al.*, 1981) is an enlargement of the merging region in which the magnetic field in the magnetosheath is shown directed southward. It is readily verified that a northward directed field would not produce a merging field configuration on the dayside magnetopause such as is depicted here. Although a strictly southward field is most convenient for describing the merging process, it should be borne in mind that this direction is only one point in a continuum of possible orientations. The direction the field assumes in the magnetosheath is determined by the direction it had in the solar wind. The solar wind field (the IMF) takes on all orientations, but it exhibits a tendency to lie in the ecliptic plane (or more precisely to be in cones of constant heliographic latitude as predicted by elementary solar wind theory). It is observed that in its fluctuations away from the preferred orientation, the IMF is as likely to point northward as southward (King, 1977).

As the figure indicates, the merging hypothesis predicts that geomagnetic activity should be greater when the IMF is directed southward compared to when it is directed northward. The prediction was confirmed by Fairfield and Cahill (1966) five years after it was made. A large number of studies subsequently also demonstrated the strong correlation between a southward IMF and geomagnetic activity, auroral activity, and magnetospheric activity generally. In the opinion of many, the numerous independent confirmations of the predicted global consequences of magnetic merging have effectively established the validity of the merging hypothesis (e.g. Burch, 1974). There are, however, several specific merging models that predict local consequences. The data necessary to assess the validity of even some of these specific models have only recently been acquired. The main differences between them concern where on the boundary the merging region is located, the number and the spatial extent of merging regions, and whether merging is an intrinsically steady or an intrinsically nonsteady process. Since certain of the local

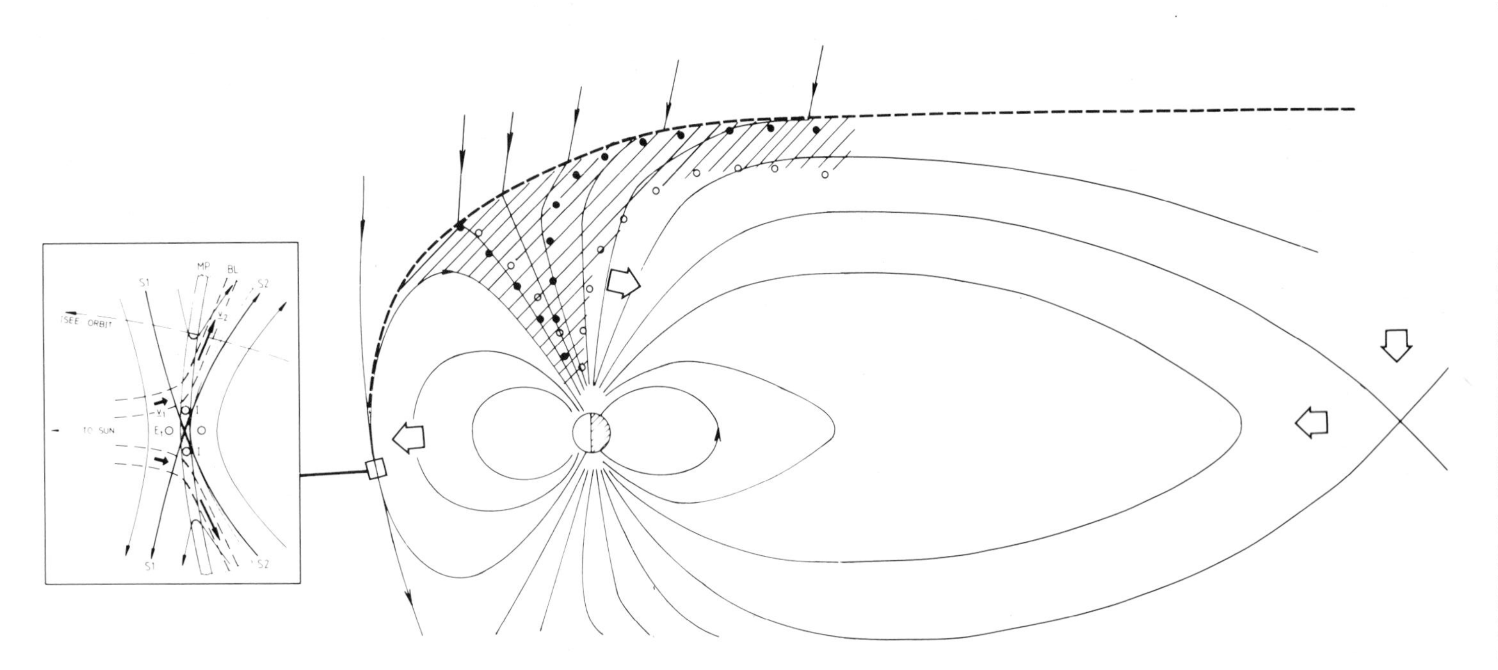

Fig. 11. Noon–midnight meridian cross-section of the magnetosphere illustrating merging of the interplanetary and geomagnetic fields and boundary layer formation. The circles trace the paths of higher (solid) and lower (open) energy magnetosheath particles as they gain access along open magnetic field lines, mirror in the converging field geometry of the cusp, and form a dispersive boundary layer inside the magnetopause. The large open arrows indicate the sense of magnetospheric convection (after Rosenbauer *et al.*, 1975). The inset shows an enlargement of the merging region. The merging electric field E_t is aligned with current I in the magnetopause MP. The plasma crossing the magnetopause is accelerated from speed $\mathbf{V}_1$ in the magnetosheath to speed $\mathbf{V}_2$ in the boundary layer BL. The separatrices S1 and S2 separate magnetosheath, merged, and Earth magnetic field lines (Sonnerup *et al.*, 1981).

and global properties of merging depicted in Figure 11 are common to most if not all of the merging models, it is useful to discuss these before describing the models.

The flowing plasma indicated in the inset carries the magnetosheath field to the magnetopause where it merges with the northward directed geomagnetic field. The merging point is marked by the intersection of the field lines labelled S1 and S2. These special field lines divide the topology of the magnetic field into three parts, one in which field lines do not connect to Earth at either end (i.e. IMF lines), another in which field lines connect to Earth at one end (the merged field lines), and a third in which field lines connect to Earth at both ends (geomagnetic field lines). The merging region where the 'freezing' law of ideal MHD is violated surrounds the merging point, although it is not specifically indicated in the figure. The calculated scale size of the merging region itself ranges roughly between 1 and 100 km, depending on which of several theoretically conceivable dissipation processes actually operates there (Vasyliunas, 1975; Sonnerup, 1979).

The dissipation region joins continuously onto an intermediate mode MHD wave (an Alfvén wave), which is identified as the magnetopause (MP) in the figure. The magnetic field penetrates the wave with a finite normal component providing an important diagnostic feature of merging. The field rotates by 180° across the wave. In the case of an isotropic plasma, the magnetic field strength and the plasma density do not change across an intermediate mode MHD wave. To match the low density–high field strength boundary condition on the magnetospheric side, a MHD slow-mode wave expansion fan is invoked to reduce the density and increase the field as the flow moves away from the magnetopause. The post-magnetopause flow is thereby confined to a boundary layer, labelled BL in the figure.

The flow speed into the merging region (labelled V_1 in the figure) is limited by theory to lie between zero and a speed of the order of the upstream (magnetosheath) Alfvén speed. The actual value depends on external boundary conditions and must be determined by matching the local solution to the global flow in which it is embedded. As yet no theoretical treatment of the self-consistent magnetosheath flow problem with merging has been presented. No calculation of a merging rate therefore exists, although it has been shown that dayside merging need proceed at only 0.1–0.2 of the maximum rate allowed by local dynamics to account for the magnetospheric phenomena driven by solar wind coupling (Burch, 1974).

Regardless of the speed of the inflowing plasma, the outflow speed (V_2 in the figure) is predicted to be of the order of twice the upstream Alfvén speed. The marked deflection and jetting of the flow as it passes through the wave are diagnostic features of this merging model. Spacecraft observations in the vicinity of the dayside boundary had failed to detect evidence of strong jetting of the plasma until recently. This failure led to the suggestion that merging as depicted in Figure 11 is inapplicable (Heikkila, 1978). However the higher time resolution and greater angular coverage for sensing plasma flow of the ISEE satellites has made possible the detection of the predicted jetting. Sonnerup *et al.* (1981) report eleven instances in the ISEE data in which the plasma velocity in the boundary layer was substantially higher than it was in the magnetosheath. The average of the ratio of the observed jetting velocity to that predicted by theory was 0.8 for the eleven events, and the average of the difference between the observed and predicted flow directions was 10°. They conclude after considering also the magnetic field and energetic particle data that the results support the merging model shown here in Figure 11.

The combination of an intermediate mode wave followed by a slow-mode expansion fan was proposed by Levy *et al.* (1964) as a model MHD merging configuration to solve the problem of matching the flow to the different plasma regimes that exist on the two sides of the dayside magnetopause. In situations where the plasma conditions on the two sides of the merging region are envisioned to be the same, for example in solar flares and in the geomagnetic tail, the MHD structure takes the form of symmetrically positioned slow-mode shock waves attached to the merging region, as proposed by Petschek (1964). The symmetrical configuration is the subject of the reviews of Vasyliunas (1975) and Sonnerup (1979); but the properties of externally controlled merging rates, namely sharply deflected flow at the MHD wave(s) attached to the merging region, a jetting of plasma away from the merging region and a normal component of the magnetic field to the attached wave(s), are common to both configurations.

As noted above, the geometry, site, and rate of merging at the dayside magnetopause are still subjects of speculation. The uncertainty results in part because the dynamics of the merging process in the presence of nonuniform field geometries and inhomogeneous flow fields, such as characterize the magnetosheath, are not fully understood, and in part because the data are not yet complete enough to decide the issues empirically. The two main parameters which are expected to govern these factors are the orientation of the magnetic field relative to Earth's field at the magnetopause and the local magnetosheath flow velocity. Merging should proceed at a maximum rate in those regions where the magnetic field is nearest antiparallel to Earth's field and where the flow is directed towards the boundary and is slow enough to allow merging to proceed. The most favorable flow conditions are in the subsolar region of the magnetosheath, near the stagnation point at the intersection of the Earth–Sun line and the magnetopause. In the case of an IMF pointing directly southward, both field and flow conditions at the stagnation point are ideal for merging. It is in this configuration that dayside merging first was proposed by Dungey (1961).

Progressively more complicated merging geometries have been proposed as more features of the process have been considered. (See the review by Crooker, 1980.) In the case of a southward IMF described above, merging is expected to occur not only at the stagnation point but also at either side of this point along a line in the equatorial plane formed by the locus of points where the IMF is antiparallel to Earth's northward field. The length of this 'merging line' depends upon how far from the stagnation point flow conditions are still suitable for merging to proceed.

If the IMF is rotated away from a strictly southward direction, it has been suggested that the merging line also rotates about the stagnation point, as illustrated in Figure 12(a) (Nishida and Maezawa, 1971). The angle of rotation is determined by the requirement that the components of the IMF and Earth's field along the merging line are equal (Sonnerup, 1974; Gonzalez and Mozer, 1974). This requirement reduces the geometry to merging of antiparallel fields in the presence of an orthogonal uniform field. Since only antiparallel components of the total fields participate in merging, the process is called 'component merging'.

An alternative geometry for merging in the case of an arbitrarily rotated IMF is shown in Figure 12(b) (Crooker, 1979). The schematic view is of the entire face of the dayside

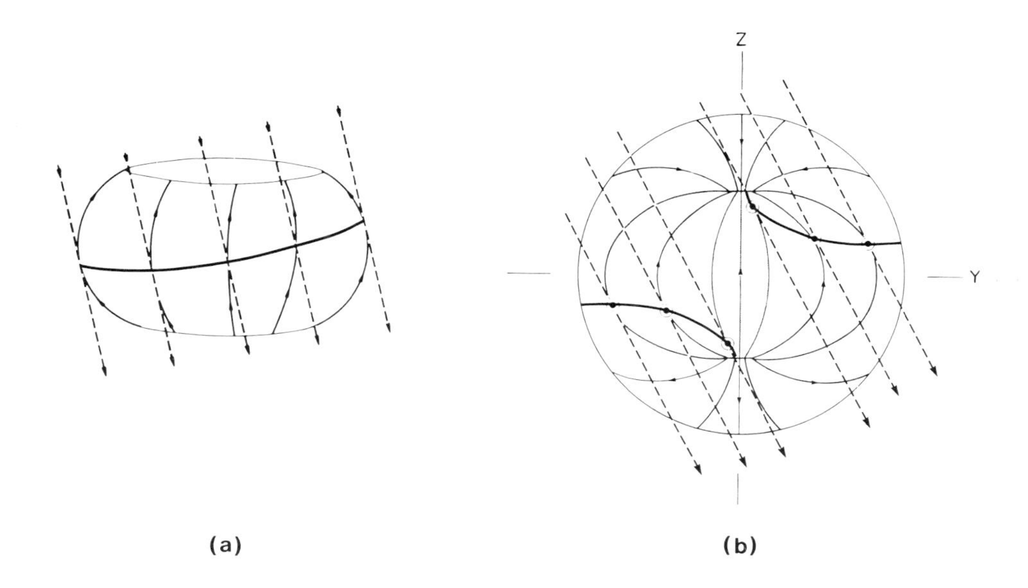

Fig. 12. Schematic illustrations of merging lines on the dayside magnetopause, as viewed from the sun, for (a) component merging (adapted from Nishida and Maezawa, 1971) and (b) antiparallel merging (Crooker, 1979). The merging lines are the heavy solid lines, the thinner curved lines are Earth field lines, and the dashed straight lines are IMF lines.

magnetopause rather than just the subsolar region as in Figure 12(a). The curved merging lines are formed by the locus of points where the total IMF and Earth fields are antiparallel. These merging lines do not pass through the stagnation point but rather lead into the northern and southern polar cusps (see Figure 1). Since the plasma jetting described above is directed away from merging lines, it would not be observed equatorward of the lines near the cusps, where the magnetosheath flow opposes the jetting. In contrast to the configuration in Figure 12(a), there is no merging at the stagnation point in Figure 12(b), except in the limiting case of a purely southward IMF, when the merging lines revert to the equatorial line described above.

In a sense, the configurations in Figure 12(a) and (b) represent two extremes: component merging emphasizes favorable flow conditions in the subsolar region, and antiparallel merging emphasizes favorable magnetic field conditions at higher latitudes. Recent theoretical work on the tearing mode instability as the cause of the dissipation responsible for merging shows that fields which are not antiparallel, even by a small rotation, are much less likely to merge than exactly antiparallel fields (Quest and Coroniti, 1981). But, on the other hand, the residence time of the magnetosheath plasma at the dayside magnetopause also is a critical factor in determining whether or not merging proceeds. Observations as yet have not been definitive in identifying the site of merging, although Sonnerup *et al.* (1981) have shown that their accelerated plasma events are consistent with merging in the subsolar region. It may be that both configurations in Figure 12 are applicable at different times, depending upon external conditions (Haerendel, 1980; Cowley, 1982).

There is increasing evidence that merging at the dayside magnetopause often proceeds intermittently in patchy regions, as suggested by Schindler (1979), rather than as a quasi-steady-state phenomenon over large regions. A picture of how intermittent merging might occur has been developed on the basis of signatures first identified in magnetometer data and called 'flux transfer events' (Russell and Elphic, 1979). The signatures are attributed to the passage of tubes of interplanetary magnetic flux which are connected to Earth's field at the magnetopause. A schematic drawing of a flux transfer event is shown in Figure 13. A magnetic flux tube protruding from the surface of the magnetopause is being convected away by the magnetosheath flow. The bend in the tube marks the juncture between the two fields. The inferred diameter of such tubes is a few Earth radii.

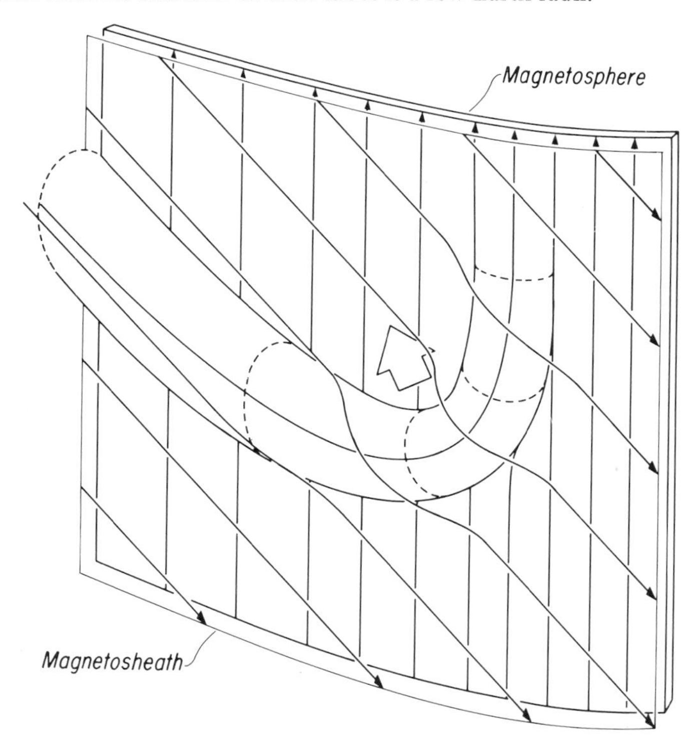

Fig. 13. Schematic illustration of a flux transfer event. Slanted magnetosheath field lines have connected with vertical Earth field lines somewhere off the lower edge of the figure, towards the right. The connected flux tube is then carried in the direction of the large arrow by the magnetosheath flow. Magnetosheath field lines not connected to the magnetosphere drape over the flux tube (Russell and Elphic, 1979).

Both quasi-steady-state and intermittent merging may occur. The accelerated plasma observations (Paschmann *et al.*, 1979; Sonnerup *et al.*, 1981) have been interpreted as signatures of the former, and the flux transfer event observations (Russell and Elphic, 1979; Paschmann *et al.*, 1982) have been interpreted as signatures of the latter (Cowley, 1982).

The large-scale view of the magnetosphere shown in Figure 11 (adapted from Rosenbauer *et al.*, 1975) illustrates how field lines and plasma ions penetrate the magnetopause as a result of merging. The act of merging changes field lines on the earthward side from a closed to an open configuration. The open field lines are then caught up in the general, tailward magnetosheath flow. As the field is convected tailward, the particles that constitute the convecting plasma can flow freely along it and thereby gain access to the magnetosphere. The circles in the figure trace paths of higher (solid) and lower (open) energy particles as they mirror in the converging magnetic field geometry of the cusp. The reflected particles form a dispersive boundary layer along the downstream magnetopause. Eventually the open field lines, which are convected into the tail lobes, reconnect with their counterparts from the opposite hemisphere at some location in the midplane of the tail where the plasma sheet thins to a point, as illustrated. The reunited closed field lines then convect sunward, completing the circulation pattern of magnetospheric convection envisioned by Dungey (1961). Open arrows in the figure indicate the sense of convective flow. The boundary layer plasma on the reconnecting field lines becomes a source for the plasma sheet (e.g. Pilipp and Morfill, 1976).

The essential feature of the merging process from the point of view of energy transfer is the electrical connection it provides between the magnetosphere and the solar wind (Atkinson, 1978). The motional electric field of the solar wind ($-\mathbf{V} \times \mathbf{B}$, where the velocity $\mathbf{V}$ and the magnetic field $\mathbf{B}$ can be evaluated anywhere along the field line) imposes an electrical potential across the length of the merging region. This potential is mapped into the magnetosphere along equipotential field lines to the ionosphere. The ionosphere is electrically conducting and is in electrical contact with all of the magnetosphere through the magnetic field. The ionosphere–magnetosphere system therefore presents an impedance to the merged field lines which draws a current from the solar wind. The product of the current and the potential gives the power input to the ionosphere–magnetosphere system.

The strength of the currents that link the magnetosphere to the solar wind are typically one to several million amps (Iijima and Potemra, 1978; Stern, 1980). The average voltage as inferred from low-altitude, polar orbiting spacecraft is approximately 50 kV (Mozer *et al.*, 1974). During substorms the polar cap potential reaches about three times its average value (Harel *et al.*, 1981b; Reiff *et al.*, 1981). The resulting power transferred to the magnetosphere from the solar wind is typically of the order of 10^{11} W but might be an order of magnitude larger during substorms.

For comparison, a representative value for the potential obtained by applying the motional electric field of the solar wind over a distance equal to the width of the magnetosphere is 350 kV, or seven times the average polar cap potential. A comparison value for the total solar wind energy flux through an area equal to the cross-sectional area of the magnetosphere is 1.4×10^{13} W (Stern, 1980), approximately two orders of magnitude greater than the typical power transferred to the magnetosphere.

Energy tranfer resulting from magnetic merging has been discussed here in terms of the associated potential and the linking currents, because this approach is so direct. Equivalent descriptions exist also in terms of the associated Poynting vector (Gonzales and Mozer, 1974; Siscoe and Crooker, 1974), and in terms of the stress (also known as tangential drag) at the boundary (Siscoe and Cummings, 1969). The latter description focuses on the ponderomotive force that slows the circumfluent solar wind, thereby extracting the energy that is transferred to the magnetosphere.

4.2. Other Mechanisms

To extract energy from the solar wind, it is sufficient that a mechanism provide a force that acts to retard it. In the language of aerodynamics, the force a moving fluid exerts tangentially on a unit area of the surface of an object is a boundary stress. The net stress over the entire surface is the drag imposed on the object by the flow, and, conversely, on the flow by the object. Magnetic merging produces a Maxwell stress that acts against the flow along the tail boundary and results in magnetic energy being added to the tail.

Attempts have also been made to evaluate the contributions viscous and wave stresses make to the energy transfer to the magnetosphere (Axford, 1964; Eviatar and Wolf, 1968, Southwood, 1979). The contribution from magnetic merging should be minimum and the contributions of other mechanisms are therefore most noticeable when the IMF is close to being strictly northward for an extended period, to allow time to dissipate energy accumulated under the Maxwell stress. A transpolar potential of 12 kV observed under such conditions has been reported by Wygant *et al.* (1982), who suggest that this is only an upper limit on the potential obtainable through nonmagnetic stresses, and that it could be less. A recent calculation by Sonnerup (1980) also found that the potential attainable under a viscous stress is a small fraction (10–15%) of the average observed value of the transpolar potential.

The flow of the solar wind along the flexible boundary of the magnetosphere should be unstable to the growth of Kelvin–Helmholtz boundary waves at least in some regions (Southwood, 1968). Multiple crossings of the magnetopause are commonly observed by spacecraft that are in transit to or from the magnetosphere, and these could be a consequence of the expected Kelvin–Helmholtz waves (see the review by Fairfield, 1976). In a case study of multiple boundary crossings interpreted as boundary waves, Southwood (1979) concluded that wave momentum was dynamically insignificant. This, again, is consistent with the small transpolar potentials observed during intervals of northward IMF – a condition which suppresses magnetic merging, as noted, but which does not suppress boundary waves.

There is a distinctive plasma feature in the form of a boundary layer inside the magnetopause lying on closed field lines at low latitudes that is apparently formed through diffusion of magnetosheath plasma (Eastman and Hones, 1979, and references therein). It is not known whether the diffusion is local or if plasma fills the boundary layer by flowing equatorward along field lines from a diffusive source in the vicinity of the polar cusps, where low field strength and an observed high level of MHD turbulence could provide favorable conditions for efficient diffusion (Haerendel *et al.*, 1978). The thickness of the boundary layer is typically a few thousand kilometres. The density and temperature of boundary layer plasma are intermediate between those of the relatively high-density, lower-temperature plasma in the adjacent magnetosheath and the relatively low-density, high-temperature plasma in the adjacent magnetosphere. Boundary layer plasma as a rule flows in the same direction as the plasma in the adjacent magnetosphere, but with a lesser speed.

4.3. Composite Model

Estimates of mass and energy transfer rates (Hill, 1979), considerations of boundary layer morphology (Crooker, 1977), correlations with solar wind parameters (Section 3.2), and inferred magnetospheric convection patterns (Section 5.1) are consistent with a composite model of the magnetosphere in which magnetic merging is the primary transfer mechanism and a viscous/diffusive interaction is a secondary but nevertheless significant additional mechanism. The three-dimensional geometry for such a model is pictured in Figure 14.

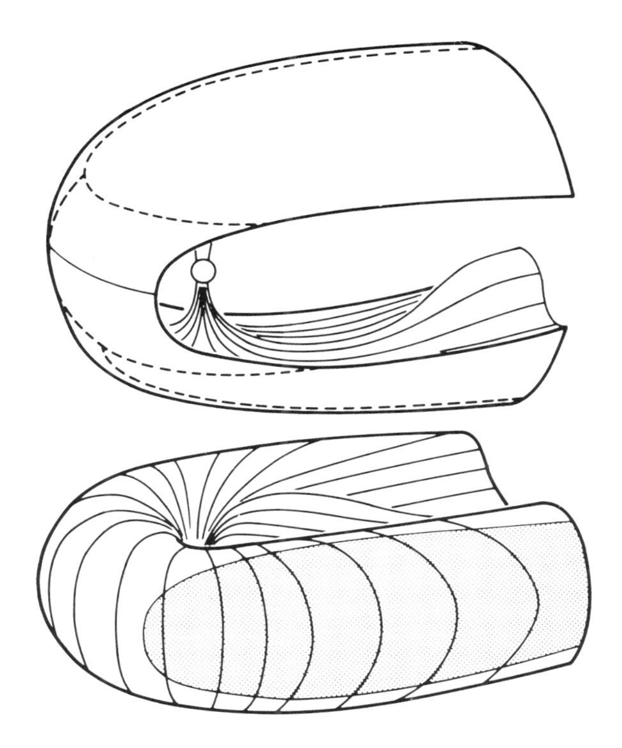

Fig. 14. Two sections of a composite model of the magnetosphere. The top section contains open field lines in a layer across the dayside and in the lobes of the tail. The bottom section contains closed field lines and encompasses the plasma sheet. When the two sections are combined to form a complete magnetosphere, the shaded area on the bottom section is exposed to the magnetosheath (Crooker, 1977).

The top volume contains the open magnetic field lines in the tail lobes and in a thin layer over the dayside. The bottom volume contains the closed field lines from the central portion of the magnetosphere encompassing the plasma sheet. Its tail cross-section is butterfly-shaped, in keeping with observations. When the two volumes are put together like pieces of a puzzle, the shaded region along the flanks of the closed volume is exposed to magnetosheath flow. It is in this region that the mass diffusion and momentum diffusion (i.e. viscosity) transfer processes operate. The noon–midnight meridian cross-section of the model is the same as Figure 11, where magnetic merging is the operative transfer process. The dashed curves near the boundary in the volume of open field lines in Figure 14 outline the boundary layer which in Figure 11 is indicated by shading. Most of the surface of the closed field line volume in Figure 14 also is covered with boundary layer plasma, since any plasma that gains access in the shaded region is free to flow along the field lines down to the ionosphere. Thus the plasma sheet boundary layer, as pictured in Figure 1, is formed.

5. Magnetospheric Convection

5.1. Convection Morphology

When momentum is transferred across the magnetopause, under the assumption that the magnetic field is 'frozen in' the plasma, the magnetic field lines (or flux tubes) inside the boundary are dragged tailward. Since no gaps of magnetic flux may be left as tubes move away from a particular position, a general circulation begins, with sunward return flow of magnetic flux in the inner regions of the tail. This is the same circulation referred to earlier as magnetospheric convection. In terms of the model in Figure 14, the merging transfer process causes tailward convection of the open field lines in the top volume and sunward convection of reconnected field lines in the lower volume. The viscous/diffusion transfer process causes convection only in the lower volume of closed fields; the tailward convection occurs in the boundary layer and the sunward convection within the volume circumscribed by the boundary layer.

Magnetic flux tubes convect through the magnetosphere as intact units along their length down to the ionosphere. At this level they are decoupled from their earthward extensions because requirements for the frozen-in condition to apply are not met there. It is useful to visualize convection patterns in the ionosphere as maps of the paths followed by the 'feet' of the flux tubes. Figure 15 is an example of such a map. The view is

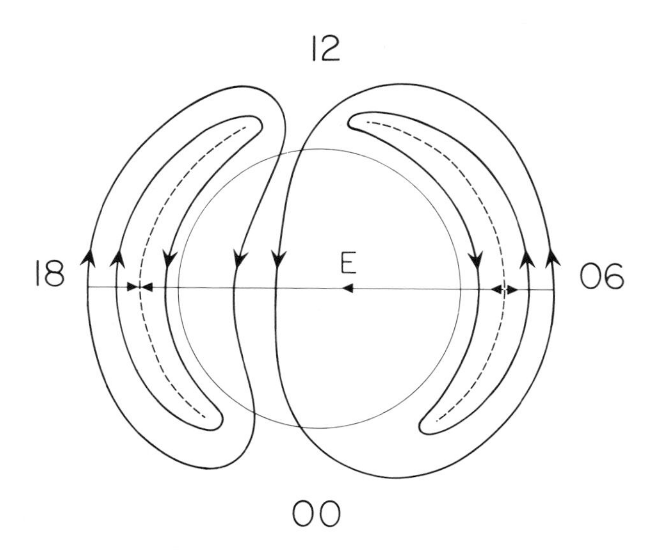

Fig. 15. View from above the northern hemisphere of the pattern of convecting magnetic flux just above the ionosphere in a composite model as in Figure 14. The circle is the polar cap, which forms the boundary between open and closed field lines. The dashed curves mark the flow reversal boundaries, and the direction of the electric field E is as indicated.

of the north polar ionosphere from above. In general convection patterns have two circulation cells. The antisunward flow across the pole in the center of the pattern maps out to field lines on the boundary, and the sunward flow at lower latitudes maps out to field lines well within the boundary. This pattern applies both to open-field-line models where the only transfer process is magnetic merging (Dungey, 1961) and to models with no open field lines where the only transfer process is a viscous/diffusive one (Axford and Hines, 1961). In an open model the polar cap boundary, defined here as the boundary between the open flux in the lobes and the closed flux in the plasma sheet, separates the sunward from antisunward flow. In the composite model in Figure 14, the polar cap boundary is the interface between the upper and lower volumes when they are fit together to form the magnetosphere. Since antisunward convection occurs on flux tubes along the flanks of the model, the flow reversal boundary maps to field lines in the ionosphere which are equatorward of the polar cap boundary, consistent with the observations of Heelis *et al.* (1980) showing the reversal boundary embedded in a region of precipitation from the plasma sheet boundary layer. The model configuration is pictured in Figure 15. The circle is the polar cap boundary, and the dashed curves mark the flow reversal boundary. The crescent-shaped flow contours which close equatorward of the polar cap boundary represent convection from viscous interaction on the volume of closed field lines, and the outer flow contours which cross the polar cap boundary represent convection from merging. The intersections of the flow contours with the polar cap boundary map out to the merging sites on the dayside and in the tail.

The rate at which magnetic flux is transferred in magnetospheric convection is expressed as the electric potential Φ between the centers of the two convection cells, where $\Phi = Ed$, d is the distance between the two cells, E is the electric field deriving from the frozen-in condition $\mathbf{E} = -\mathbf{V} \times \mathbf{B}$, and $\mathbf{V}$ is the velocity of the convecting plasma carrying Earth's magnetic field $\mathbf{B}$. This potential is often referred to as the driving potential. As noted earlier, its value is related to the rate of energy transfer from the solar wind. In Figure 15 the electric field is directed from dawn-to-dusk across the potential drop between the two cells and dusk-to-dawn equatorward of the cell centers.

In the ionosphere at about 100 km altitude, ion-neutral collisions prevent the ions from convecting with the electrons, and the frozen-in condition no longer applies. The convecting electrons then become a negative flow of Hall current. The Hall current flows perpendicular to the electric field in the same pattern as the convective flow but in a direction opposite to the arrows in Figure 15. Current flow in the ionosphere also has a component parallel to the electric field. This Pedersen current is carried by the ions as they collide with neutrals and become progressively displaced in the direction of the electric field.

The flow configuration in Figure 15 is not symmetric about the noon–midnight meridian. The asymmetry is drawn in accordance with observed electric field asymmetries which are correlated with the azimuthal component B_y of the IMF (e.g. Heppner, 1972). When B_y is negative, the electric field is strongest on the dusk side of the northern polar cap, as in Figure 15. When B_y is positive, the electric fields are strongest on the dawn side. This effect has been interpreted as an azimuthal stress on Earth's magnetic field lines when they merge with an IMF with a finite azimuthal component (Jorgensen *et al.*, 1972; Russell and Atkinson, 1973). Some dawn–dusk asymmetry also has been attributed to the day–night conductivity contrast across the polar cap (Lyatsky *et al.*, 1974; Atkinson and Hutchison, 1978). This latter asymmetry is always in the sense of a stronger electric field on the dawn side.

Convection patterns as in Figure 15 have been deduced from electric field measurements (e.g. Heppner, 1972; Cauffman and Gurnett, 1972; Mozer *et al.*, 1974), ground-based magnetometer measurements (e.g. Friis-Christensen and Wilhjelm, 1975; Maezawa, 1976), and more directly from measurements of vector ion drift velocities (e.g. Heelis *et al.*, 1976, 1980). Patterns different from Figure 15 have also been observed for times when the southward component of the IMF is zero. Single convection cells in the polar cap occur, with clockwise circulation (viewed from above) for positive B_y and anticlockwise circulation for negative B_y in the northern hemisphere, and with opposite circulation in the southern hemisphere (Friis-Christensen and Wilhjelm, 1975). This pattern is known as the Svalgaard–Mansurov effect (e.g. Svalgaard, 1973). When the IMF has a northward component and the effects of B_y are subtracted from the data, two small cells with convection in a sense reversed from that shown in Figure 15 appear wholly within the polar cap (Maezawa, 1976). These patterns are consistent with the merging morphology. For a composite model, as in Figure 14, an IMF with a northward component should produce both a convection cell or cells wholly within the polar cap and two additional, separate cells equatorward of the polar cap boundary, from viscous interaction. Burke *et al.* (1979) show electric field observations during times of northward IMF consistent with one or two cells in the polar cap and two additional cells at lower latitudes, as predicted by the composite model. (See review by Crooker, 1980, for schematic drawings and more discussion of convection patterns for northward IMF.)

5.2. Birkeland Currents, Alfvén Layers, and Shielding

One of the most important features in the magnetosphere, because of the central role it plays in convection, is the observed set of field-aligned or Birkeland currents flowing into and out of the ionosphere at auroral latitudes (e.g. Zmuda and Armstrong, 1974; Iijima and Potemra, 1976). Figure 16 (from Iijima and Potemra, 1978) shows statistical patterns based on low-altitude satellite magnetometer data for geomagnetically quiet and disturbed conditions as viewed from above the northern polar ionosphere. Both patterns show current flow at the statistical location of the auroral oval, which moves to lower latitudes during disturbed periods. The general sense of flow is into the ionosphere at dawn and out at dusk at the poleward edge (Region 1) and in the opposite sense at the equatorward edge (Region 2).

The Region 1 Birkeland current system is a necessary consequence of magnetospheric convection. The diverging electric field in the ionosphere, described above with respect to the convection pattern in Figure 15, implies diverging Pedersen currents at the flow reversal boundaries. In order to enforce continuity of current, these must continue as Birkeland currents flowing in the sense of the Region 1 system.

The Region 2 Birkeland currents also flow in response to the driving potential, but in a less direct manner. Earthward convection, driven by the potential, occurs in the plasma sheet and its earthward extension, the ring current. The general circulation is carried by cool plasma, to which the 'frozen-in' condition applies. But the hot plasma particles, which dominate the plasma sheet and ring current populations, follow paths which diverge from the general circulation. As the hot plasma approaches stronger magnetic

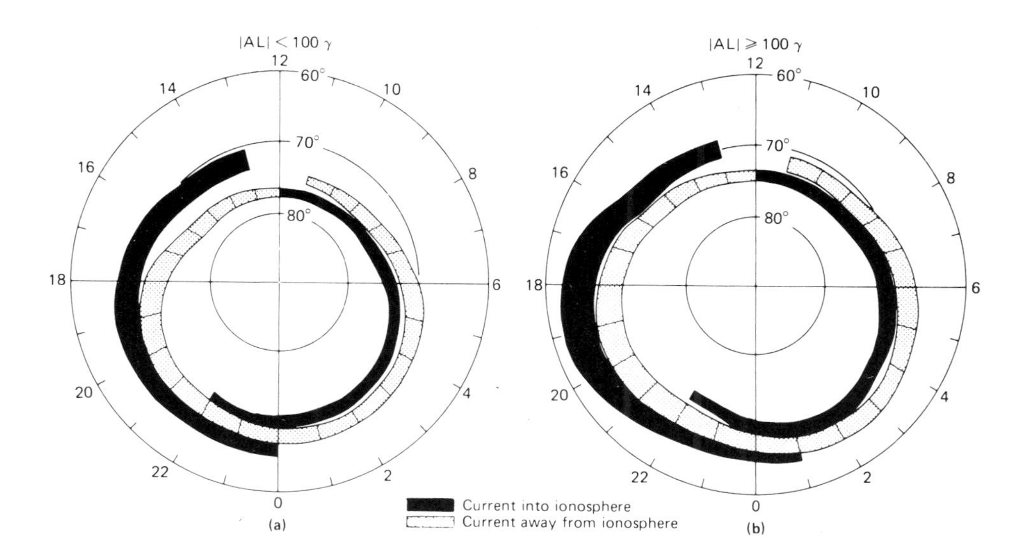

Fig. 16. View from above the northern hemisphere of the distribution and direction of Birkeland current flow into and away from the ionosphere: (a) during quiet periods when the westward auroral electrojet index AL of geomagnetic activity was less than 100γ; (b) during more active periods when AL was greater than 100γ. (From Iijima and Potemra, 1978.)

fields near Earth, the field gradient and curvature cause electrons and ions to drift in opposite directions and form a westward current. Because the drift paths for ions come closer to Earth on the duskside than on the dawnside, and conversely for the electrons, regions of charge separation, called Alfvén layers, form and draw neutralizing current from the ionosphere (Schield *et al.*, 1969). These neutralizing currents are identified with the Region 2 Birkeland current system (Harel *et al.*, 1981b).

In steady state, the Alfvén layers effectively shield latitudes equatorward of Region 2 from the driving potential (e.g. Vasyliunas, 1972; Jaggi and Wolf, 1973; Southwood, 1977). The electric field there associated with Region 2 Birkeland currents is directed opposite to and approximately cancels that associated with Region 1 currents. Thus, in steady state, little ionospheric Pedersen and Hall currents associated with convection flow equatorward of Region 2, and sunward convection in the magnetosheath is confined to the plasma sheet, which maps into the auroral oval in the ionosphere.

The complete circuit of the Birkeland current system is believed to be connected as follows: Most of Region 2 currents are connected to Region 1 currents across the width of the auroral oval (e.g. Yasuhara *et al.*, 1975). Some of the Region 1 current closes across the polar cap. In the magnetosphere the Region 2 currents flow along the inner edge of the plasma sheet and close as westward current in the Alfvén layers. The Region 1 currents flow along the outer boundary of the plasma sheet to the magnetopause. This junction in Figure 14 is marked by the outline of the shaded region on the lower volume. At the magnetopause the Region 1 Birkeland currents close over the tail lobes, joining the shielding and tail current flow on the magnetopause (e.g. Atkinson, 1978).

5.3. THE PLASMASPHERE: A CONVECTION/COROTATION FORBIDDEN ZONE

In addition to the electric fields associated with the driving potential and the Alfvén layers, or, equivalently, with the Regions 1 and 2 Birkeland current systems, the rotation of Earth's dipolar magnetic field also imposes a motional electric field throughout the magnetosphere. The rotational electric field $\mathbf{E}_\Omega$ is $-(\mathbf{r} \times \boldsymbol{\Omega}) \times \mathbf{B}$, where $\mathbf{r}$ is the geocentric radius vector and $\boldsymbol{\Omega}$ is the angular velocity vector of Earth. To a good approximation the electric field $\mathbf{E}_\Omega$ is directed radially inward in the equatorial plane and its strength decreases from Earth as r^{-2}. It is desirable to compare $\mathbf{E}_\Omega$ with the strength of the convection electric field. To simplify the discussion leading to the main point of this section, it is useful to ignore the contribution of the Alfvén layers to the electric field. Then convection can be represented by a uniform, constant electric field $\mathbf{E}_c$ directed from the dawnside to the duskside of the magnetosphere.

Close to Earth, corotation dominates over convection. As an important consequence of this condition, magnetic flux tubes near Earth tend to corotate with it, moving in closed, Earth encircling paths. Electrons and ions escaping from the ionosphere fill these tubes until an equilibrium is reached in which the flux out of the ionosphere is matched by a return flux from the magnetosphere. The resulting plasma body, the plasmasphere, is in effect a continuous extension of the ionosphere into the magnetosphere.

The outer boundary of the plasmasphere, the plasmapause, is usually well marked by a substantical drop in density within a relatively narrow radial distance (see Figure 17,

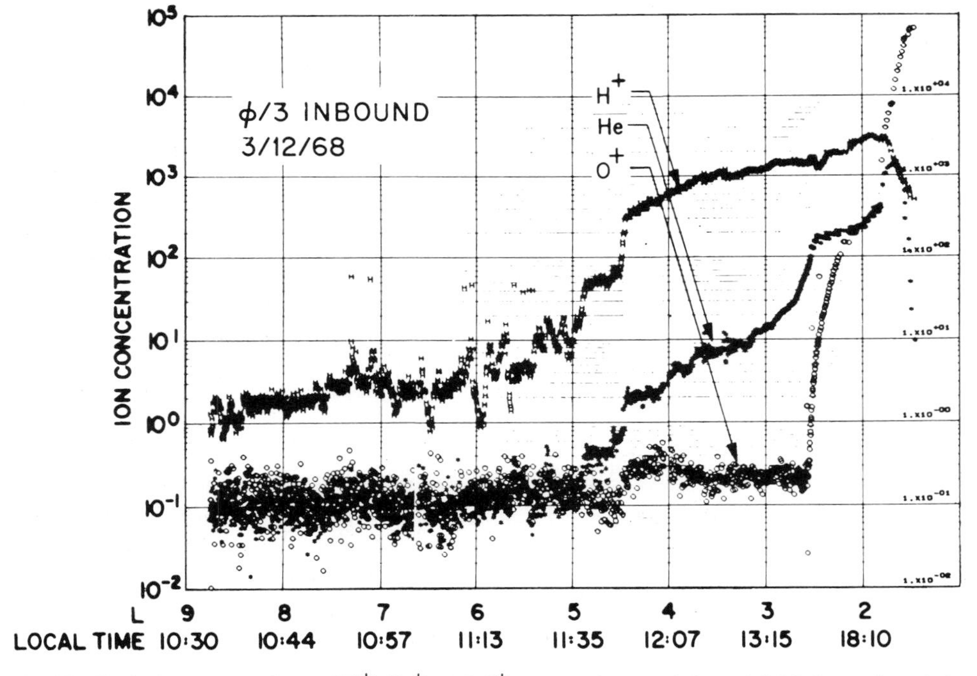

Fig. 17. Typical concentrations of H^+, He^+, and O^+ measured on an inbound OGO 5 crossing of the plasmapause. The parameter L along the abscissa is the distance in Earth radii between Earth's center and the equatorial point along the magnetic flux tube on which the spacecraft is located. (From Chappell *et al.*, 1970.)

from Chappell *et al.*, 1970). Beyond the plasmasphere, convection dominates over corotation. Instead of circling Earth in closed paths, flux tubes in this region move with the convective flow. In their general sunward course, they stream around the dawn and dusk sides of the plasmasphere, sweeping away plasma that has escaped from the ionosphere. Since the magnetospheric circulation time associated with convection is short compared to the time required for the ionosphere to come into equilibrium with its plasmaspheric extension, the region dominated by convection is relatively depleted in ionospheric plasma (Nishida, 1966; Brice, 1967).

In the mathematical formulation of this phenomenon (e.g. Kavanaugh *et al.*, 1968; Chen, 1970; Kivelson *et al.*, 1980), the convective and quasi-corotational flow regimes are separated by a singular flow streamline, the separatrix. With respect to its accessibility by low energy (i.e. ionospheric) ions and electrons in the convective flow, the region of quasi-corotating flow on the other side of the separatrix is a forbidden zone, and of course, the converse is also true. It is in this sense that the plasmasphere is a forbidden zone to the convective flow.

Since the strength of the convection electric field is modulated by variations of solar wind parameters, the position of the separatrix changes in time. Correspondingly, the density drop marking the plasmapause is observed at different positions from one observing opportunity to the next.

5.4. Time Dependent Convection: The Substorm Cycle

For the sake of simplifying the introduction of a complex phenomenon, convection was discussed in the previous sections as if it were a steady process. However, usually the magnetosphere is not in a steady state. The rates of dayside and tail magnetic merging at any instant are usually not the same. Dayside merging may be relatively steady, depending on the steadiness of solar wind conditions. The prevailing assumption for fixed solar wind conditions is that the total rate of dayside merging is reasonably steady, even if it occurs in intermittent patches. In contrast, a substantial portion of merging in the tail is thought to be nonsteady, occurring in temporally separated, well-marked substorm intervals. Prior to a substorm, dayside merging is largely uncompensated by tail merging, and the magnetic flux in the tail lobes increases. The imbalance is reversed during the ensuing substorm.

A listing of the major events in the lifecycle of a typical substorm begins with a southward turning of the IMF. This initiates the buildup phase in which the strength of the magnetic field in the tail increases, the magnetic field configuration in the near-Earth plasma sheet becomes less dipolar and more tail-like, the flare of the tail boundary increases, the boundary on the dayside contracts earthward, and the ionsophere impressions of the dayside polar cusps move equatorward. About 40 min after the IMF turns southward, the explosive phase of the substorm occurs. At this time, the magnetospheric changes that characterize the buildup phase reverse. The time-scale for the restoration of pre-buildup conditions in the magnetosphere is again approximately 40 min. However, the associated ionospheric phenomena develop more rapidly, as suggested by the phrase 'explosive phase'. The intensities of auroral activity, the auroral electrojets and the accompanying magnetic disturbances peak within 10 min of the onset of the explosive phase of the substorm. (For recent reviews of the details of substorm phenomenology, see Akasofu, 1977, and McPherron, 1979.)

The changes in the boundary morphology and in the intensity of the field in the tail that take place during a substorm cycle can be understood simply in terms of the seesawing flow of magnetic flux to and from the tail lobes, as already noted. However, the role of the Birkeland currents needs to be acknowledged to account quantitatively for these changes and to understand even qualitatively the morphological changes and the dynamical process that occur deeper in the magnetosphere.

When the IMF turns southward, convection begins and magnetic flux is transferred from the dayside to the tail lobes, causing the observed increase in tail field magnitude. The tailward convection activates the Region 1 Birkeland current system, or increases the strength of the pre-existing system. The disturbance field from the Birkeland currents and their closure on the boundary causes the inward movement of the dayside boundary and contributes to the flaring of the tail boundary. At lower latitudes the convection proceeds with no significant merging in the tail, so that there is a depletion of closed field lines as they convect sunward to replace the flux merged at the dayside. The depletion of the closed field lines and the addition of open flux to the tail lobes result in a deflation of the reservoir of dipolar field lines on the nightside of the inner magnetosphere, which gives a more tail-like configuration of the near-Earth plasma sheet. The tail current system encircling the two lobes is then closer to Earth.

At the explosive phase of the substorm there is a rapid increase in the transpolar driving potential and the associated Birkeland currents. There is some controversy as to whether the potential increase is supplied directly by the solar wind, in which case the observed 40 min lag must be interpreted as an inductance time, with the tail as an inductor in series with the ionosphere, or whether it is supplied by the discharge of the energy stored in the tail (e.g. Akasofu, 1980). It may be that both the direct and stored energy processes contribute to the potential increase.

The increase in the Region 1 Birkeland current at the onset of the explosive phase appears to result from a diversion into the ionosphere of the cross-tail portion of the tail current system encircling the lobes (Figure 18). That is, the near-Earth portion of the tail current system, which moves earthward prior to onset, converts to a Birkeland current system, which closes over the boundary, as described at the end of Section 5.1. The consequent dissipation of the near-Earth portion of the tail current restores the magnetic field in this region to its dipolar configuration.

The impulsive increase in the Region 1 Birkeland current system correspondingly intensifies convection at lower latitudes and deepens its earthward penetration. The equilibrium position of the inner edge of the plasma sheet is suddenly shifted earthward. To reach the new position, ions and electrons radiate from the old inner edge by moving in new drift trajectories determined by the new convection electric field. A spatially coherent front of drifting particles is formed in this way. A particle detector on a spacecraft in the path of this front records a signature known as a plasma injection event, which is characteristic of substorms (Kivelson *et al.*, 1980). The ultimate dispersal of these particles results in an approximately azimuthally homogeneous plasma torus around the Earth, generally known as the ring current.

5.5. Computer Modeling of Convection

Viewed as a problem in mechanics in which one would like to predict the state and

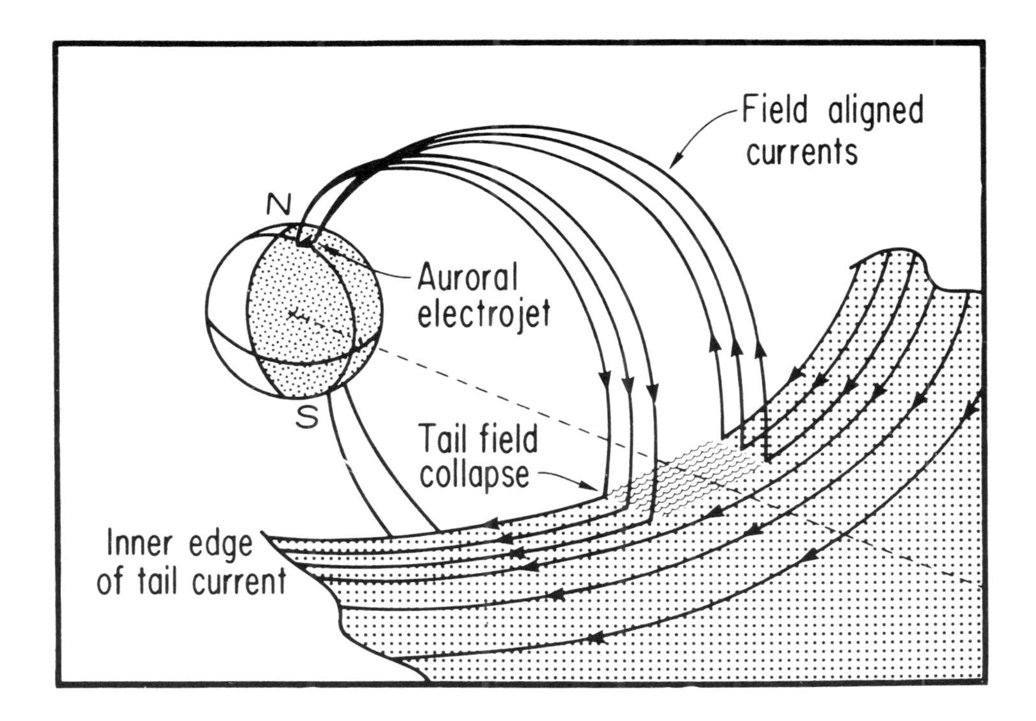

Fig. 18. Schematic drawing of current flow on the nightside of the inner magnetosphere during a substorm. Current flowing across the tail has been diverted into the ionosphere along field lines to form the auroral electrojet (McPherron *et al.*, 1973).

the behavior of a system from given initial conditions and boundary conditions, magnetospheric convection is notably complex. The three-dimensional geometry of the magnetosphere alone precludes an analytic approach to a realistic model. Electric fields and currents are projected from one part of the system to another along magnetic field lines, which are dipolar close to Earth but which are greatly distorted and essentially three-dimensional in the outer magnetosphere. Large and localized variations in the conductivity of the ionosphere contribute importantly to the spatial complexity of the problem.

The system is not only inherently three-dimensional, it is inherently time dependent. The global morphology changes through a substorm cycle. Current systems wax and wane, shift their positions and even their paths of closure. Ionospheric conductivity changes in localized regions in response to the changing patterns and intensities of precipitating particles.

The various components of the system can not be treated in isolation and the results then combined to produce a functioning model of magnetospheric convection, because the components are linked interactively through feedback loops. This point is perhaps not immediately obvious, and it is sufficiently important to warrant demonstration with an example. The feedback nature of the coupling between the plasma sheet/ring current particle population and the ionosphere under the influence of the convection electric field was formulated explicitly by Vasyliunas (1970, 1972). It can be reformulated in terms of the relation between the Regions 1 and 2 Birkeland current systems. The first represents the direct response of the ionosphere to an imposed transpolar potential; that is, the current drawn when a potential is applied across a resistor. The fringing field of this potential extends to lower latitudes in the ionosphere, where it is mapped along equipotential magnetic field lines into the magnetosphere. In this way the convection electric field reaches the inner magnetosphere from a source connected to the polar regions of Earth. The resulting differential drift of the ions and electrons in the inner magnetosphere creates charge separation layers (the Alfvén layers) which act as sources and sinks for the Region 2 Birkeland current system. The Region 2 currents close through the ionosphere and in so doing create a Region 2 potential which is virtually coextensive with the Region 1 potential. It can now be seen that the feedback problem is to find the Region 2 potential that when superimposed on the given Region 1 potential produces Region 2 currents that, through their closure in the ionosphere, recreate the Region 2 potential with which one began. The system is then self-consistently coupled. Low-latitude shielding is one of the more pronounced consequences of this coupling.

A different but related, triangular feedback loop has as one leg the modification of ionospheric conductivity by particle precipitation, as a second leg the effect of ionospheric conductivity on the distribution of ionospheric potential, and as the closing leg the control of particle precipitation by the ionospheric potential mapped into the magnetosphere. Southwood and Wolf (1978) showed that this coupling can give rise to rapid convective flows in a narrow latitudinal zone just equatorward of the auroral oval, and such flows are observed.

From this brief recapitulation it is evident that the approach of computer simulation is needed to model global magnetospheric convection. The most sophisticated and comprehensive of such computer models has been developed at Rice University by R. A. Wolf and colleagues (e.g. Jaggi and Wolf, 1973; Harel *et al.*, 1979, 1981a). The model calculates the evolution in space and energy of plasma sheet particles as they convect inward from an initial edge at $10\,R_e$ with a given initial density and initial energy distribution. The resulting magnetospheric electric current is obtained by integrating the drift trajectories of the ions and electrons separately. The total magnetospheric current is kept divergenceless by allowing the convergence or divergence of the net drift current to act as a source or sink of Birkeland currents. This couples the magnetospheric particle population to the ionosphere.

In the ionosphere the Pedersen and Hall currents are calculated as solutions to a potential problem on a spherical surface with a boundary at 72° latitude (corresponding to the $10\,R_e$ plasma sheet edge in the magnetosphere) on which the transpolar potential is imposed. The Birkeland currents from the magnetosphere are incorporated in the potential problem as sources and sinks of currents. The solution gives the potential throughout the ionosphere, which is then mapped into the magnetosphere to calculate the next step in the particle drift trajectories. The loop is now closed self-consistently and can be repeated to develop the evolution of the system.

The model possesses specially programmed capabilities and features to achieve a close simulation of the natural system. Ionospheric conductivity is specified to match models based on observations. The conductivity changes in response to electron precipitation as calculated by the model itself. Electrons are assumed to precipitate at the maximum rate allowed by pitch-angle scattering. The three-dimensional distortions of the magnetic field are embodied through the use of the magnetic field model of the magnetosphere developed by Olson and Pfitzer (1974). The shunting of the cross tail current into the ionosphere during the explosive phase of the substorm is handled with a 'hard wire' circuit, the 'substorm current loop', linking the ionosphere and the tail current sheet in a way that approximates the geometry that is inferred for the actual current system (e.g. McPherron *et al.*, 1973).

A logic diagram of the Rice magnetospheric convection model is shown in Figure 19.

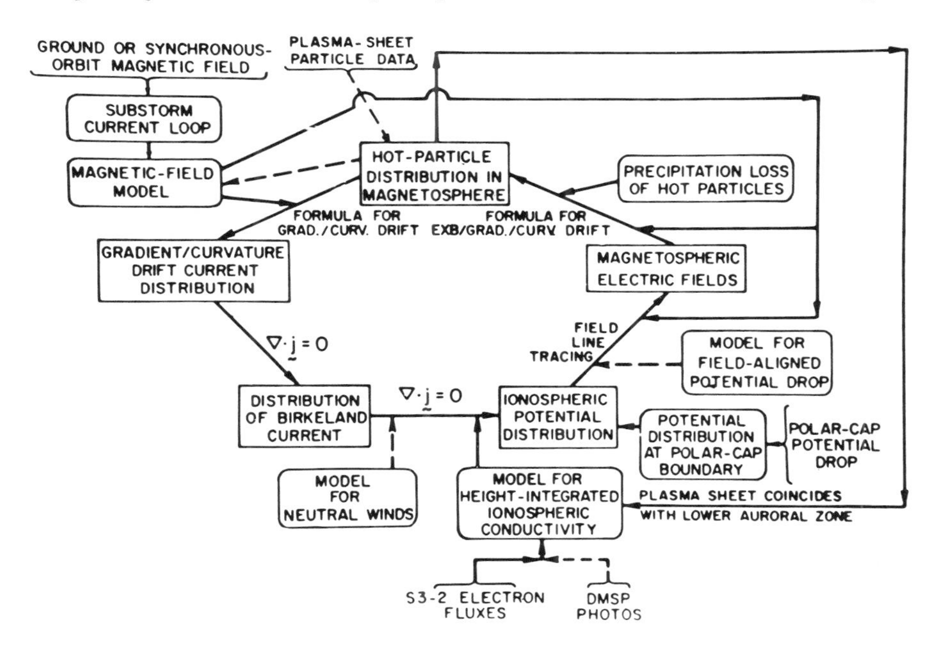

Fig. 19. Logic diagram of the substorm computer simulation model of Harel *et al.* (1981a). Dashed lines indicate features to be incorporated in the future. The rectangles at the corners of the central pentagon represent basic computed parameters. Input models are indicated by rectangles with round corners; input data are indicated by curly brackets. The program cycles through the main loop approximately every 30 s. The basic equations solved by the computer are described next to the logic flow lines.

The ionosphere–magnetosphere feedback loop is at the center of the diagram. A number of inputs connect into the loop. The initial plasma sheet particle population, the driving potential, and the strength of the current in the substorm current loop must be specified. The last two are required as given inputs at each step of the calculation, and may therefore be allowed to vary as the system evolves to simulate the time dependence of specific events.

The model was tested using as input data polar cap potential drops and substorm current intensities measured during a particular substorm-type event (Harel *et al.*, 1981a, b; Spiro *et al.*, 1981; Karty *et al.*, 1982). The model produced satisfactory agreement with a number of observations, especially with observations of the high-latitude electric field, the degree of low-latitude shielding, and a dawn–dusk asymmetry in the electric

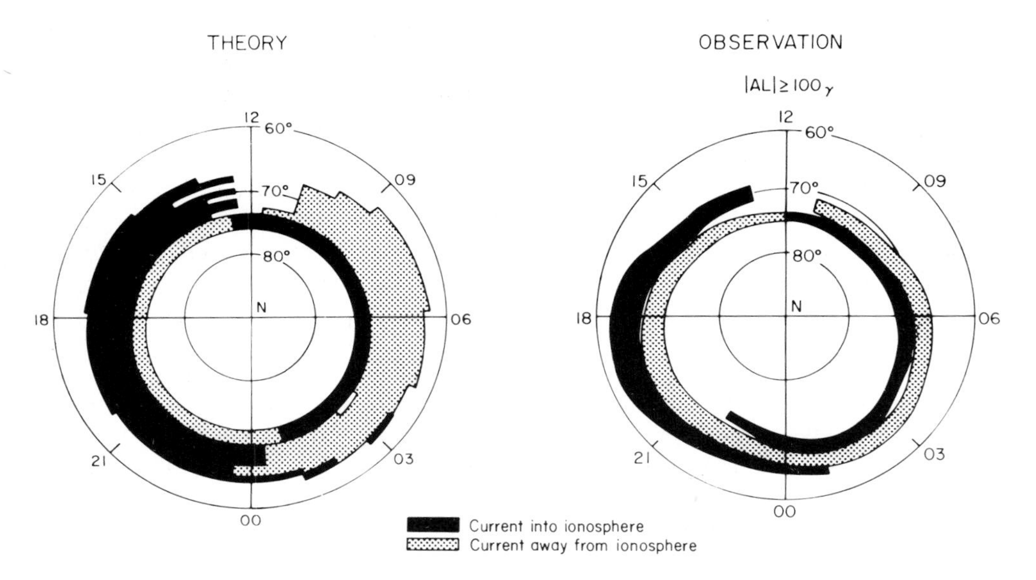

Fig. 20. Comparison of computed and observed Birkeland current patterns. (From Harel *et al.*, 1981b.)

field. To illustrate the level of detail achievable with this model, Figure 20 shows a comparison between a computed Birkeland current pattern which was typical for this event and the statistical Birkeland current systems compiled by Iijima and Potemra (1978) from data acquired during active periods. The computed Region 1 currents lie just poleward of the polar boundary of the model. The pattern was derived by a Birkeland current continuation of the computed horizontal currents that flowed into or out of the boundary. The thickness of the computed Region 1 current was not determined and was therefore set arbitrarily. The overall agreement between the two patterns is apparent, even to the overlapping of the systems near midnight. The most notable disagreement – a counterclockwise rotation of the whole pattern – could indicate that the driving potential is not strictly dawn-to-dusk, but has also a night-to-day component.

The model still neglects a number of important factors. These are indicated by input boxes connected by dashed lines to the feedback loop in the logic diagram in Figure 19. Enough is known about the behavior of these items and about their dependence on the model variables which are already incorporated that their inclusion in the model is 'simply' a programming problem that can in principle be carried out. By contrast, the way the solar wind couples to the magnetosphere and the involvement of the tail in the convection process are at present not well enough understood to permit extending even in principle a convection model to the point where it could accept solar wind parameters as the primary input data.

5.6. Convection and Magnetic Merging in the Magnetotail

Although the workings of the magnetotail are not as well understood as are the processes that occur in the near-Earth or inner magnetosphere, much is known about magnetotail phenomenology and considerable theoretical progress has been made on the macrophysics and microphysics of plasmas in tail-like configurations. This section reviews representative work aimed at elucidating and testing the buildup-and-release model of magnetotail convection, which is the most generally adopted working hypothesis. (For an alternative model, however, see Akasofu, 1980.)

In the buildup-and-release model of magnetotail convection, magnetic flux which has merged with the IMF at the dayside magnetopause is carried by the flow of the plasma in the magnetosheath to the tail lobes where it accumulates until the threshold of some instability is passed, resulting in reconnection and return of the flux to the inner region of sunward convection. It follows as an immediate prediction of the model that the magnetotail should expand during the buildup phase and contract during the release phase. As a further immediate consequence the field strength should synchronously increase and decrease. The first prediction was confirmed by Maezawa (1975) in an analysis of the correlation between substorm activity and the locations of crossings of the magnetotail boundary by different spacecraft. That the expected concomitant variation of field strength also occurs has been reported by Meng and Colburn (1974) and Caan *et al.*, (1975). Figure 21, taken from the latter work, illustrates the behavior of the field energy density in the tail lobes through the two phases of a substorm cycle. The transition between the phases is identified as the onset of the explosive phase, which is marked in the mid-latitude, surface level magnetic field data by a sudden increase in the horizontal component. Energy densities from 20 events, each normalized to the value at the onset of the explosive phase, have been averaged to provide a statistically representative picture. The result shows a clearly evident buildup phase beginning approximately two hours before the transition to a relaxation phase, which persists for approximately an additional hour.

During the buildup phase, the increase of the flux in the tail at the expense of the flux in the inner magnetosphere should effect an earthward encroachment on the nightside of tail-like characteristics (McPherron, 1974). This predicted behavior agrees with spacecraft observations and is an especially regular occurrence in data from satellites in synchronous orbit, at 6.6 R_e (McPherron *et al.*, 1973). An unusually comprehensive confirmation of the direct predictions of the buildup–release model of magnetotail convection was presented by Baker *et al.*, (1981b). Because of the availability of well-timed auroral images from polar orbiting spacecraft, and extensive riometer data from fortuitously positioned sites in Greenland and Scandinavia, they were able to obtain an unambiguous chronology of events during the life cycle of a particular substorm. They verified that signatures attributed to the buildup phase in fact occurred prior to the explosive phase onset. They demonstrated also that prior to onset, the magnetic field and pitch-angle distribution of electrons observed at synchronous orbit evolved in precisely the way expected from an earthward development of the magnetotail.

A pronounced change in the thickness of the plasma sheet is observed that is correlated with specific phases of the substorm cycle (McPherron *et al.*, 1973). Although this phenomenon is not an obviously necessary consequence of the buildup–release model, it synchronizes with the two phases, and therefore supports the proposal of a two-phase process. During the buildup phase the near-Earth part of the plasma sheet thins (Pytte and West, 1978). In the explosive phase, the thinning is more pronounced and more extensive. During the recovery from the explosive phase, it thickens to typically 10 R_e

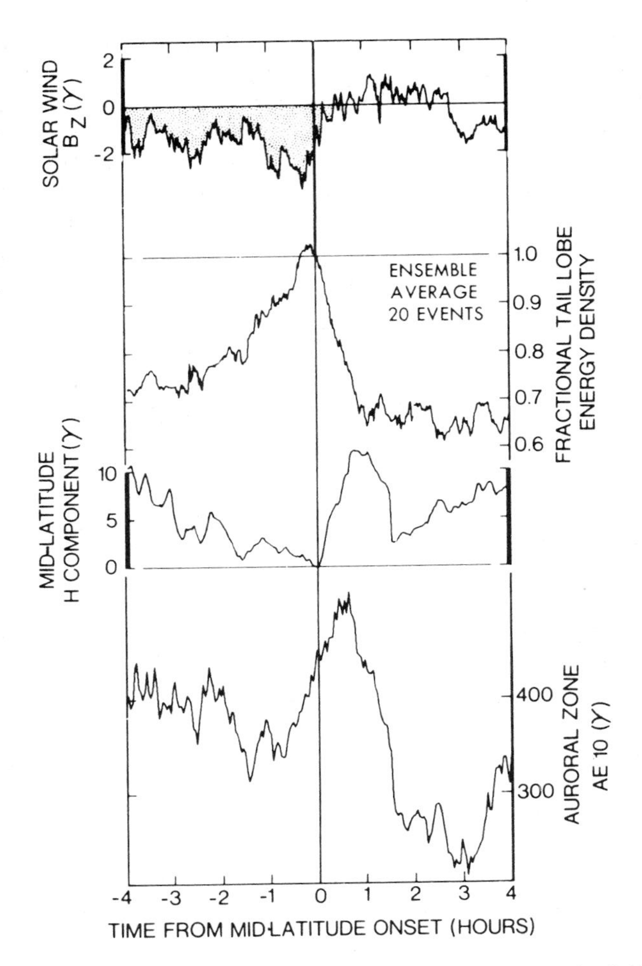

Fig. 21. Superimposed epoch averages of solar wind north–south magnetic field component B_Z (GSM), tail lobe magnetic energy density, disturbance in horizontal component of geomagnetic field at mid-latitudes, and AE index during a substorm. (From Caan *et al.*, 1975.)

(Hones *et al.*, 1970). The density of the particles in the plasma sheet does not change appreciably in the course of these variations in thickness, implying that they are not principally the result of compressions and rarefactions. The observed phenomenology of the plasma sheet through the substorm cycle provides an important constraint on models of convection in the magnetotail.

While magnetic flux is accumulating in the buildup phase, convective flow in the plasma sheet should be weak compared to that which characterizes the explosive phase. Highest flow speeds in the plasma sheet are in fact observed in the explosive phases of substorms. Synoptic studies of these flows should provide important information on the nature of the dynamics driving them. Since the early days of the space age (e.g. Dungey, 1961; Axford *et al.*, 1965; Dessler, 1968), it has been generally held to be a reasonable working hypothesis that magnetic merging is the basic physical process underlying the substorm explosive phase. (For a general review of magnetic merging models, especially in tail-like geometries, see Vasyliunas, 1975). In the context of the merging model, the direction of the flow in the plasma sheet indicates the direction of the merging region from which it came. It is observed that when a spacecraft is positioned near the midplane of the plasma sheet and in the central portion of the tail, the strongest flows beyond about 18 R_e are directed away from Earth (Hones, 1980; Hayakawa *et al.*, 1982). The implication that a merging region exists earthward of the spacecraft at these times is reinforced by the behavior of the magnetic field, which must reverse its orientation across a merging region. Since Earth's field points northward in the plasma sheet, an interposed merging region would create a southward pointing field at the spacecraft. The rapid flows away from Earth are found to carry southward directed fields (Hayakawa *et al.*, 1982).

Weaker flows, which are more common in the plasma sheet, in some cases clearly do not exhibit the simultaneous field and flow signatures diagnostic of a merging region (Lui, 1980, and references therein; Hayakawa *et al.*, 1982). It is speculated that some other process operates to produce more pervasive and more frequent weaker flows in the plasma sheet than those that are driven by magnetic merging during the explosive phase of the substorm cycle.

Certain features in the data obtained from energetic particle detectors in the magnetotail also seem to require an explanation in terms of magnetic merging. Events occur in association with the explosive phase of substorms in which energetic electrons (> 0.2 MeV) stream along field lines as if away from a more earthward merging region in which they are generated (Bieber and Stone, 1980). These events are found to correlate in detail with rapid flows in the plasma sheet which are directed away from Earth and which carry southward pointing magnetic field (Bieber *et al.*, 1982). The onsets of energetic particle (> 0.3 MeV proton) events are sudden, implying acceleration times as short as one minute or less (Baker *et al.*, 1979; Krimigis and Sarris, 1980). The energy achieved by some particles exceeds by a factor of 20 the typical potential available from the coupling to the solar wind, that is from the equivalent steady state convection potential. The action of a powerful induction electric field in the tail seems to be required (Pellinen and Heikkila, 1978; Baker *et al.*, 1979).

The buildup-and-release model of magnetotail convection is inherently inductive (Boström, 1974). The release process must return magnetic flux at a rate faster than it accumulates under steady solar wind convection in order to discharge the accumulated excess while tailward transport continues. The ratio of the return EMF to the equivalent steady EMF is equal to the ratio of the full substorm cycle period to the duration of the discharge interval of the explosive phase. Since the characteristic lifetime of a substorm cycle is somewhat greater than 1 h, a discharge interval lasting only 1 min would provide an EMF amplification factor of the order of 10^2. The temporal profile of the changes that occur during the discharge interval, however, appears to be longer and more characteristic of an actual inductive discharge, that is a rapid rise and a slower decay (e.g. Boström, 1974). Thus, the highest EMFs would occur nearly simultaneously with the explosive phase onset and might persist for a time of the order of 1 min, but a substantial part of the return would occur over a longer interval at lesser rates.

There remains the problem of identifying the mechanism that can generate a discharge EMF of the order of 10^6 V, which brings the discussion back to magnetic merging. A speculative answer to this question has been given by Galeev *et al.* (1978) in an analysis of time dependent merging based on the assumption of collisionless tearing mode reconnection, which has long been the dissipation process favored for merging in the magnetotail (Coppi *et al.*, 1966; Schindler, 1974). They find that once the mode has attained a sufficient development, explosive nonlinear growth ensues. They obtained 400 kV as a rough estimate of the magnitude of the EMF achievable by this mechanism.

Figure 22 shows a phenomenologically based picture of the flows in the magnetotail and the changes in geometry that occur in the discharge and recovery intervals of the explosive phase of the substorm cycle (from Hones, 1977). Merging is envisioned to begin in the near-Earth part of the plasma sheet (stage 2), proceeding from a state which has already become a much distended tail-like configuration (stage 1). Merging continues until the plasma sheet has been cut through (stage 5), resulting in a detached plasma bubble (a 'plasmoid'). Subsequent merging reconnects and closes lobe field lines that were formerly open. The plasmoid is now propelled tailward in the manner of a slingshot by the field lines that drape around it and stretch tailward to some remote point where they become anchored in the solar wind (stage 8). Energetic ions and electrons, created in the merging EMF, stream away from the merging region on open field lines and form quasi-trapped populations in closed field line regions, including the plasmoid. When the discharge episode has ended and the recovery period begins, the plasma sheet thickens and develops in tailward extent (stage 9). The cycle is now complete, and the system could evolve from stage 9 through a subsequent buildup phase directly to stage 1 to initiate a new cycle.

The major features of the phenomenological model were reproduced in a computer simulation of merging in a tail-like geometry (Birn and Hones, 1981). A three-dimensional, time dependent, resistive MHD code was used. The computation begins with a configuration that is in equilibrium in the absence of resistivity. The initial state is meant to represent the stage in the evolution of the buildup phase where anomalous resistivity is generated by current driven instabilities. (The subsequent development is expected to be independent of whether the dissipation is resistive or collisionless.) The introduction of resistance initiates the resistive tearing mode instability. Merging at a near-Earth merging point ensues, rapid tailward jetting develops together with a large EMF across the merging region, and a plasmoid forms which proceeds to progress down the tail.

A fundamental question concerning convection in the magnetotail remains to be addressed. Why is it inherently time dependent? There are instances of what appear to be intervals of steady-state convection (Pytte *et al.*, 1978) and clear examples of external triggering of the explosive phase of the substorm cycle (e.g. McPherron, 1980). Nevertheless, removal of these cases still leaves the bulk of the events requiring an explanation in terms of an intrinsic internal instability. Cowley and Southwood (1980) showed that if ions are cycled in steady state through a given tail configuration, they can reproduce typical plasma sheet parameters. However, theirs is not a dynamically self-consistent solution, for which an MHD approach would be better suited. Erickson and Wolf (1980) in a qualitative analysis discovered that MHD solutions to the steady-state convection problem have the property of producing much too large a plasma pressure in the near-Earth plasma sheet. The plasma sheet could not be confined by the magnetic field in these solutions. They postulated that time-dependent convection with merging is necessary to avoid the hyperpressure catastrophe. Schindler and Birn (1982) treated the problem

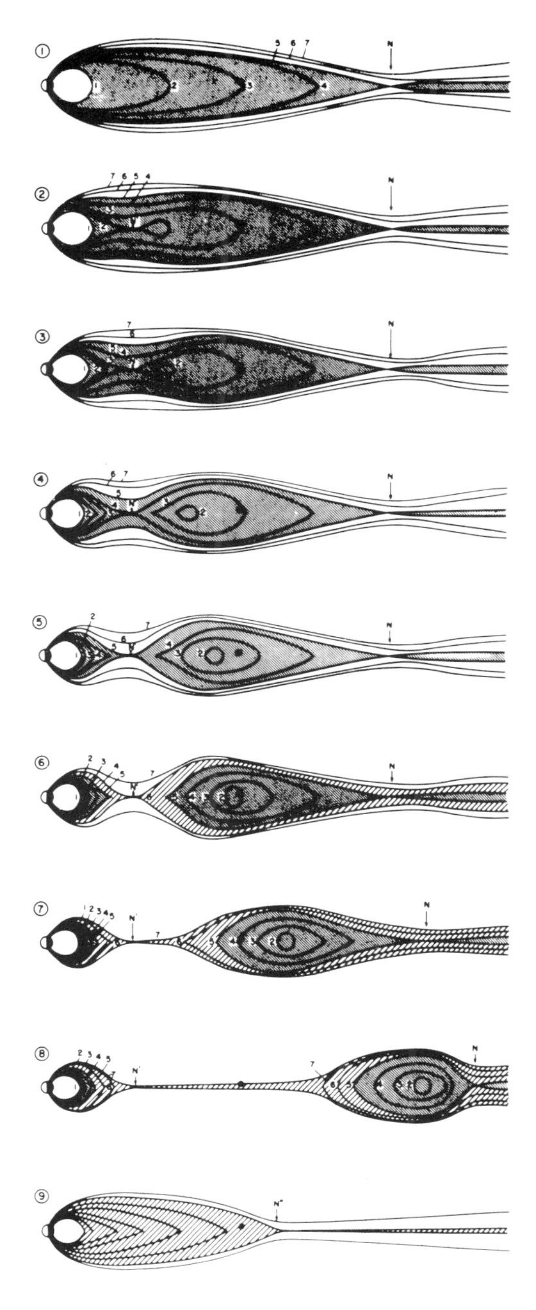

Fig. 22. Sequence of changes of magnetic and plasma configuration of the plasma sheet during a substorm. Five closed field lines (1, 2, 3, 4, and 5) of the presubstorm plasma sheet are depicted as well as two open field lines (6 and 7) that were in the tail lobe before the substorm. Fine hatching delineates the plasma of the presubstorm neutral line N. Coarse hatching delineates plasma populating the newly merged previously open field lines that have entered the merging region at N' from the north and south tail lobes. (From Hones, 1977.)

in a fully self-consistent, two-dimensional MHD calculation. They found that a singular class of steady-state solutions is in principle possible, but these solutions do not match any reasonable boundary conditions with the solar wind. They concluded that time dependent evolution is inevitable. They showed in addition that the evolution proceeds in the manner specified for the buildup phase of the substorm cycle, and that it would eventually bring about the tearing mode instability, regardless of whether the collisionless or the resistive mode is responsible.

6. Magnetosphere Effects on the Ionosphere and Thermosphere

In the previous sections the atmosphere was represented solely by the ionosphere which played the role of an anisotropic electrical conductor carrying currents in response to electric fields presented to it from the magnetosphere. Additionally it played the important role of providing electrical contact between the outer and inner regions of the magnetosphere. This section looks in greater detail at some of the properties of the atmosphere–magnetosphere interface that are affected by or determined by magnetospheric processes. The interested reader will find other aspects of this large subject treated in the review by Banks (1979).

6.1. Low-Latitude Electric Fields and Currents

Some geomagnetic disturbances in the polar regions occur almost simultaneously with disturbances at low latitudes. A prime example is the DP2 disturbance (Nishida *et al.*, 1966). Its pattern appears to be the result of ionospheric Hall currents in the usual convection pattern, as in Figure 15, except that the regions of sunward convection are not confined to the auroral oval but extend through low latitudes to the equator. The disturbance at the equator on the dayside is pronounced, suggesting that the electric field impressed across the polar cap has penetrated to the equatorial ionosphere to drive ionospheric currents there, which form an equatorial electrojet because of the special geometry of the geomagnetic field at the equator. That is, the DP2 disturbance at low latitudes appears to result mainly from local ionospheric currents driven by electric fields in the low-latitude ionosphere rather than from remote magnetospheric Birkeland currents with associated ionospheric Pedersen and Hall currents confined to the polar regions. Other forms of polar or auroral disturbances also are accompanied by nearly simultaneous disturbances from the equatorial electrojet (e.g. Akasofu and Chapman, 1963; Onwumechili *et al.*, 1973; Araki, 1977).

Ionospheric electric fields at low latitudes have been sensed indirectly by incoherent-scatter radar measurements. These measurements are supportive of penetration of auroral electric fields, since the low-latitude fields are correlated both with the auroral fields themselves and with the z-component of the IMF, which controls the potential which drives auroral fields (e.g. Blanc, 1978; Fejer *et al.*, 1979; Gonzales *et al.*, 1979). Also consistent with penetration of polar or auroral electric fields to low latitudes are the correlations between the z-component of the IMF and magnetic disturbances from the equatorial electrojet (e.g. Nishida, 1968; Patel, 1978).

What is surprising about the penetration of polar electric fields to low latitudes is, first, that if the signal propagates as a magnetohydrodynamic wave through the ionized medium, it should take a time of the order of several tens of minutes to reach the equator, whereas the polar and low-latitude disturbances are instead observed to occur almost simultaneously. Second, the process violates the shielding of Region 1 electric fields by Region 2 Birkeland currents, as discussed in Section 5.2. The first point has been addressed by Kikuchi *et al.* (1978) and Kikuchi and Araki (1979a, b). These authors show that penetration can be virtually instantaneous if the space between Earth's surface and the ionosphere is treated as a waveguide. The second point has been addressed by many authors (e.g. Vasyliunas, 1972; Jaggi and Wolf, 1973; Southwood, 1977; Harel *et al.*, 1981b). The condition of perfect shielding applies for steady-state convection. In a time-dependent situation, the response time of the Region 2 current to an increase in Region 1 current is expected to be about 1 h. Thus the DP2 disturbance, for example, which occurs on a time-scale of about 1 h, may be interpreted as penetration of the polar cap electric field before the buildup of shielding currents (e.g. Lyatsky *et al.*, 1974). Kelley *et al.* (1979) present observations that are consistent with the occurrence of a reverse situation in which the polar cap electric field decreases in response to a northward turning of the IMF, and the oppositely directed electric field associated with the Region 2 currents, which decreases on a slower time-scale, then penetrates to low latitudes.

It is noted here that although some electric field perturbations are observed at low latitudes in conjunction with substorms (e.g. Gonzales *et al.*, 1979), the well-known substorm magnetic disturbance field at low latitudes (Silsbee and Vestine, 1942) is not interpreted as the result of overhead ionospheric currents. During substorms the polar cap electric field increases rapidly, so that temporarily there is no shielding, as in the case of DP2. However, because the auroral oval conductivity is high, the ionospheric currents concentrate there rather than flow down to low latitudes. The magnetic disturbance then results from the more remote Region 1 Birkeland currents and associated ionospheric currents in the auroral region (Vasyliunas, 1970; Crooker and Siscoe, 1981).

6.2. Parallel Electric Fields

On magnetic field lines where auroral particle precipitation occurs, at altitudes of several thousand kilometres, the electric field is often observed to have a component parallel to the magnetic field (Mozer *et al.*, 1977). A statistical study by Mozer and Torbert (1980) showed that the average potential drop along magnetic field lines at altitudes below 1.5 R_e in the auroral zone is 3 kV. A conceptual model proposed by Swift *et al.* (1976) for the electrical potential associated with the region of parallel electric fields is shown in Figure 23, from Swift (1979). The diagram is a vertical cross-section perpendicular to an auroral arc. It extends above the ionosphere along a vertical magnetic field. The electric fields, directed perpendicular to the equipotentials, are strongest at the highest altitudes. Between the heights of z_0 and z_1, the electric field is parallel to the geomagnetic field along the dashed line in the center of the figure at $x = 0$, then gradually changes direction away from the center until it is directed perpendicular to the geomagnetic field at $|x| > a$. The potential is such that electrons are accelerated downward and ions upward. This gives rise to the characteristic 'inverted V' signatures in the charged particle data obtained from spacecraft traversing such features.

The existence of parallel potential drops of the order of 10% of the convection

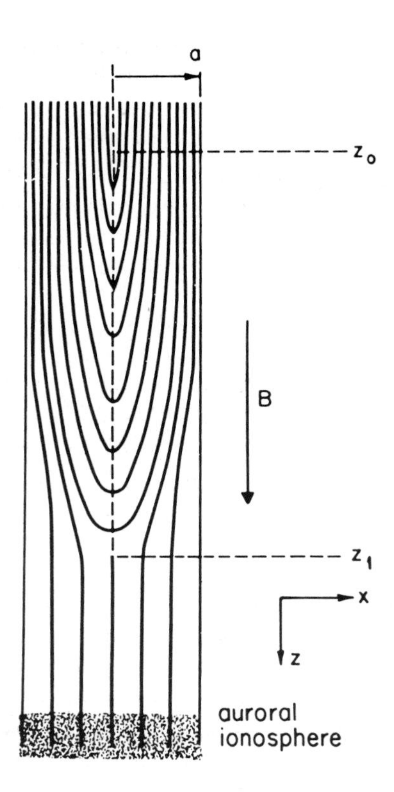

Fig. 23. Vertical cross-section perpendicular to an auroral arc of model equipotential surfaces above the arc. The electric field is perpendicular to the magnetic field B at $|x| > a$ but has a component parallel to B between heights z_0 and z_1 at $|x| < a$. (From Swift, 1979.)

potential is of great interest because of its theoretical implications and because it causes electrons to precipitate with elevated energies, which has important consequences for the upper atmosphere. Parallel electric fields are not an immediate consequence of magnetohydrodynamic convection, in which the electric field $\mathbf{E} = -\mathbf{V} \times \mathbf{B}$ yields $\mathbf{E} \cdot \mathbf{B} = 0$. However, as discussed earlier, convection in the presence of the conducting ionosphere gives rise to parallel electric currents, the strength of which is governed by the convection potential and the ionospheric conductivity. Electron precipitation from the magnetosphere to the ionosphere and electron upflow from the ionosphere to the magnetosphere are invoked to provide the parallel currents required by convection.

Electron precipitation must act against the magnetic mirror force, which confines the solid angle available for precipitating electrons to a cone around the magnetic field line with a half-angle α given by

$$\sin^2 \alpha = \frac{B(o)}{B(i)} \left[1 + \frac{e\Delta\phi}{W_o} \right]$$

where $B(o)$ is the field strength at any point (labelled o) along the current-carrying field line, $B(i)$ is the field strength in the ionosphere, $\Delta\phi$ is the potential drop between o and i, and W_o is the kinetic energy of the electrons at point o. At auroral latitudes the ratio $B(o)/B(i)$ is typically 1/2000, where $B(o)$ is evaluated in the equatorial plane.

Parallel currents are required by convection in order to maintain charge balance. If they are larger than can be provided by precipitation with $\Delta\phi = 0$, the resulting charge imbalance creates a $\Delta\phi$ parallel to the magnetic field, in the right sense to increase the size of α and, hence, to increase the current that can be carried by precipitation. Explicit models using convection electric fields predict parallel potential drops and the characteristic size of inverted V features which are in agreement with observations (Lyons, 1981; Chiu *et al.*, 1981).

6.3. Ionospheric Outflow

As described earlier, the ions and electrons that make up the ionosphere expand outward along the geomagnetic field. At low and middle latitudes, where the field lines are closed and relatively static, the plasma flow produces the quasipermanent plasma body, the plasmasphere. At higher latitudes convection carries upflowing ions away, and at even higher latitudes the outflow is unconstrained and continues as a polar wind to remote regions down the tail, presumably eventually to merge into the circumfluent solar wind. The cause of the outflow is solar UV heating of the ionosphere and upper atmosphere. The outflow is a manifestation of a pressure gradient between the ionospheric and magnetospheric portions of a given magnetic flux tube. There are interesting complications relating to the physics of this flow, such as the role of the ambipolar electric field in accelerating primarily the lighter ions and the saturation of the outflow by H^+–O^+ collisions (Banks, 1979), but these are not magnetospheric effects.

The magnetosphere affects the density of the ionosphere at outflow altitudes in the ways already mentioned – namely, through providing convection and through determining topology. With regard to the first of these, one further effect should be mentioned. The convection and rotational motions combine to produce a region of nearly stagnated flow in the dusk local time sector at latitudes between the plasmapause and the auroral zone. The ionosphere in this region spends a relatively long interval of time in a condition of low ion production and high ion loss. Thus a quasipermanent region of low ion density called the mid-latitude trough is produced.

The parallel electric fields and the plasma waves associated with the auroral zones also give rise to outflow of ionospheric ions. There is an important distinction between the auroral zone outflow and the ubiquitous polar wind outflow. The former involves heavy ions, principally O^+, as well as light ions, and the energy of the ions can reach kilovolt values, corresponding to the magnitude of the parallel potential drop. (See the recent article by Gorney *et al.*, 1981, and references therein.) Two types of auroral zone ion outflow, referred to as ion beams and ion conics, are observed. Ion beams appear to result simply from the ionospheric reaction to the parallel potential drop to which it is subjected in the auroral regions. Ion beams are observed mainly above 5000 km altitude, which suggests that most of the potential drop occurs above that altitude. Also, they are observed primarily during times of magnetic disturbance, when they are largely confined to the dusk sector, in conformity with the location of the region of large parallel potentials.

Ion conics are characterized by a conical distribution relative to the magnetic field direction of the intensity of the upflowing ion flux. Ion beams by contrast have a maximum intensity in the direction parallel to the field. Ion conics are observed at lower altitudes than are ion beams, and during both quiet and disturbed times. They occur at all local times, but preferentially in the noon local time sector during quiet times. Ion conics are thought to result from ion heating by the absorption of ion cyclotron waves in the ionosphere. This process gives an ion a large velocity perpendicular to the magnetic field direction in the ionosphere. The component of elevated velocity becomes more field-aligned as the ion moves outward into regions of reduced field strength, thereby producing the characteristic conic signature.

The outflow of moderate energy heavy ions into the magnetosphere provides particles and energy for the ring current and alters its composition. An important, still unanswered question concerns the relative amounts of heavy ions from the ionosphere and light ions from the solar wind that make up the ring current during magnetically disturbed times (Hultqvist, 1982).

6.4. Effects on Thermosphere

The thermosphere receives energy from the magnetosphere in several ways. The imposition of electric fields causes ionospheric Pedersen currents (Section 5.1) which result in Joule heating. As the ions respond to electric fields and collide with neutral particles, in addition to carrying current, they also transfer momentum to the neutrals and cause ion drag winds and consequent viscous dissipation. At high altitudes where the convection electric field is strongest, ion drag winds clearly dominate over winds driven by radiative heating. Particle precipitation adds energy to the thermosphere mostly by heating. Some energy from precipitating particles is used to ionize and excite neutral particles, which, in turn, radiate the energy as the aurora. Recent observations suggest that an additional means of thermospheric heating may be through unstable plasma waves (Schlegel and St-Maurice, 1981). Waves also may act to heat the ionosphere in the vicinity of the plasmapause. At high latitudes the total amount of energy deposited in the thermosphere from magnetospheric sources is comparable to that supplied by extreme ultraviolet radiation, which is the global principal source of heating and ionization in the thermosphere. (See review by Mayr and Harris, 1979.)

Heat from magnetospheric sources produces global effects in the thermosphere, especially during geomagnetic storms. Figure 24, from Roble (1977), shows model thermospheric meridional wind patterns for quiet, average, and active periods. During quiet times the circulation is determined by EUV heating. During substorms the concentrated heating at high latitudes causes a complete reversal of the circulation pattern.

Heating and winds from magnetospheric energy deposition at high latitudes produce global changes in thermospheric composition, and these changes are intimately related to ionospheric storms, which are patterned changes in density structure. (See reviews by Mayr *et al.*, 1978; Prölss, 1980). More locally at high latitudes, observed ion and electron density troughs are ascribed to the effects of compositional changes caused by heating from ion drag winds in regions where the convection electric field is strong (e.g. Schunk *et al.*, 1976; Knudsen *et al.*, 1977). Magnetospheric energy deposition at high latitudes also generates large-scale gravity waves which propagate to lower latitudes; these waves interact with the ionosphere and cause traveling ionospheric disturbances. (See review by Richmond and Roble, 1979.)

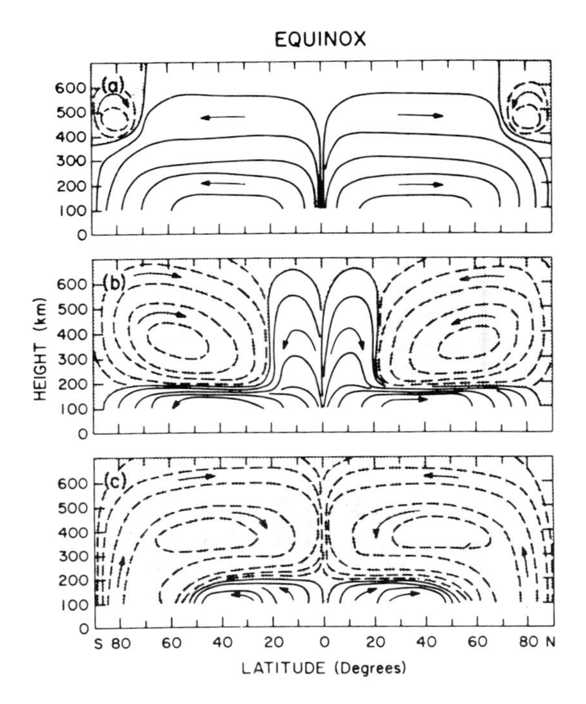

Fig. 24. Vertical cross-section of model meridional circulation in the thermosphere during equinox for various levels of geomagnetic activity: (a) extremely quiet; (b) average; (c) active. (From Roble, 1977.)

7. Middle and Lower Atmosphere

Although there is no known mechanism by which energy can flow from the thermosphere and ionosphere downward to the middle and lower atmosphere, highly energetic electrons in the outer radiation belt in the magnetosphere precipitate through the thermosphere into the middle atmosphere during geomagnetically active periods. It appears that these trapped electrons gain their relativistic energies and precipitate as the result of solar wind driven magnetospheric convection and related wave processes and thus are a form of solar wind energy deposition in the middle atmosphere. Their total energy input is small compared to the radiative energy input; but, like energetic solar protons discussed by Reid in Chapter 22, they may cause a significant amount of ozone depletion. (See review by Thorne, 1980.) Also, like solar protons, it is possible that relativistic electron precipitation may have some effect on weather in the lower atmosphere, although such possibilities are highly speculative.

Among the many Sun–weather correlations reported in the literature (see reviews by King, 1975; Pittock, 1978; Siscoe, 1978), there is one which is directly related to the solar wind. Wilcox *et al.* (1974, 1976) have found that decreases in a vorticity area index correlate with times when the heliospheric current sheet or sector boundary (Section 3.1) is carried past Earth by the solar wind. The vorticity area index measures the intensity and extent of low-pressure troughs or 'bad weather' systems in the lower atmosphere. The correlation appears to be firm throughout the 1963–1973 period, although it seems to have disappeared in the succeeding 1974–1977 period (Williams and Gerety, 1978). Since energy transfer from solar wind plasma to the lower atmosphere is highly unlikely to occur, explanations for the correlation are sought in less direct mechanisms. Specifically, explanations in terms of the modulation of galactic cosmic rays by the solar wind have been proposed. Sector boundaries are associated with the leading edges of solar wind streams where the magnetic field is highly compressed and cosmic ray intensity is depressed (Barouch and Burlaga, 1975). Also, there is a maximum in cosmic ray intensity one day prior to sector boundary crossings (Lethbridge, 1981). Thus, since cosmic rays are ordered by solar wind structure and penetrate directly to the lower atmosphere, it is possible that they may alter some feature of the lower atmosphere, such as the fair weather electric field, which, in turn, may modulate weather systems and lead to the observed correlation (Markson and Muir, 1980; Lethbridge, 1981).

A related correlation between surface pressure in the lower atmosphere near the Gulf of Alaska and the polarity of the IMF has been found by Rostoker and Sharma (1980). As discussed in Section 5.1, the y-component and, hence, the polarity of the IMF controls the location of concentrated convection electric fields on the ionosphere at high latitudes. Following the approach of Hines (1974), these authors suggest that the electric fields may cause small wind perturbations which may trigger instabilities leading to changes in the planetary wave configuration and the observed correlation.

The suggested mechanisms mentioned above for coupling between weather and relativistic electron precipitation, solar sector structure, and IMF polarity, are discussed by Reid (Sections 5.2 and 5.3) relative to solar particles.

Acknowledgements

This work was supported in part by the National Science Foundation under grants ATM 79-15251 and ATM 81-20455.

References

Akasofu, S.-I.: 1977, *Physics of Magnetospheric Substorms*, D. Reidel, Dordrecht.
Akasofu, S.-I.: 1979, *Planet. Space Sci.* **27**, 425.
Akasofu, S.-I.: 1980, in S.-I. Akasofu (ed.), *Dynamics of the Magnetosphere*, D. Reidel, Dordrecht, p. 447.
Akasofu, S.-I. and Chapman, S.: 1963, *J. Geophys. Res.* **68**, 2375.
Araki, T.: 1977, *Planet. Space Sci.* **25**, 373.
Arthur, C. W., McPherron, R. L., and Hughes, W. J.: 1977, *J. Geophys. Res.* **82**, 1149.
Atkinson, G.: 1978, *J. Geophys. Res.* **83**, 1089.
Atkinson, G. and Hutchison, D.: 1978, *J. Geophys. Res.* **83**, 725.
Axford, W. I.: 1964, *Planet. Space Sci.* **12**, 45.
Axford, W. I. and Hines, C. O.: 1961, *Can. J. Phys.* **39**, 1433.
Axford, W. I., Petschek, H. E., and Siscoe, G. L.: 1965, *J. Geophys. Res.* **70**, 1231.
Baker, D. N., Belian, R. D., Higbie, P. R., and Hones, E. W., Jr: 1979, *J. Gelphys. Res.* **84**, 7138.
Baker, D. N., Hones, E. W., Jr, Payne, J. B., and Feldman, W. C.: 1981a, *Geophys. Res. Lett.* **8**, 179.
Baker, D. N., Hones, E. W., Jr, Higbie, P. R., Belian, R. D., and Stauning, P.: 1981b, *J. Geophys. Res.* **86**, 8941.

Ballif, J. R., Jones, D. E., Coleman, P. J., Jr, Davis, L., Jr, and Smith, E. J.: 1967, *J. Geophys. Res.* **72**, 4357.
Ballif, J. R., Jones, D. E., and Coleman, P. J., Jr.: 1969, *J. Geophys. Res.* **74**, 2289.
Bame, S. J., Asbridge, J. R., Feldman, W. C., Felthauser, H. E., and Gosling, J. T.: 1977, *J. Geophys. Res.* **82**, 173.
Banks, P. M.: 1979, in C. F. Kennel, L. J. Lanzerotti and E. N. Parker (eds), *Solar System Plasma Physics*, Vol. II, North-Holland, New York, p. 57.
Barouch, E. and Burlaga, L. F.: 1975, *J. Geophys. Res.* **80**, 449.
Berthelier, A.: 1976, *J. Geophys. Res.* **81**, 4546.
Bieber, J. W. and Stone, E. C.: 1980, *Geophys. Res. Lett.* 7, 945.
Bieber, J. W., Stone, E. C., Hones, E. W., Jr, Baker, D. N., and Bame, S. J.: 1982, *Geophys. Res. Lett.* 9, 664.
Birn, J. and Hones, E. W., Jr,: 1981, *J. Geophys. Res.* **86**, 6802.
Blanc, M.: 1978, *Geophys. Res. Lett.* **5**, 203.
Boller, B. R. and Stolov, H. L.: 1970, *J. Geophys. Res.* **75**, 6073.
Boström, R.: 1974, in B. M. McCormac (ed.), *Magnetospheric Physics*, D. Reidel, Dordrecht, p. 45.
Brice, N. M.: 1967, *J. Geophys. Res.* **72**, 5193.
Burch, J. L.: 1973, *J. Geophys. Res.* **78**, 1047.
Burch, J. L.: 1974, *Rev. Geophys. Space Phys.* **12**, 363.
Burke, W. J., Kelley, M. C., Sagalyn, R. C., Smiddy, M., and Lai, S. T.: 1979, *Geophys. Res. Lett.* **6**, 21.
Burlaga, L. F.: 1975, *Space Sci. Rev.* **17**, 327.
Burlaga, L. F. and Lepping, R. P.: 1977, *Planet. Space Sci.* **25**, 1151.
Burton, R. K., McPherron, R. L., and Russell, C. T.: 1975, *Science* **189**, 717.
Caan, M. N., McPherron, R. L., and Russell, C. T.: 1975, *J. Geophys. Res.* **80**, 191.
Cauffman, D. P. and Gurnett, D. A.: 1972, *Space Sci. Rev.* **13**, 369.
Chapman, S. and Ferraro, V. C. A.: 1931, *Terr. Mag. Atmos. Elec.* **36**, 171.
Chappell, C. R., Harris, K. K., and Sharp, G. W.: 1970, *J. Geophys. Res.* **75**, 50.
Chen, A. J.: 1970, *J. Geophys. Res.* **75**, 2458.
Chernosky, E. J.: 1966, *J. Geophys. Res.* **71**, 965.
Chiu, Y. T., Newman, A. L., and Cornwall, J. M.: 1981, *J. Geophys. Res.* **86**, 10029.
Clauer, C. R., McPherron, R. L., Searls, C., and Kivelson, M. G.: 1981, *Geophys. Res. Lett.* **8**, 915.
Coleman, P. J., Jr: 1970, *Cosmic Electrodyn.* **1**, 145.
Coppi, B., Laval, G., and Pellat, R.: 1966, *Phys. Rev. Lett.* **16**, 1207.
Cowley, S. W. H.: 1982, *Rev. Geophys. Space Phys.* **20**, 531.
Cowley, S. W. H. and Southwood, D. J.: 1980, *Geophys. Res. Lett.* 7, 833.
Crooker, N. U.: 1977, *J. Geophys. Res.* **82**, 3629.
Crooker, N. U.: 1979, *J. Geophys. Res.* **84**, 951.
Crooker, N. U.: 1980, in S.-I. Akasofu (ed.), *Dynamics of the Magnetosphere*, D. Reidel, Dordrecht, p. 101.
Crooker, N. U. and Siscoe, G. L.: 1981, *J. Geophys. Res.* **86**, 11201.
Crooker, N. U., Feynman, J., and Gosling, J. T.: 1977, *J. Geophys. Res.* **82**, 1933.
Dessler, A. J.: 1968, *J. Geophys. Res.* **73**, 209.
Dungey, J. W.: 1961, *Phys. Rev. Lett.* **6**, 47.
Eastman, T. E. and Hones, E. W., Jr: 1979, *J. Geophys. Res.* **84**, 2019.
Erickson, G. W. and Wolf, R. A.: 1980, *Geophys. Res. Lett.* 7, 897.
Eviatar, A. and Wolf, R. A.: 1968, *J. Geophys. Res.* **73**, 5561.
Fairfield, D. H.: 1976, in B. M. McCormac (ed.), *Magnetospheric Particles and Fields*, D. Reidel, Dordrecht, p. 67.
Fairfield, D. H. and Cahill, L. J., Jr: 1966, *J. Geophys. Res.* **71**, 155.
Fejer, B. G., Gonzales, C. A., Farley, D. T., Kelley, M. C., and Woodman, R. F.: 1979, *J. Geophys. Res.* **84**, 5797.
Feynman, J.: 1980, *Geophys. Res. Lett.* 7, 971.
Feynman, J. and Crooker, N. U.: 1978, *Nature* **275**, 626.
Feynman, J. and Silverman, S. M.: 1980, *J. Geophys. Res.* **85**, 2991.

Fraser-Smith, A. C.: 1972, *J. Geophys. Res.* 77, 4209.
Friis-Christensen, E. and Wilhjelm, J.: 1975, *J. Geophys. Res.* **80**, 1248.
Galeev, A. A., Coroniti, F. V., and Ashour-Abdalla, M.: 1978, *Geophys. Res. Lett.* **5**, 707.
Garrett, H. B.: 1974, *Planet. Space Sci.* **22**, 111.
Garrett, H. B., Dessler, A. J., and Hill, T. W.: 1974, *J. Geophys. Res.* **79**, 4603.
Gonzales, C. A., Kelley, M. C., Fejer, B. G., Vickrey, J. F., and Woodman, R. F.: 1979, *J. Geophys. Res.* **84**, 5803.
Gonzalez, W. D. and Mozer, F. S.: 1974, *J. Geophys. Res.* **79**, 4186.
Gorney, D. J., Clarke, A., Croley, D., Fennell, J., Luhmann, J., and Mizera, P.: 1981, *J. Geophys. Res.* **86**, 83.
Gosling, J. T., Asbridge, J. R., and Bame, S. J.: 1977, *J. Geophys. Res.* **82**, 3311.
Gosling, J. T., Asbridge, J. R., Bame, S. J., and Feldman, W. C.: 1976b, *J. Geophys. Res.* **81**, 5061.
Gosling, J. T., Asbridge, J. R., Bame, S. J., Feldman, W. C., and Zwickl, R. D.: 1980, *J. Geophys. Res.* **85**, 3431.
Gosling, J. T., Hildner, E., MacQueen, R. M., Munro, R. H., Poland, A. I., and Ross, C. L.: 1974, *J. Geophys. Res.* **79**, 4581.
Gosling, J. T., Hundhausen, A. J., and Bame, S. J.: 1976a, *J. Geophys. Res.* **81**, 2111.
Greenstadt, E. W., McPherron, R. L., and Takahashi, K.: 1980, *J. Geomag, Geoelectr.* **32**, SII89.
Greenstadt, E. W., Singer, H. J., Russell, C. T., and Olson, J. V.: 1979, *J. Geophys. Res.* **84**, 527.
Gul'elmi, A. V.: 1974, *Space Sci. Rev.* **16**, 331.
Gul'yel'mi, A. V. and Bol'shakova: 1973, *Geomag. Aeron.* **13**, 459.
Haerendel, G., Paschmann, G., Sckopke, N., Rosenbauer, H., and Hedgecock, P. C.: 1978, *J. Geophys. Res.* **83**, 3195.
Haerendel, G.: 1980, in C. S. Deehr and J. A. Holtet (eds), *Exploration of the Polar Upper Atmosphere*, Reidel, Dordrecht, p. 219.
Hansen, R. T., Hansen, S. F., and Sawyer, C.: 1976, *Planet. Space Sci.* **24**, 381.
Harel, M., Wolf, R. A., and Reiff, P. H.: 1979, in W. P. Olson (ed.), *Quantitative Modeling of Magnetospheric Processes*, AGU, Washington, D.C., p. 499.
Harel, M., Wolf, R. A., Reiff, P. H., Spiro, R. W., Burke, W. J., Rich, F. J., and Smiddy, M.: 1981a, *J. Geophys. Res.* **86**, 2217.
Harel, M., Wolf, R. A., Spiro, R. W., Reiff, P. H., and Chen, C.-K.: 1981b, *J. Geophys. Res.* **86**, 2242.
Hayakawa, H., Nishida, A., Hones, E. W., Jr, and Bame, S. J.: 1982, *J. Geophys. Res.* **87**, 277.
Hedgecock, P. C.: 1975, *Solar Phys.* **42**, 497.
Heelis, R. A., Hanson, W. B., and Burch, J. L.: 1976, *J. Geophys. Res.* **81**, 3803.
Heelis, R. A., Winningham, J. D., Hanson, W. B., and Burch, J. L.: 1980, *J. Geophys. Res.* **85**, 3315.
Heikkila, W. J.: 1978, *J. Geophys. Res.* **83**, 1071.
Heppner, J. P.: 1972, in E. R. Dyer (ed.), *Critical Problems of Magnetospheric Physics*, National Academy of Sciences, Washington, D. C., p. 107.
Hill, T. W.: 1979, in B. Battrick (ed.), *Magnetospheric Boundary Layers*, ESA SP-148, European Space Agency, Paris, p. 325.
Hines, C. O.: 1974, *J. Atmos. Sci.* **31**, 589.
Hirshberg, J. and Colburn, D. S.: 1969, *Planet. Space Sci.* **17**, 1183.
Hones, E. W., Jr: 1977, *J. Geophys. Res.* **82**, 5633.
Hones, E. W., Jr.: 1980: in S.-I. Akasofu (ed.), *Dynamics of the Magnetosphere*, D. Reidel, Dordrecht, p. 545.
Hones, E. W., Jr, Akasofu, S.-I., Perreault, P., Bame, S. J., and Singer, S.: 1970, *J. Geophys. Res.* **75**, 7060.
Hultqvist, B.: 1982, *Rev. Geophys. Space Phys.* **20**, 589.
Hundhausen, A. J.: 1972, *Coronal Expansion and Solar Wind*, Springer-Verlag, New York, p. 135.
Hundhausen, A. J.: 1977, in J. B. Zirker (ed.), *Coronal Holes and High Speed Streams*, Colorado Assoc. Univ. Press, Boulder, p. 225.
Hundhausen, A. J., Bame, S. J., and Montgomery, M. D.: 1971, *J. Geophys. Res.* **76**, 5145.
Iijima, T. and Potemra, T. A.: 1976, *J. Geophys. Res.* **81**, 2165.
Iijima, T. and Potemra, T. A.: 1978, *J. Geophys. Res.* **83**, 599.
Jaggi, R. K. and Wolf, R. A.: 1973, *J. Geophys. Res.* 78, 2852.

Jorgensen, T. S., Friis-Christensen, E., and Wilhjelm, J.: 1972, *J. Geophys. Res.* **77**, 1976.
Joselyn, J. A. and McIntosh, P. S.: 1981, *J. Geophys. Res.* **86**, 4555.
Karty, J. L., Chen, C.-K., Wolf, R. A., Harel, M., and Spiro, R. W.: 1982, *J. Geophys. Res.* **87**, 777.
Kavanaugh, L. D., Jr, Freeman, J. W., Jr, and Chen, A. J.: 1968, *J. Geophys. Res.* **73**, 5511.
Kelley, M. C., Fejer, B. G., and Gonzales, C. A.: 1979, *Geophys. Res. Lett.* **6**, 301.
Kikuchi, T. and Araki, T.: 1979a, *J. Atmos. Terr. Phys.* **41**, 917.
Kikuchi, T. and Araki, T.: 1979b. *J. Atmos. Terr. Phys.* **41**, 927.
Kikuchi, T., Araki, T., Maeda, H., and Maekawa, K.: 1978, *Nature* **273**, 650.
King, J. H.: 1977, *Interplanetary Medium Data Book*, Rept NSSDC 77-04, NASA/Goddard Space Flight Center, Greenbelt, MD.
King, J. H.: 1979, *J. Geophys. Res.* **84**, 5938.
King, J. W.: 1975, *Astronaut. Aeronaut.* **13**, 10.
Kivelson, M. G., Kaye, S. M., and Southwood, D. J.: 1980, in S.-I. Akasofu (ed.), *Dynamics of the Magnetosphere*, D. Reidel, Dordrecht, p. 385.
Knudsen, W. C., Banks, P. M., Winningham, J. D., and Klumpar, D. M.: 1977, *J. Geophys. Res.* **82**, 4784.
Krimigis, S. M. and Sarris, E. T.: 1980, in S.-I. Akasofu (ed.), *Dynamics of the Magnetosphere*, D. Reidel, Dordrecht, p. 599.
Lethbridge, M. D.: 1981, *Geophys. Res. Lett.* 8, 521.
Levy, R. H., Petschek, H. E., and Siscoe, G. L.: 1964, *AIAA J.* **2**, 2065.
Lui, A. T. Y.: 1980, in S.-I. Akasofu (ed.), *Dynamics of the Magnetosphere*, D. Reidel, Dordrecht, p. 563.
Lyatsky, W. B., Maltsev, Yu. P., and Leontyev, S. V.: 1974, *Planet. Space Sci.* **22**. 1231.
Lyons, L. R.: 1981, in S.-I. Akasofu and J. R. Kan (eds), *Physics of Auroral Arc Formation*, AGU, Washington, D. C., p. 252.
Maezawa, K.: 1975, *J. Geophys. Res.* **80**, 3543.
Maezawa, K.: 1976, *J. Geophys. Res.* **81**, 2289.
Maezawa, K.: 1979, in W. P. Olson (ed.), *Quantitative Modeling of Magnetospheric Processes*, AGU, Washington, D.C., p. 436.
Markson, R. and Muir, M.: 1980, *Science* **208**, 979.
Mayaud, P. N.: 1970, *Ann. Geophys.* **26**, 313.
Mayaud, P. N.: 1975, *J. Geophys. Res.* **80**, 111.
Mayr, H. G. and Harris, I.: 1979, *Rev. Geophys. Space Phys.* **17**, 492.
Mayr, H. G., Harris, I., and Spencer, N. W.: 1978, *Rev. Geophys. Space Phys.* **16**, 539.
McIntosh, D. H.: 1959, *Phil. Trans. Roy. Soc. London* **A251**, 525.
McPherron, R. L.: 1974, *EOS* **55**, 994.
McPherron, R. L.: 1979, *Rev. Geophys. Space Phys.* **17**, 657.
McPherron, R. L.: 1980, in S.-I. Akasofu (ed.), *Dynamics of the Magnetosphere*, D. Reidel, Dordrecht, p. 631.
McPherron, R. L., Russell, C. T., and Aubry, M. P.: 1973, *J. Geophys. Res.* **78**, 3131.
Meng, C.-I. and Colburn, D. S.: 1974, *J. Geophys. Res.* **79**, 1831.
Mozer, F. S. and Torbert, R. B.: 1980, *Geophys. Res. Lett.* **7**, 219.
Mozer, F. S., Carlson, C. N., Hudson, M. K., Torbert, R. B., Parady, B., Yatteau, J., and Kelley, M. C.: 1977, *Phys. Rev. Lett.* **38**, 292.
Mozer, F. S., Gonzalez, W. D., Bogott, F., Kelley, M. C., and Schutz, S.: 1974, *J. Geophys. Res.* **79**, 56.
Munro, R. H., Gosling, J. T., Hildner, E., MacQueen, R. M., Poland, A. I., and Ross, C. L.: 1979, *Solar Phys.* **61**, 201.
Murayama, T.: 1974, *J. Geophys. Res.* **79**, 297.
Murayama, T., Aoki, T., Nakai, H., and Hakamada, K.: 1980, *Planet. Space Sci.* **28**, 803.
Neupert, W. M. and Pizzo, V.: 1974, *J. Geophys. Res.* **79**, 3701.
Nishida, A.: 1966, *J. Geophys. Res.* **71**, 5669.
Nishida, A.: 1968, *J. Geophys. Res.* **73**, 5549.
Nishida, A. and Maezawa, K.: 1971, *J. Geophys. Res.* **76**, 2254.
Nishida, A., Iwasaki, N., and Nagata, T.: 1966, *Ann. Geophys.* **22**, 478.

Olson, W. P. and Pfitzer, K. A.: 1974, *J. Geophys. Res.* **79**, 3739.

Onwumechili, A., Kawasaki, K., and Akasofu, S.-I.: 1973, *Planet. Space Sci.* **21**, 1.

Parker, E. N.: 1958, *Astrophys. J.* **128**, 664.

Paschmann, G., Haerendel, G., Papamastorakis, I., Sckopke, N., Bame, S. J., Gosling, J. T., and Russell, C. T.: 1982, *J. Geophys. Res.* **87**, 2159.

Paschmann, G., Sonnerup, B. U. Ö., Papamastorakis, I., Sckopke, N., Haerendel, G., Bame, S. J., Asbridge, J. R., Gosling, J. T., Russell, C. T., and Elphic, R. C.: 1979, *Nature* **282**, 243.

Patel, V. L.: 1978, *J. Geophys. Res.* **83**, 2137.

Pellinen, R. J. and Heikkila, W. J.: 1978, *J. Geophys. Res.* **83**, 1544.

Petschek, H. E.: 1964, in W. N. Hess (ed.), *The Physics of Solar Flares*, NASA SP-50, Washington, D.C., p. 425.

Piddington, J. H.: 1960: *J. Geophys. Res.* **65**, 83.

Pilipp, W. and Morfill, G.: 1976, in B. M. McCormac (ed.), *Magnetospheric Particles and Fields*, D. Reidel, Dordrecht, p. 55.

Pittock, A. B.: 1978, *Rev. Geophys. Space Phys.* **16**, 400.

Pizzo, V. J.: 1980, *J. Geophys. Res.* **85**, 727.

Pizzo, V. J.: 1982, *J. Geophys. Res.* **87**, 4374.

Prölss, G. W.: 1980, *Rev. Geophys. Space Phys.* **18**, 183.

Pytte, T. and West, H. I., Jr: 1978, *J. Geophys. Res.* **83**, 3791.

Pytte, T., McPherron, R. L., Hones, E. W., Jr, and West, H. I., Jr: 1978, *J. Geophys. Res.* **83**, 663.

Quest, K. B. and Coroniti, F. V.: 1981, *J. Geophys. Res.* **86**, 3289.

Reiff, P. H., Spiro, R. W., and Hill, T. W.: 1981, *J. Geophys. Res.* **86**, 7639.

Richmond, A. D. and Roble, R. G.: 1979, *J. Atmos. Terr. Phys.* **41**, 841.

Roble, R. G.: 1977, in *Studies in Geophysics, The Upper Atmosphere and Magnetosphere*, National Academy of Sciences, Washington, D.C., p. 57.

Rosenbauer, H., Grunwaldt, H., Montgomery, M. D., Paschmann, G., and Sckopke, N.: 1975, *J. Geophys. Res.* **80**, 2723.

Rosenberg, R. L. and Coleman, P. J., Jr: 1969, *J. Geophys. Res.* **74**, 5611.

Rostoker, G. and Sharma, R. P.: 1980, *Can. J. Phys.* **58**, 255.

Russell, C. T.: 1974, *Geophys. Res. Lett.* **1**, 11.

Russell, C. T.: 1975, *Solar Phys.* **42**, 259.

Russell, C. T. and Atkinson, G.: 1973, *J. Geophys. Res.* **78**, 4001.

Russell, C. T. and Elphic, R. C.: 1979, *Geophys. Res. Lett.* **6**, 33.

Russell, C. T. and Fleming, B. K.: 1976, *J. Geophys. Res.* **81**, 5882.

Russell, C. T. and McPherron, R. L.: 1973, *J. Geophys. Res.* **78**, 92.

Russell, C. T., McPherron, R. L., and Burton, R. K: 1974, *J. Geophys. Res.* **79**, 1105.

Schield, M. A., Freeman, J. W., and Dessler, A. J.: 1969, *J. Geophys. Res.* **74**, 247.

Schindler, K.: 1974, *J. Geophys. Res.* **79**, 2803.

Schindler, K.: 1979, *J. Geophys. Res.* **84**, 7257.

Schindler, K. and Birn, J.: 1982, *J. Geophys. Res.* **87**, 2263.

Schlegel, K. and St-Maurice, J. P.: 1981, *J. Geophys. Res.* **86**, 1447.

Schreiber, H.: 1981, *J. Geophys.* **49**, 169.

Schunk, R. W., Banks, P. M., and Raitt, W. J.: 1976, *J. Geophys. Res.* **81**, 3271.

Schwenn, R., Rosenbauer, H., and Mülhäuser, K.-H.: 1980, *Geophys. Res. Lett.* **7**, 201.

Sheeley, N. R., Jr and Harvey, J. W.: 1981, *Solar Phys.* **70**, 237.

Sheeley, N. R., Jr, Asbridge, J. R., Bame, S. J., and Harvey, J. W.: 1977, *Solar Phys.* **52**, 485.

Silsbee, H. C. and Vestine, E. H.: 1942, *Terr. Mag. Atmos. Elec.* **47**, 195.

Siscoe, G. L.: 1978, *Nature* **276**, 348.

Siscoe, G. L.: 1980, *Rev. Geophys. Space Phys.* **18**, 647.

Siscoe, G. L. and Crooker, N.: 1974, *Geophys. Res. Lett.* **1**, 17.

Siscoe, G. L. and Cummings, W. D.: 1969, *Planet. Space Sci.* **17**, 1795.

Siscoe, G. L., Crooker, N. U., and Christopher, L.: 1978, *Solar Phys.* **56**, 449.

Smith, E. J. and Wolfe, J. H.: 1976, *Geophys. Res. Lett.* **3**, 137.

Smith, E. J., Tsurutani, B. T., and Rosenberg, R. L.: 1978, *J. Geophys. Res.* **83**, 717.

Sonnerup, B. U. Ö.: 1974, *J. Geophys. Res.* **79**, 1546.

Sonnerup, B. U. Ö.: 1979, in C. F. Kennel, L. J. Lanzerotti and E. N. Parker (eds), *Solar System Plasma Physics*, Vol. III, North-Holland, New York, p. 45.

Sonnerup, B. U. Ö.: 1980, *J. Geophys. Res.* **85**, 2017.
Sonnerup, B. U. Ö., Paschmann, G., Papamastorakis, I., Sckopke, N., Haerendel, G., Bame, S. J., Asbridge, J. R., Gosling, J. T., and Russell, C. T.: 1981, *J. Geophys. Res.* **96**, 10049.
Southwood, D. J.: 1968, *Planet. Space Sci.* **16**, 587.
Southwood, D. J.: 1977, *J. Geophys. Res.* **82**, 5512.
Southwood, D. J.: 1979, in B. Battrick (ed.), *Magnetospheric Boundary Layers*, ESA SP-148, European Space Agency, Paris, p. 357.
Southwood, D. J. and Wolf, R. A.: 1978, *J. Geophys. Res.* **83**, 5227.
Spiro, R. W., Harel, M., Wolf, R. A., and Reiff, P. H.: 1981, *J. Geophys. Res.* **86**, 2261.
Stern, D. P.: 1980, *Energetics of the Magnetosphere*, Tech. Mem. 82039, NASA, Goddard Space Flight Center, Greenbelt, Maryland.
Suess, S. T.: 1979, *Planet. Space Sci.* **27**, 1001.
Svalgaard, L.: 1973, *J. Geophys. Res.* **78**, 2064.
Svalgaard, L.: 1975, SUIPR Rept 646, Institute for Plasma Research, Stanford Univ., Stanford.
Svalgaard, L.: 1977, in J. B. Zirker (ed.), *Coronal Holes and High Speed Wind Streams*, Colorado Assoc. Univ. Press, Boulder, p. 371.
Svalgaard, L. and Wilcox, J. M.: 1975, *Solar Phys.* **41**, 461.
Swift, D. W., Stenbaek-Nielson, H. C., and Hallinan, T. J.: 1976, *J. Geophys. Res.* **81**, 3931.
Swift, D. W.: 1979, *Rev. Geophys. Space Phys.* **17**, 681.
Thorne, R. M.: 1980, *Pure Applied Geophys.* **118**, 128.
Vasyliunas, V. M.: 1970, in B. M. McCormac (ed.), *Particles and Fields in the Magnetosphere*, D. Reidel, Dordrecht, p. 60.
Vasyliunas, V. M.: 1972, in B. M. McCormac (ed.), *Earth's Magnetospheric Processes*, D. Reidel, Dordrecht, p. 29.
Vasyliunas, V. M.: 1975, *Rev. Geophys. Space Phys.* **13**, 303.
Wilcox, J. M. and Scherrer, P. H.: 1972, *J. Geophys. Res.* 77, 5385.
Wilcox, J. M., Scherrer, P. H., Svalgaard, L., Roberts, W. O., Olson, R. H., and Jenne, R. L.: 1974, *J. Atmos. Sci.* **31**, 581.
Wilcox, J. M., Svalgaard, L., and Scherrer, P. H.: 1976, *J. Atmos. Sci.* **33**, 1113.
Williams, R. G. and Gerety, E. J.: 1978, *Nature* **275**, 200.
Wygant, J. R., Torbert, R. B., and Mozer, F. S.: 1982 in *Origins of Plasmas and Electric Fields in the Magnetosphere*, (abstract) Yosemite, p. 40.
Yasuhara, F., Kamide, Y., and Akasofu, S.-I.: 1975, *Planet. Space Sci.* **23**, 1355.
Zmuda, A. J. and Armstrong, J. C.: 1974, *J. Geophys. Res.* 79, 4611.

Dept of Atmospheric Sciences,
Univ. of California.
Los Angeles, CA 90024,
U.S.A.

CHAPTER 22

SOLAR ENERGETIC PARTICLES AND THEIR EFFECTS ON THE TERRESTRIAL ENVIRONMENT

GEORGE C. REID

1. Introduction

Evidence that the Sun could produce significant fluxes of energetic (i.e. $10^6 - 10^9$ eV) particles was first presented by Forbush (1946), who reported transient increases in the counting rate of ground-based cosmic-ray detectors occurring in association with solar flares. Little further development took place for the next decade, until the intense relativistic energy event of February 23, 1956, created widespread interest through its effects on the Earth's ionosphere, particularly at polar latitudes. The succeeding years of the late 1950s and early 1960s saw an explosion of interest in the near-Earth space environment brought about by the cooperative scientific efforts of the International Geophysical Year and by the development of artificial Earth satellites. Ground-based radio measurements revealed the existence of energetic-particle events that could not be seen by cosmic-ray detectors, while satellites detected events that were yet weaker, and provided direct measurements of the flux and composition of the particles.

The total number of events recorded now reaches several hundred, and they have given birth to an extensive scientific literature. The particles themselves are now known to be mainly protons, with a variable admixture of alpha-particles and heavier nuclei, and often with an accompanying flux of electrons with energies ranging from tens of keV to a few MeV. The energies of the protons and other heavier nuclei extend over at least five decades on occasion, from less than 100 keV to more than 10 BeV, and the fluxes range over several orders of magnitude. Their arrival in the vicinity of the Earth is determined by such factors as the Earth's location relative to the primary solar flare and the magnetic structure in the intervening solar wind. Once they arrive at the Earth, the particles (referred to often as simply 'solar protons') are influenced by the geomagnetic field, which effectively shields all but the polar caps from the direct effects, and whose outer structure can be inferred to some extent from the ways in which the particles behave. Entering the atmosphere, the solar particles lose energy chiefly by ionization, but partly also by nuclear interactions that lead to the production of neutrons and such unstable isotopes as C^{14}. The ionization is mainly produced in the 'middle' atmosphere, between heights of 20 and 100 km, and leads to marked changes in the ion chemistry and minor-constituent neutral composition.

In this chapter we shall trace the solar particles through these stages, and describe some of the practical consequences of their presence. Near the peak of solar activity in the late 1950s these would have seemed of immediate concern, since major events were

Peter A. Sturrock (ed.), Physics of the Sun, Vol. III, pp. 251–278.

occurring several times per year. The two succeeding solar cycles, however, have been much less active in terms of intense solar particle events, and the effects have thus turned out to be of less practical concern than would have been forecast. The reasons for such marked differences between individual solar cycles is not obvious, and poses an important question for solar physics.

2. Solar Energetic Particles and the Magnetosphere

Access of solar energetic particles to the vicinity of the Earth is governed by the properties of the outer magnetosphere. Even before the existence of a magnetosphere was recognized, the potential usefulness of solar particles as 'probes' of the distant geomagnetic field was pointed out (Reid and Leinbach, 1959). In particular, it appeared that observations of solar protons at the Earth might provide a definitive answer to the question of whether the geomagnetic field is directly connected to the interplanetary field, as suggested by Dungey (1961), or is dragged downstream by the solar wind to form a tail many thousands of Earth radii in length (Michel and Dessler, 1965). If the former were true, one would expect to see time variations in solar particle fluxes in interplanetary space fairly rapidly duplicated in the fluxes seen at the Earth, since the particles could simply travel along the connected field lines. In the latter case, however, one would probably see a substantially delayed response at the Earth, with a blurring of detailed structure, since the particles would have to diffuse across the interface between the solar wind and the magnetosphere.

The observations have not yielded a clear-cut answer to this question, which has generated a considerable amount of controversy over the years. The observations prior to the mid-1970s have been reviewed by Lanzerotti (1972) and by Paulikas (1974). The general conclusion, if one exists, is that the so-called 'open' and 'closed' models of the magnetosphere represent two extreme cases, and that in reality some regions of the magnetosphere are generally open, while others are closed. The controversy remains alive, however (Michel and Dessler, 1975; Hynds and Morfill, 1976).

The classical Størmer theory of the motion of a charged particle in a dipole magnetic field predicts the existence of a 'cutoff' magnetic latitude for a given particle rigidity, or of a corresponding 'cutoff' rigidity for a given latitude. Particles of a given rigidity are able to reach the surface of a given sphere surrounding the dipole only at latitudes higher than the corresponding cutoff latitude. Størmer theory has been extensively developed as a means of explaining the latitude variation of galactic cosmic rays, and it meets this objective very well, particularly when the departures of the Earth's internally generated magnetic field from the dipolar form are taken into account. Solar energetic particles, however, have characteristic rigidities that are much less than those of galactic cosmic rays, with correspondingly higher cutoff latitudes. Earth observations showed that the behavior predicted by Størmer theory did not take place, and the solar protons had access to latitudes considerably lower than the Størmer cutoffs. In fact, there appeared to exist a polar 'plateau' surrounding the geomagnetic pole, to which essentially all solar protons had access, regardless of energy, and an abrupt transition region at the edge of the plateau in which the geomagnetic cutoff climbed rapidly towards the Størmer value (Pieper *et al.*, 1962). Furthermore, the extent of the polar plateau varied with geomagnetic disturbance (Leinbach *et al.*, 1965), reaching middle geographic latitudes over North America during severe disturbances (Freier *et al.*, 1959).

The discovery of the Earth's magnetotail (Ness, 1965) pointed the way to an explanation of many of the anomalous cutoff effects, since the tail is a broad open magnetic structure containing weak fields and is presumably accessible to solar particles either immediately via field lines connected to the interplanetary magnetic field or more slowly via pitch-angle scattering. Once in the tail, the particles are free to travel along the field lines to the polar caps, thereby explaining the lack of a persistent magnetic cutoff in the polar caps (Michel, 1965; Reid and Sauer, 1967a). Several theoretical calculations of solar proton trajectories in model magnetospheres of varying degrees of sophistication have been carried out (Gall *et al.*, 1968; Smart *et al.*, 1969), and predict a diurnal variation in cutoff at latitudes below the polar plateau, in general agreement with observations (Stone, 1964; Imhof *et al.*, 1971; Fanselow and Stone, 1972), although the observed cutoffs appear to be displaced to lower latitudes than the predicted values by a few degrees.

Pronounced north–south asymmetry in solar particle fluxes is often observed, with one polar cap showing substantially larger fluxes than the other, or with one showing marked inhomogeneity while the other displays a smooth uniform flux (Van Allen *et al.*, 1971; Evans and Stone, 1969; Domingo and Page, 1971). These effects were explained in terms of the 'open' model of the magnetosphere (Reid and Sauer, 1967b; Van Allen *et al.*, 1971): if the geomagnetic field is directly connected to the interplanetary field one polar cap must be linked magnetically to the incoming solar wind, while the other is linked to the departing solar wind. Any anisotropy in the solar particle flux will then be revealed by the difference in flux seen in the two polar caps. In a sense the Earth acts as a bidirectional detector for the particles, with the sense of the directional response determined by that of the radial component of the interplanetary magnetic field.

Many observations of solar energetic particles have been made with spacecraft in the outer magnetosphere, particularly from the synchronous orbit (Lanzerotti, 1968; Paulikas and Blake, 1969; Blake *et al.*, 1974). These have shown again that solar protons of medium and low energies can penetrate the geomagnetic field to considerably greater depths than theory would predict, and that entry takes place partly through direct trajectories connecting with the dawn flank of the magnetosphere and partly through radial diffusion. The solar particles at synchronous altitude show evidence of transient trapping in the geomagnetic field, completing a few circuits of the Earth before disappearing (Blake *et al.*, 1974). The phenomenon of transient trapping may have a bearing on the 'midday recovery' often seen by high-latitude riometers during solar particle events (Leinbach, 1967). The ionospheric radio-wave absorption (polar cap absorption) recorded close to the edge of the polar plateau frequently shows a substantial decrease lasting for a few hours centered near local noon, indicating a temporary weakening of solar-particle precipitation. This may be related to the known diurnal variation in geomagnetic cutoff or to the development of a pitch-angle anisotropy in pseudotrapped solar particle fluxes as they drift around towards the dayside of the magnetosphere (Leinbach, 1967).

The topic of magnetospheric influences on solar energetic particles now has an extensive literature of its own, and many intriguing aspects of the interaction remain unexplained. Only a few of the highlights have been mentioned here, and the interested reader can consult the references listed for further details. Most of the studies mentioned were carried out in the 1960s and early 1970s, with an apparent decrease in activity in recent years. This is partly a result of the scarcity of major solar particle events and partly due to a growing preoccupation on the part of spacecraft experimenters with local magnetospheric processes as opposed to the large questions of overall magnetospheric topology and dynamics. The same questions remain largely unanswered, however, and pose a significant challenge for future research.

3. Energy Loss Processes

3.1. Polar Cap Absorption

The dominant energy loss process for solar particles in the atmosphere is ionization, since the energies generally lie below those for which losses by nuclear interactions are competitive. The stopping power of air for protons and alpha-particles is such that the bulk of the ionization produced by 1–100 MeV solar particles is at heights between about 30 and 90 km, well below the conventional ionospheric layers responsible for long-distance radio-wave propagation. Since the frequency of collisions of free electrons with air molecules in this region of the atmosphere is comparable with the frequencies of ionospherically reflected radio waves, the electrons can efficiently pick up ordered energy from the radio waves and transform it into random heat energy, resulting in severe attenuation of radio signals in the lower VHF band and below (i.e. frequencies below about 50 MHz). Because the enhanced ionization tends to be confined to high magnetic latitudes on the Earth due to the shielding effect of the geomagnetic field, the phenomenon became known as 'polar cap absorption', or simply PCA (Reid and Leinbach, 1959), a term that has frequently been extended to cover the entire range of effects associated with solar energetic particles in the terrestrial environment.

Historically, the radio-wave absorption effects provided the first direct evidence that the Sun emitted particles with energies intermediate between those of cosmic rays and those of the hot solar wind plasma (Bailey, 1957, 1959; Reid and Collins, 1959; Leinbach and Reid, 1959; Hultqvist and Ortner, 1959). Initially the radio measurements were used to provide information on the fluxes and spectra of the solar particles themselves, but as direct spacecraft measurements of these parameters became available, more attention was paid to the complementary approach of using the measurements to provide information on the properties of the relatively unknown region of the atmosphere in which the ionization occurs. Such efforts have continued for the past 20 years, but the relative scarcity of large solar particle events has caused progress to be slow and highly intermittent. Furthermore, it has become clear, as we shall see later, that the heavy ionization produced during these events substantially alters the ion chemistry and even the minor-constituent neutral composition of the atmosphere, so that conclusions drawn from PCA measurements cannot necessarily be carried over to normal conditions. The practical consequences of the events, however, have stimulated a great deal of interest in such studies, and have revealed much about the response of the middle atmosphere to intense ionization events.

Range–energy relations for energetic particles in air are well known (e.g. Sternheimer, 1959), and are directly applicable to the height region of concern here. Figure 1 shows

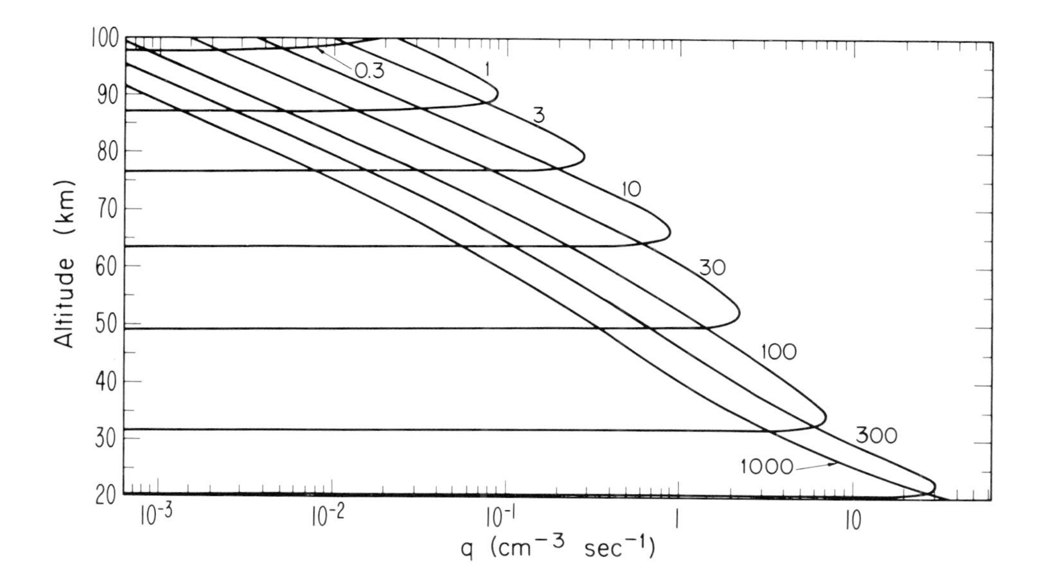

Fig. 1. Rate of production of ion pairs in the atmosphere by monoenergetic fluxes of protons incident isotropically over the upward-looking hemisphere with a flux of 1 proton cm^{-2} s^{-1} $ster^{-1}$. Energies are in MeV.

the rate of ionization by monoenergetic protons with a flux of 1 cm^{-2} s^{-1} $ster^{-1}$ isotropic over the upward-looking hemisphere. The integrated ionization rate profile at any time during a PCA event can be found by folding in the actual particle spectrum and by using the fact that alpha-particles and protons of the same velocity have approximately the same range, with the alpha-particle producing four times as much ionization as the proton. Heavier nuclei make a negligible contribution, as do solar electrons in most cases. Figure 2 shows some examples of ionization rates calculated in this way for specific events. Since normal daytime production rates are less than 10 ion cm^{-3} s^{-1} for most of the height range shown, the perturbation in ionization brought about by solar particles can be enormous. The effects are usually geographically confined, however, as can be seen from Figure 3, which shows the regions of the Earth's polar caps that are affected. Reductions of geomagnetic cutoffs caused by magnetic storms can extend these regions to lower latitudes by several degrees.

3.2. Ion Chemistry of the Middle Atmosphere: Influence of Solar Energetic Particles

The steady-state ion composition of the middle atmosphere during a solar particle event is determined by a balance between the rate of production discussed above and the rate of loss. The latter involves the complex and still poorly understood details of atmospheric ion chemistry, which will be outlined here as they apply to these events. A general survey of the ion chemistry of the mesosphere can be found in Reid (1976).

The three primary charged species produced in the atmosphere are free electrons and

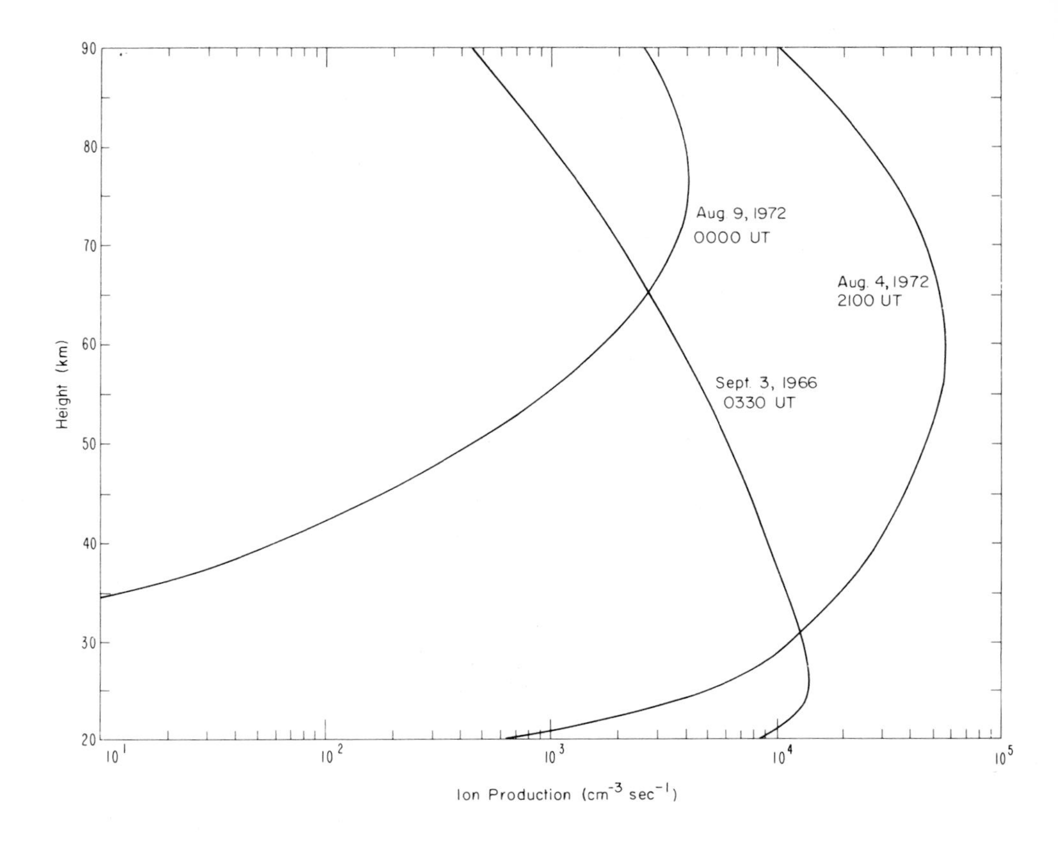

Fig. 2. Rate of production of ion pairs in the atmosphere in actual solar particle events at the times indicated.

the two positive-ion species N_2^+ and O_2^+. Much smaller quantities of such species as Ar^+, CO_2^+, O_3^+, O^+, and N^+ are also formed, but play minor roles in the ion chemistry. The free electrons are lost either through recombination with a positive ion or through attachment to form a negative ion. The latter process occurs almost entirely through three-body attachment to O_2, forming O_2^- in the reaction

$$e + O_2 + O_2 \longrightarrow O_2^- + O_2. \quad (1)$$

The corresponding reaction with nitrogen does not take place, since N_2 does not form a stable negative ion, but a small additional negative-ion contribution probably comes from dissociative attachment to ozone via the reaction

$$e + O_3 \longrightarrow O^- + O_2. \quad (2)$$

If there were no mechanism for releasing the electrons, attachment would compete successfully with recombination as a loss mechanism up to 90 km or higher. Both O_2^-

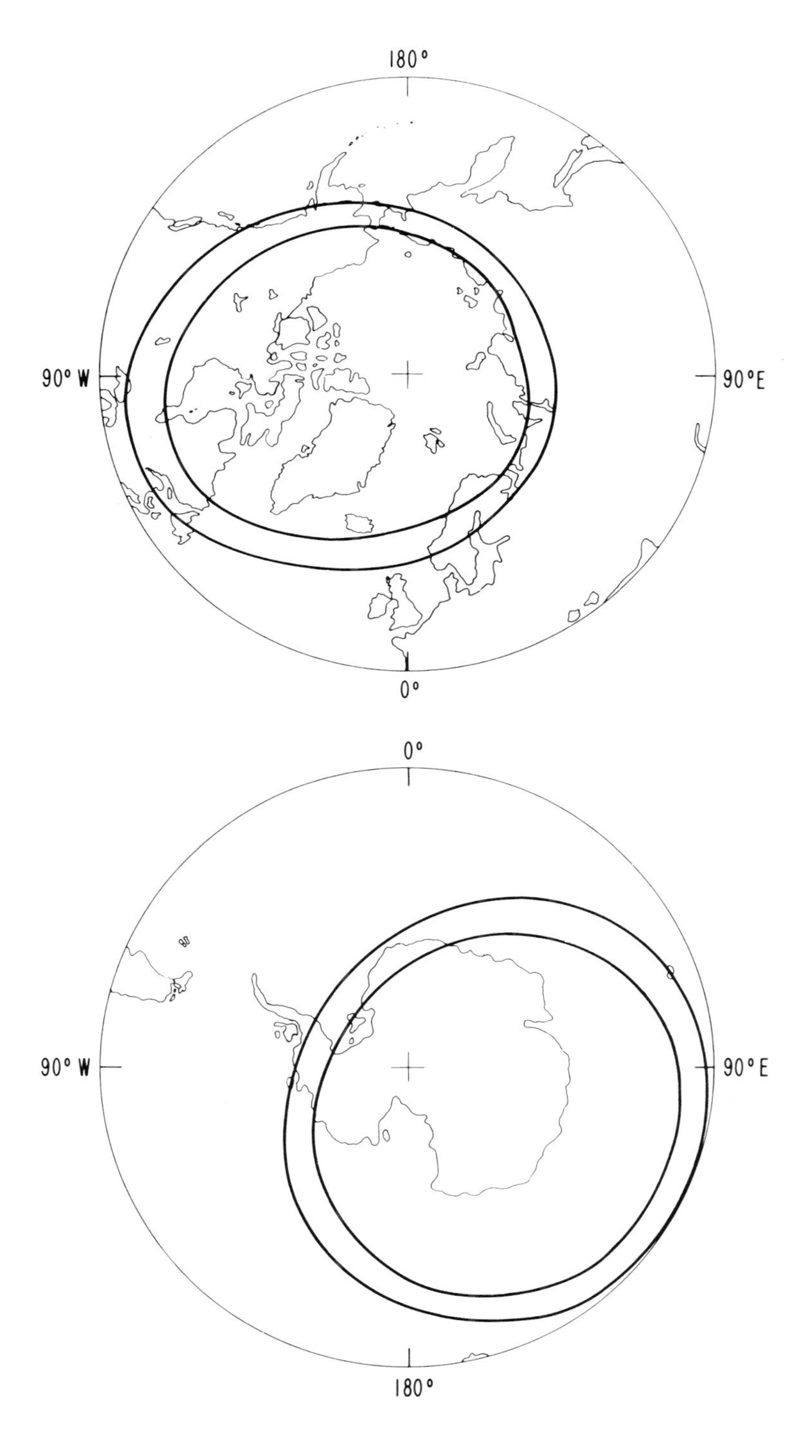

Fig. 3. The normal areas affected by polar cap absorption. The regions inside the inner curves represent the polar 'plateaux', while the regions outside the outer curves are usually unaffected, except during severe geomagnetic disturbances.

and O^-, however, are attacked by atomic oxygen in associative detachment reactions of the form

$$O_2^- + O \longrightarrow O_3 + e \tag{3}$$

$$O^- + O \longrightarrow O_2 + e \tag{4}$$

that keep electrons cycling through loops formed by reactions (1) and (3) and by (2) and (4). Loss of an electron from the system occurs only when one of the negative ions reacts with a neutral molecule to form a more stable negative ion that is immune to attack by O. The principal such reaction is

$$O_2^- + O_3 \longrightarrow O_3^- + O_2, \tag{5}$$

forming the stable negative ion of ozone.

Since atomic oxygen tends to destroy negative ions and release free electrons, while ozone tends to stabilize the negative ions, the ratio of the concentrations of O and O_3 is an important parameter in determining the equilibrium electron concentration. Model calculations show that negative-ion concentrations should be small at heights above 70–75 km in daytime when the O/O_3 ratio is large, and above 80–85 km at night when atomic oxygen virtually disappears below these levels.

Electrons in the middle atmosphere thus go through a fairly sharp transition between an attachment-dominated lower region and a recombination-dominated upper region. One of the most important features of solar energetic particle events from the ionospheric point of view is the fact that they 'illuminate' this transition region with a strong source of ionization that is independent of sunlight, unlike the normal ionization sources in this height range. Time dependence of chemical effects can thus be separated from time dependence of ionization rate.

Once formed, the negative ions can undergo a complicated series of reactions with minor atmospheric constituents such as NO, CO_2, O_3, and H, ultimately forming such stable species as the bicarbonate and nitrate ions, HCO_3^- and NO_3^-. The latter is particularly stable, having an electron affinity of about 3.9 eV (Ferguson *et al.*, 1972), and is eventually lost through a recombination reaction with a positive ion or possibly through photodetachment by solar radiation with wavelength less than about 300 nm. The importance of photodetachment and photodissociation of atmospheric negative ions remains obscure, but quantitative estimates are now becoming possible through laboratory measurements (Peterson, 1976).

The 'terminal' negative-ion species are also expected to become hydrated, forming relatively heavy clusters of several molecules of water, and at the lower heights of such alternative polar molecules as HNO_3 and H_2SO_4, around the core ion. Heavy negative-ion species have been observed by mass spectrometers flown in rockets during a PCA event (Narcisi *et al.*, 1972b) and have been tentatively identified as hydrated clusters of NO_3^- and CO_3^-. The same observations, however, have revealed surprisingly large concentrations of light ions identified as O_2^- and O^-, which should exist only at very low concentrations because of the rapidity of reactions (3) and (4).

At any given height in the atmosphere there is a certain equilibrium negative-ion composition that depends on the local neutral composition. Whether or not that equilibrium state is reached depends on the lifetime of a negative ion against loss by ion–ion recombination which always competes against the chemical reactions that lead towards equilibrium. This lifetime is itself inversely proportional to the positive-ion concentration, which depends on the ionization rate. As an event increases in intensity, then, the ambient negative-ion composition should show progressively greater departures from equilibrium, and the nature of the departures should reveal the details of the ion chemistry that is taking place. Only PCA events can provide the calibrated ionization source that such an experiment would require.

Turning to positive ions, PCA events have provided important information on the atmospheric sequence of chemical reactions. Early mass spectrometer observations of positive-ion composition in the mesosphere had revealed the presence of hydrated protons – the so-called water cluster ions – as the dominant species below 80 km (Narcisi and Bailey, 1965). The problem of identifying a reaction sequence that would lead from the primary species to the water clusters immediately became important, and such a sequence was soon found in the laboratory by Febsenfeld and Ferguson (1969) and Good *et al.* (1970), starting from O_2^+ as the precursor. Since N_2^+ is quickly converted into O_2^+ by charge exchange in the mesosphere, this is the appropriate sequence for PCA events, and mass spectrometer measurements (Narcisi *et al.*, 1972a) have shown that it is consistent with the observations. The case of the normal undisturbed mesosphere, however, where the dominant primary species is NO^+, has not yet been completely elucidated (Reid, 1977a).

The ionization rate, q, and the steady-state electron concentration, n_e, are related by the simple expression

$$q = \Psi n_e^2, \tag{9}$$

where Ψ is a height-dependent parameter often called the electron loss coefficient. In the upper mesosphere, where recombination is the dominant loss mechanism, Ψ is simply the weighted average recombination coefficient of the various positive-ion species present. At lower heights, however, Ψ depends on the negative-ion concentration and is not completely independent of q, as well as having a pronounced diurnal variation.

PCA events have long been recognized as presenting excellent opportunities to measure Ψ directly, since n_e is often large enough to be measured fairly easily and q can readily be calculated if the energetic particle spectrum is known. Many such coordinated measurements have been made, using rocket- or satellite-based spectrum measurements and electron concentrations derived from rocket (Megill *et al.*, 1971; Larsen *et al.*, 1972; Ulwick, 1973; Swider and Dean, 1975; Sellers and Stroscio, 1975) and radar (Reagan and Watt, 1976) measurements. The latter technique has obvious advantages over rocket techniques, since it can yield essentially continuous data, but for reasons of sensitivity it is likely to be applicable only to the most intense events. With a few exceptions, the measurements shows fairly reasonable agreement among themselves, despite a fairly wide range of ionization rates. Below 80 km nighttime values are considerably larger than daytime values, as expected from the dominance of negative-ion processes at these altitudes. The twilight transition is well defined and strongly height dependent (Reagan and Watt, 1976).

To summarize, PCA events are of considerable importance in the context of the

Earth's ionosphere. They provide a strong and measurable source of ionization in a region of the atmosphere where normal ionization rates are very low, thereby allowing studies of atmospheric ion chemistry that would not otherwise be possible. Partially counteracting this, however, is the fact that the ion chemistry is itself modified during these events, particularly during the more intense ones, so that application of results derived from PCA events to the undisturbed atmosphere must be made cautiously. As we shall see later, PCA events also have important practical consequences through the effect they have on ionospheric radio propagation.

3.3. Polar Glow Aurora: Optical Effects of Solar Energetic Particle Precipitation

Part of the energy lost by solar particles in the atmosphere goes into the production of radiation at optical wavelengths. The resultant 'polar glow aurora' was first reported by Sandford (1961) using patrol-spectrograph measurements of the intensity of the first negative bands of N_2^+ at the South Pole and at Thule, Greenland. In contrast to the normal aurora, which is often highly structured and variable on short time-scales, polar glow aurora is a diffuse featureless glow that covers the entire sky at polar cap locations, and its lack of contrast makes it difficult to observe even when its intensity is above the threshold of visibility. In principle, the high spatial resolution that is possible with optical measurements ought to provide a means for detecting spatial inhomogeneities in precipitation of the particles using polar glow observations. Events with a high enough emission rate are rare, however, and reports of intensity measurements are few. Argemi (1964) reported a systematic variation in intensity across the sky at Dumont d'Urville, Antarctica, during the major events of the IGY period. More recently, Weber *et al.* (1976) have presented the results of measurements made during the event of August 1972 at the South Pole. In terms of total energy input to the middle atmosphere, this was the most intense event yet recorded, and was accompanied by major auroral activity that dominated the optical emissions most of the time even at the high magnetic latitude of the South Pole.

The theoretical intensities of the various optical emissions have been calculated by a number of authors (Sandford, 1963; Brown, 1964; Alcaydé, 1968; Edgar *et al.*, 1973, 1975; Singh and Singhal, 1978), with emphasis on the N_2^+ first negative bands and the 557.7 nm emission of atomic oxygen. The latter is a forbidden transition and is strongly quenched below 90 km. The allowed transitions are much less affected, but quenching becomes significant even for them at stratospheric heights. Volume emission rates have been calculated by Edgar *et al.* (1975) for several emissions of N_2^+, O_2, O_2^+, and O, and were found to vary strongly with the assumed geomagnetic cutoff for the particles and with the shape of the spectrum at low energies. The emission expected from the enhanced formation of NO_2 by reaction of NO with ozone was calculated by Brown (1967) and found to be small using data then available on the concentration of NO during PCA events.

The normal aurora is often loosely attributed to the atmospheric impact of solar particles. While it is true that the ultimate source of the aurora is the influence of solar wind plasma on the magnetosphere, it is also clear that there is a long chain of plasma effects connecting the source with the visual phenomena. With the possible exception of the 'cusp' regions on the dayside of the magnetosphere, the solar wind plasma itself does not enter the atmosphere. Polar glow aurora is thus the only direct visual effect produced by solar particles in the atmosphere. Unfortunately its occurrences have been so sporadic that it has been little more than a curiosity to date, and its potential as a diagnostic tool for particle precipitation patterns and for middle-atmosphere studies has not been realized.

4. Atmospheric Alterations and Nuclear Interactions

4.1. ALTERATIONS IN MIDDLE-ATMOSPHERE COMPOSITION

Some of the energy introduced into the middle atmosphere by solar particles is used to dissociate molecular species, thereby initiating chemical reactions that lead to significant changes in the minor-constituent chemical composition. The major effects are produced by the dissociation of nitrogen and water vapor, with oxygen dissociation probably playing an insignificant role.

The possibility that alterations in neutral composition might take place during PCA events was first raised by Herzberg (1960), who suggested that dissociative recombination of N_2^+ ions would produce N atoms. These would go on to produce nitric oxide through the reaction

$$N + O_2 \longrightarrow NO + O \tag{1}$$

and substantial increases in the NO concentration of the middle atmosphere might be expected. It is now known, however, that N_2^+ ions react rapidly with O_2 in a charge–exchange reaction, so that this particular pathway to dissociation is not available in the mesosphere, where O_2 concentrations are large. It does become important, however, in the thermosphere.

Dalgarno (1967) examined the chemistry of PCA events in some detail using kinetic data available at the time, and concluded that enhancements in NO concentration were likely. The chief sources were the dissociative ionization of N_2 into N and N^+, together with the hypothetical reaction

$$O_2^+ + N_2 \longrightarrow NO^+ + NO. \tag{2}$$

Although exothermic, this reaction has never been seen in the laboratory. Since N_2 molecules are so numerous in the atmosphere, however, the reaction would be important even if its rate were well below the upper limit of 10^{-15} cm^3 s^{-1} placed on it by Ferguson (1974). In fact, its rate is probably zero, since the reaction involves the unlikely breaking and remaking of two chemical bonds simultaneously.

Further interest in neutral composition changes was aroused by the observation by Weeks *et al.* (1972) of a substantial depletion of ozone in the mesosphere (54–67 km) during the PCA event of November 1969. The ozone concentration near the peak of the event was found to be less by a height-dependent factor of 2–4 than the corresponding values measured two days later. This effect was explained by Swider and Keneshea (1973) as a consequence of the formation of water cluster ions, involving the dissociation of water vapor. The currently accepted positive-ion reaction chain in the middle atmosphere includes the reaction of a hydrated O_2^+ ion with a water molecule, leading to dissociation:

$$O_2^+ \cdot H_2O + H_2O \longrightarrow H_3O^+ \cdot OH + O_2 \tag{3}$$

A subsequent step frees the OH radical, and the ultimate disappearance of the ion through recombination frees a hydrogen atom, either alone or in the form of nitric acid (HNO_3) if the recombination is with a nitrate ion (NO_3^-). The odd hydrogen species generated in this way enter into a complicated series of reactions, some of which lead to destruction of ozone. For example, the reaction pairs

$$\begin{aligned} H + O_3 &\longrightarrow OH + O_2 \\ OH + O &\longrightarrow H + O_2 \end{aligned} \tag{4}$$

$$\begin{aligned} OH + O_3 &\longrightarrow HO_2 + O_2 \\ HO_2 + O_3 &\longrightarrow OH + 2O_2 \end{aligned} \tag{5}$$

$$\begin{aligned} OH + O_3 &\longrightarrow HO_2 + O_2 \\ HO_2 + O &\longrightarrow OH + O_2 \end{aligned} \tag{6}$$

are equivalent to the ozone-destroying reactions

$$O + O_3 \longrightarrow 2O_2 \tag{7}$$

or

$$2O_3 \longrightarrow 3O_2 \tag{8}$$

with H or OH acting as a catalyst that is not itself removed. The odd hydrogen species eventually disappear by reacting with each other to re-form stable species, such as H_2 and H_2O. Typical time constants for this recombination process in the middle atmosphere are a few hours to a day, so that the ozone fairly quickly recovers to its normal concentration. Crutzen and Solomon (1980) have pointed out that the dissociation of water vapor during an intense event such as that of August 1972 must have been great enough to seriously deplete the mesospheric water vapor concentration. Since much of the dissociated hydrogen recombines to form H_2 rather than H_2O, the water vapor concentrations will remain depleted for some time following the event, until they are finally restored by diffusion from below. Ozone concentrations will thus be enhanced, after their initial depletion, since photodissociation of water vapor is the source of the odd hydrogen in the mesosphere under normal conditions.

From the point of view of ozone destruction, the odd nitrogen production in PCA events is probably much more serious than the odd hydrogen production, since the odd nitrogen has a photochemical lifetime of years in the stratosphere, where most of the ozone is located. As we have seen above, the production of NO by solar particles in the middle atmosphere was predicted by Herzberg (1960) and Dalgarno (1967), but the key role of odd nitrogen compounds in stratospheric photochemistry was not realized at the time.

The reactions discussed above illustrate the catalytic destruction of ozone by the dissociation products of water vapor. A similar catalytic role is played by the nitrogen dissociation products, chiefly through the reaction pair

$$NO + O_3 \longrightarrow NO_2 + O_2 \tag{9}$$

$$NO_2 + O \longrightarrow NO + O_2 \tag{10}$$

which is equivalent to reaction (7) above (Crutzen, 1970). In the undisturbed stratosphere, NO is produced by the action of metastable O atoms on N_2O (nitrous oxide), which is a stable byproduct of bacterial action in the soil, and which diffuses up into the stratosphere. NO is lost by conversion to nitric acid and eventual return to the surface of the earth in rainfall, but the balance between production and loss allows odd nitrogen concentrations large enough that the catalytic cycle accounts for a large part of the required sink for stratospheric ozone (Johnston, 1975).

Warneck (1972) and Brasseur and Nicolet (1973) pointed out that galactic cosmic rays could provide a natural source of odd nitrogen in the lower stratosphere. This suggestion was discussed in detail by Nicolet (1975), who concluded that approximately one free nitrogen atom was produced per ion pair by the energetic secondary electrons resulting from cosmic-ray ionization. The nitrogen atoms are the products of both dissociative ionization

$$N_2 + e \longrightarrow N^+ + N + 2e \tag{11}$$

whose cross-section has been measured by Rapp *et al.* (1965), and simple dissociation

$$N_2 + e \longrightarrow N + N + e \tag{12}$$

with a cross-section measured by Winters (1966) (see also Zipf and McLaughlin, 1978). The twin facts that the cosmic-ray flux undergoes a solar cycle modulation and that cosmic-ray-generated odd nitrogen participates in ozone destruction would suggest that the stratospheric concentration of ozone might show a solar cycle variation, a possibility that was investigated by Ruderman and Chamberlain (1975). Although its existence is controversial, there are indications of such a solar cycle variation in the long series of total ozone measurements at Arosa, Switzerland, and Tromso, Norway, with a peak ozone content occurring some three years after solar maximum. If the variation is real, its cause may lie in some factor other than the cosmic-ray variation (such as a solar cycle variation in solar ultraviolet radiation), but Ruderman and Chamberlain (1975) showed that the sense of the variation was consistent with the theory, which would predict maximum ozone following the minimum in cosmic-ray flux, i.e. following solar maximum.

Crutzen *et al.* (1975) calculated the odd nitrogen production that must have occurred during some intense PCA events as a result of the same mechanism, i.e. dissociation of nitrogen by fast secondary electrons. They estimated that the ratio of N-atom to ion-pair production should be about 1.5, some 50% higher than Nicolet's (1975) value, but a subsequent theoretical study of energy loss by relativistic particles in nitrogen by Porter *et al.* (1976) indicated a value of 1.27. The results of Crutzen *et al.* (1975) indicated that one or two major PCA events, each lasting only 2–3 days, could create as much odd nitrogen as an entire year's integrated cosmic-ray flux. Figure 4 shows their estimated

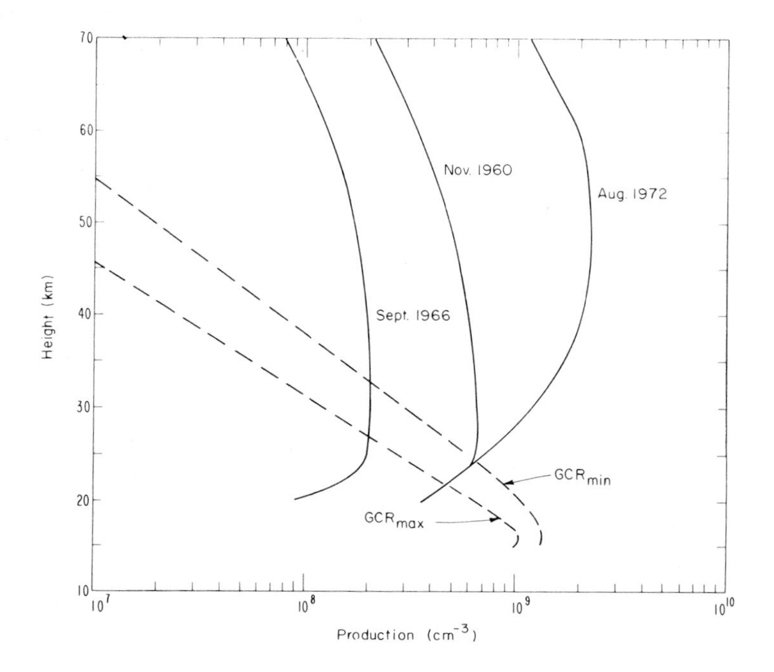

Fig. 4. Calculated total production of NO during the PCA events indicated. Broken curves show the production by galactic cosmic rays during one year at solar maximum (GCR_{max}) and solar minimum (GCR_{min}).

production by the events of November 1960, September 1966, and August 1972, together with the estimated annual production by cosmic rays at solar maximum and minimum. Since the solar particles have lower energies than the cosmic rays, the bulk of their odd nitrogen is produced in the upper stratosphere and mesosphere, where the only significant loss is through photodissociation of NO followed by chemical destruction

$$NO + h\nu \longrightarrow N + O \tag{13}$$

$$N + NO \longrightarrow N_2 + O. \tag{14}$$

Removal by this reaction pair is slow, whereas the cosmic-ray-produced NO, being close to the tropopause, can be washed out by precipitation processes more rapidly following conversion to nitric acid. The effective lifetime of odd nitrogen produced by solar particles should thus be significantly longer than that of cosmic-ray-produced odd nitrogen, and solar particle events could represent an important sink for stratospheric ozone if they occurred with a frequency of more than one major event per year.

In fact, the frequency of occurrence of PCA events has been highly variable in the recent past. When they were first studied following the peak solar activity of 1957–1958, they occurred at a rate of several per year; in more recent years their occurrence has been quite sporadic, but with a tendency to maximize in the years immediately following solar maximum. Their destructive effect on stratospheric ozone should thus maximize near solar maximum, leading to an out-of-phase relationship between ozone and solar activity. If anything, ozone appears to be in phase with solar activity, so it does not appear that solar particle events can explain the variation.

The effect of solar-particle-produced nitrogen oxides on upper stratospheric ozone was clearly seen following the very intense event of August 1972. Figure 5 (Heath *et al.*, 1977) shows satellite measurements of the total ozone column above the 4-millibar pressure level, corresponding approximately to 40 km altitude, in three latitude bands in the northern hemisphere. The event occurred on day 217 (August 4), and was followed quickly by a drop of about 15% in ozone content in the polar latitude band (75°–80° N). A sudden drop followed by a rapid recovery and then a slower decrease occurred in the 55°–65° band, while there was no uniquely identifiable effect near the equator (the slow decrease following the event is at least partially attributable to other factors). The magnitude of the ozone decrease near the pole is consistent with model predictions (Crutzen *et al.*, 1975; Heath *et al.*, 1977; Reagan *et al*;. 1981) for the August 1972 event. The behavior in the 55°–65° latitude band is probably explainable on the basis of the advection of polar air into and out of this region during the event, while the lack of an identifiable response at the equator within the month or so following the event is expected.

Oxygen dissociation takes place with an efficiency comparable to that of nitrogen, and was studied by Maeda and Aikin (1968) for the case of energetic auroral electrons rather than solar particles. The solar particle case has been discussed by Frederick (1976) as part of a comprehensive study of particle-induced chemical effects. The nature of the primary ionizing radiation is not important since the dissociation is a product of the fast secondary electrons, and not of the primary particles themselves. In the case of oxygen, the free oxygen atoms will mostly produce ozone through three-body reaction with molecular oxygen

$$O + O_2 + M \longrightarrow O_3 + M \tag{15}$$

so that solar particles produce ozone as well as destroying it through the odd nitrogen and odd hydrogen catalytic cycle. The production, however, is negligible compared to the normal daytime production of O by solar photodissociation. Furthermore, each O atom produces only one ozone molecule, whereas each H or N atom consumes many ozone molecules through the catalytic cycles described above. Except possibly during the polar night, when there is no photodissociation, it appears that the dissociation of O_2 has only minor significance.

To summarize, solar particle events clearly have the potential for modifying the chemical composition of the middle atmosphere. When they are reasonably intense and frequent, they become an important factor in the global ozone cycle, with consequences that will be discussed later. The same mechanisms operate for any primary ionizing radiation, and Thorne (1977) has pointed out that aurorally associated relativistic electron precipitation might also be important, since such events are considerably more frequent

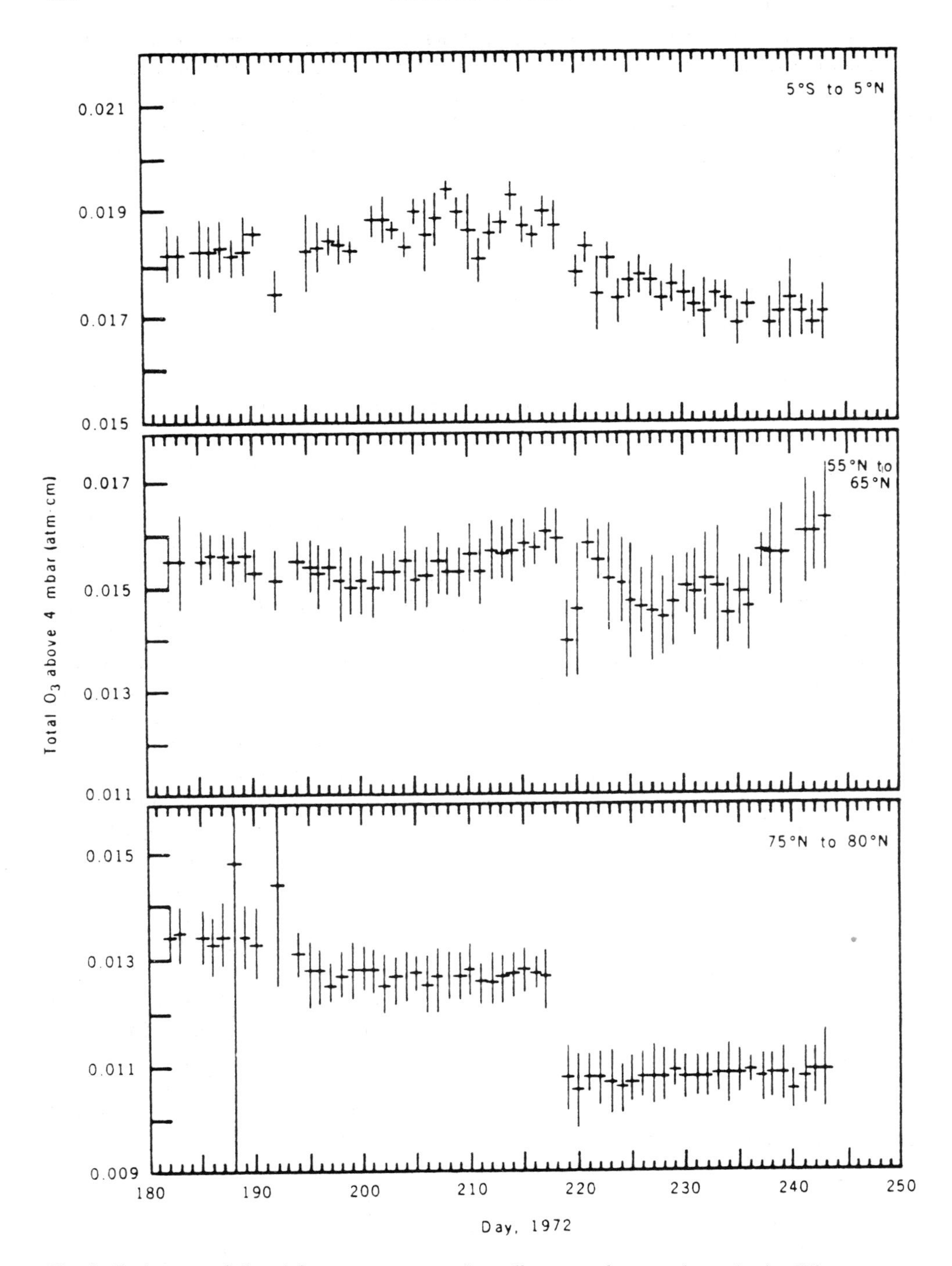

Fig. 5. Backscattered ultraviolet measurements of zonally averaged ozone above the 4 millibar pressure level (Heath *et al.*, 1977). Data are from Nimbus 4 over the latitude bands indicated. (Copyright 1977 by the American Association for the Advancement of Science.)

than major solar particle events. The current status of the topic has been reviewed by Thorne (1980), but much remains to be learned about the long-term role of solar particle events in atmospheric chemistry. Refinements in the theory are still being made (e.g. Rusch *et al.*, 1981; Solomon *et al.*, 1981), and there is a pressing need for more observations to provide tests of the theoretical predictions. These must await the cooperation of the Sun.

4.2. NUCLEAR INTERACTIONS AND C^{14} PRODUCTION

A significant component of cosmic radiation recorded at the surface of the Earth is in the form of neutrons. These neutrons are the products of interactions between the primary particles and the nuclei of air molecules, principally in the lower stratosphere. They are produced at high (million electron-volt) energies, and subsequently lose energy by elastic and inelastic collisions, eventually becoming thermalized. The atmosphere thus contains a steady flux of neutrons with a continuous energy spectrum that extends from thermal energies of a fraction of an electron-volt up to almost the full energy of the primary particles.

The cosmic-ray neutrons themselves can undergo unclear interaction, producing a wide variety of isotopes. The most important of these is C^{14}, which is a product of an (n, p) reaction with N^{14}, the most abundant isotope in the atmosphere. The C^{14} (also known simply as radiocarbon) ultimately becomes oxidized to form $C^{14}O_2$ which is chemically indistinguishable from ordinary $C^{12}O_2$, and which becomes incorporated in all the various reservoirs of the terrestrial carbon cycle. In particular, C^{14} is incorporated into the carbon of living organisms, all of which ultimately depend on photosynthetic reactions involving atmospheric CO_2. Its proportion in the cells of the organism will be identical to the proportion of C^{14} in the atmosphere during the life of the organism (with a slight deviation arising from differences in uptake of the heavier molecule). When the organism dies, however, carbon is no longer taken up from the atmosphere, and the C^{14} decays to N^{14} by beta emission with a half-life of 5730 years. If the C^{14} proportion of atmospheric carbon and the CO_2 fraction of the atmospheric composition are both assumed to be constant with time, the amount of C^{14} in a sample of organic material gives a measure of the time elapsed since the death of the organism. This is the principle of the radiocarbon dating technique (Libby, 1955), which has found a wide variety of practical applications.

In fact, it is now known that the above assumptions are not strictly valid, and that the atmospheric content of C^{14} has varied in the past, causing systematic differences between radiocarbon dates and those determined by other means. The topic has been reviewed by Damon *et al.* (1978), who also discuss the terrestrial carbon cycle in outline. According to calculations by Lingenfelter and Ramaty (1970), the global time-averaged production rate of C^{14} over the period 1937–1967 was 2.2 ± 0.4 atom cm^{-2} s^{-1}. In a steady state, this should equal the global time-averaged decay rate, and current estimates (Damon *et al.*, 1978) do in fact find reasonable agreement between the two.

Simpson (1960) first pointed out that solar protons and alpha-particles might provide a significant additional source of C^{14}. At the time, near the peak of the largest solar maximum of recent times, major solar particle events were frequent, and their contribution to the atmospheric C^{14} inventory could have created an awkward imbalance between production and loss. The solar proton contribution was first calculated by Lingenfelter and Flamm (1964a) and has been updated more recently by Mendell *et al.* (1973). According to the latter authors, the time-integrated production of C^{14} by solar proton events during solar cycle 20 (peaking in 1969) was 2–3 orders of magnitude lower than the production by galactic cosmic rays during the same period. The sporadic and highly variable nature of the solar particle source is illustrated, however, by the fact that the single event of February 23, 1956, produced roughly 150 times the amount of C^{14} produced by all of the events of solar cycle 20. It added about 10% to the total radiocarbon produced during the 11 years of solar cycle 19. This solar cycle was so active as a whole, however, that the galactic cosmic ray production of C^{14} was reduced by about 10% below that of solar cycle 20. The net result was that the two cycles, which were of greatly differing intensity, produced approximately equal amounts of atmospheric C^{14}.

The enhanced fast-neutron fluxes that result from solar particle bombardment of the atmosphere have been observed on several occasions (Chupp *et al.*, 1967; Lockwood and Friling, 1968; Greenhill *et al.*, 1970; Mendell *et al.*, 1973). Since the neutron is unstable, decaying into a proton and an electron with a radioactive half-life of about 12 min, some early observations led to the suggestion that 'albedo' neutrons from solar proton bombardment of the polar atmosphere might provide a significant source of trapped protons for the inner Van Allen radiation belt (Armstrong *et al.*, 1961). Calculations of the neutron leakage flux have since been performed (Lingenfelter and Flamm, 1964b; Dragt *et al.*, 1966; Mendell *et al.*, 1973), but the question of the contribution to the radiation belts does not appear to have been settled.

5. Effects of Solar Particle Events

5.1. Effects on Radio Communication and Navigation

The principal effect of solar particle events from the ionospheric point of view is the creation of very large amounts of ionization in a region of the atmosphere where there is normally very little ionization. As an extreme example, electron concentrations measured at Chatanika, Alaska, during the intense event of August 1972 exceeded 5×10^4 cm^{-3} at 70 km (Reagan and Watt, 1976), an altitude at which normal daytime values are less than 10^2 cm^{-3}. Such enormous increases in free-electron concentration cause marked effects on radio waves propagating through the lower ionosphere, but the precise nature of the effects depends on the radio-wave frequency.

At the frequencies for which the ionosphere acts as a reflector (the medium- and high-frequency bands) the chief effect is strong absorption of the radio-wave energy by the electrons, whose high collision frequency at mesospheric heights leads to efficient transfer of the ordered energy of the radio waves to disordered thermal energy of the neutral atmosphere. The amplitudes of incident and received waves are related by the usual expression

$$I = I_0 \exp(-\int k \, ds), \tag{1}$$

where the integration is taken along the wave path, and the absorption coefficient k is given in the original Appleton–Hartree development of magneto-ionic propagation theory as

$$k = \frac{e^2}{2\epsilon_0 mc} \cdot \frac{1}{\mu} \cdot \left(\frac{N\nu}{\nu^2 + \omega_e^2}\right) \tag{2}$$

$$\omega_e = \omega \pm \omega_L. \tag{3}$$

Here N is the electron concentration, ν the effective collision frequency, ω the radio-wave angular frequency, ω_L the component of the local gyrofrequency in the direction of propagation, and μ the real refractive index. ϵ_0 is the free-space permittivity, e and m are the charge and mass of the electron, and c is the speed of light. The positive and negative signs refer, respectively, to the ordinary and extraordinary birefringent rays produced by the geomagnetic field. For most purposes μ can be taken as 1, since the radio waves of concern are reflected at much greater heights, so that the properties of the absorption coefficient are determined by the quantity in parentheses in (2).

At higher levels, where $\nu \ll \omega_e$, $k \propto N\nu/\omega_e^2$, while at lower levels, where $\nu \gg \omega_e$, $k \propto N/\nu$. Physically, this behavior is easily interpreted. In order to absorb energy, the electrons must be able to react to the electric field of the radio wave, and to transfer the resulting kinetic energy to the molecules of the neutral gas. The first of these two steps is hindered if the collision frequency is too high, which is the situation at lower levels. The second step is impeded if the collision frequency is too low, which is the situation at the higher levels. The maximum absorption per free electron occurs where $\nu = \omega_e$.

Figure 6 shows the quantity $f = \nu/(\nu^2 + \omega_e^2)$ as a function of height in the atmosphere, using a standard approximation for the collision frequency. The curves correspond to radio frequencies ($\omega_e/2\pi$) in the MF, HF, and VHF bands that are used for communications purposes, and illustrate the enhanced sensitivity of absorption at these frequencies to the presence of free electrons in the mesosphere and upper stratosphere. Comparison with Figure 1 shows immediately the reason for the dramatic radio-wave absorption effects that accompany solar particle precipitation. From the communications point of view, the most obvious effect of a PCA event is a total 'blackout' of ionospherically propagated radio signals at high latitudes. Since ionospheric propagation is a major technique for long-distance communication at high latitudes, where satellite communications are often impossible, PCA events constitute serious potential disruptions with durations of possibly several days at a time.

Similar disruptions of ionospheric communications are associated with auroral events, but the duration is usually short. Their existence, however, coupled with the need for some means of continuous communication at high latitudes, led to the development of alternative systems that operated at higher frequencies and did not rely on ionospheric reflection. One such technique was that of VHF forward scatter, which used weak scattering from electron density irregularities to propagate the radio signals. It proved fairly immune to auroral effects, but was seriously affected by PCA events during daytime, and in fact played a major role in the discovery and early study of solar energetic particles (Bailey, 1957, 1959). Another technique utilized reflections from meteor trails at VHF. Since the reflections take place at heights of 90–95 km, well below the height of most

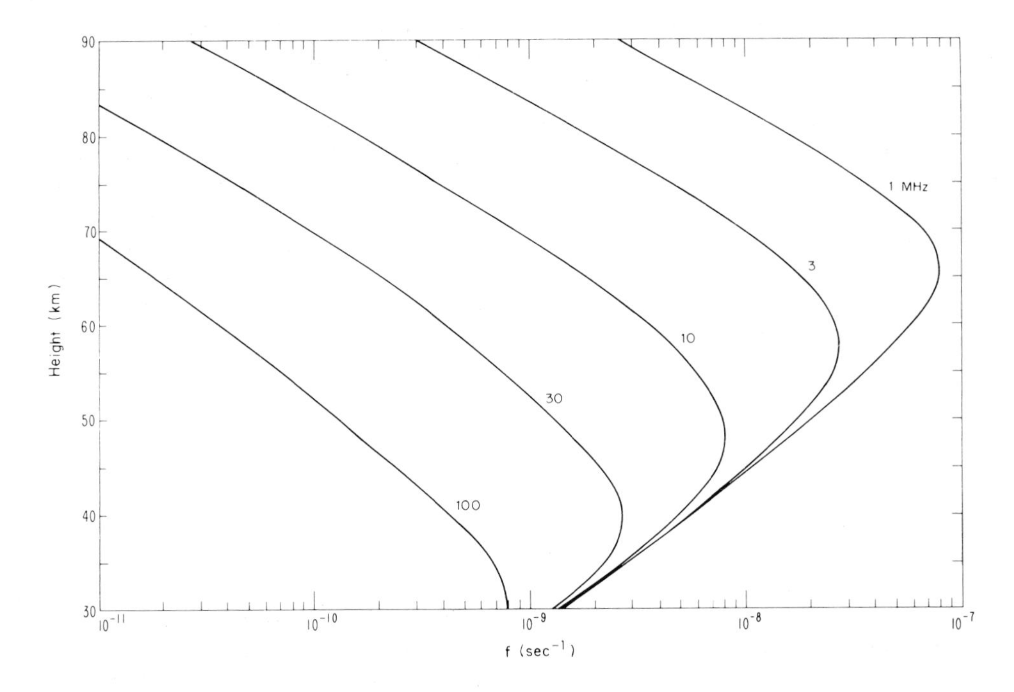

Fig. 6. The collision frequency parameter f as a function of height for various radio-wave frequencies. f is a measure of the efficiency of a free electron in producing radio-wave absorption at the frequency shown.

auroral ionization, a meteor-burst communication system should be relatively immune to auroral disturbances. The ionization produced by solar particles, however, lies below the meteor-trail ionization, and PCA events again cause major disruptions (Crysdale, 1960).

At frequencies in the LF (30–300 kHz), VLF (3–30 kHz), and ELF (< 3 kHz) bands, the mode of long-distance propagation changes to one in which the waves travel in a ducted 'waveguide' having the base of the ionosphere as one boundary and the surface of the earth as the other. PCA events also effect the propagation of these waves, and severe disturbances of both amplitude and phase of signals received on transpolar paths were noted in the early days of PCA studies (Ortner *et al.*, 1960; Eriksen and Landmark, 1961). The phase shifts that occur during PCA events are particularly important from the point of view of global navigation systems, which rely on the phase of the long waves received from known distant transmitters to determined the location of the receiver (Swanson, 1974). The reason for the phase changes lies in the reduction in width of the waveguide caused by the enhancement in conductivity of the lower ionosphere, while the additional free electrons below the waveguide boundary presumably cause the change in amplitude. The ground conductivity along the path is apparently also important, since Westerlund *et al.* (1969) have reported much larger phase advances on VLF paths that cross the Greenland ice cap, which has a low ground conductivity, than on paths that cross more normal terrain.

Many reports of VLF propagation anomalies during PCA events have appeared in the literature, but the difficulties associated with understanding propagation modes have prevented much quantitative study. VLF effects do, however, take place even for very weak particle fluxes, so that VLF observations are probably the most sensitive ground-based technique for detecting PCA events (Oelbermann, 1970).

At even lower frequencies, in the ELF band, PCA effects have again been observed (Davis, 1974). Current uses of such extremely low frequencies are restricted to global communication with nuclear submarines (Willim, 1974), but the importance of maintaining such communication at high latitudes is great. Theoretical studies (Field, 1969) have shown that at the frequencies used for this purpose ($\sim$ 50 Hz), heating of the ions becomes an important source of energy dissipation, particularly during PCA events, when ion concentrations are large.

5.2. Effects on Global Atmospheric Electricity

A vertical electric field exists throughout the lower and middle atmospheres of the Earth. In areas free from thunderstorms and other local electrical disturbances the field is directed vertically downward, with a magnitude of about 100 V m^{-1} at the surface, and is accompanied by a downward vertical current of a few times 10^{-12} A m^{-2}. According to the classical theory of atmospheric electricity, the 'fair-weather' electric field is a result of global thunderstorm activity. Each individual thunderstorm generates a net upward-directed current that charges the spherical-shell capacitor formed by the base of the ionosphere and the surface of the Earth (the same spherical shell that acts as a waveguide for propagation of low-frequency radio waves, as we saw in the last section). The capacitor in turn discharges by a continuous downward current in the areas of the Earth not affected by thunderstorm activity. The size of the discharging current, and the total potential drop across the capacitor, are determined by the columnar resistance of the atmosphere. Near the surface, several factors influence the conductivity, but throughout most of the lower and middle atmosphere the chief determining factor under normal conditions is the ionization produced by galactic cosmic rays.

In these circumstances, it would not be surprising if solar particle events were found to cause substantial changes in global electrical parameters. The most direct evidence for such a change is shown in Figure 7, which illustrates the results of a series of balloon measurements of the stratospheric electric field at high latitudes during the event of August 1972 (Holzworth and Mozer, 1979). The electric field is evidently decreased by the presence of the solar particle flux, the size of the decrease amounting to an order of magnitude at the peak of the event. The analysis of Holzworth and Mozer (1979) strongly suggests that the drop in field strength can be understood quite simply in terms of an increase in atmospheric conductivity and a constant current flow.

The well-known Forbush decreases in galactic cosmic-ray flux that follow major solar flares ought to produce an opposite effect, since the atmospheric conductivity must decrease. Increases in electric field strength at mid-latitude stations following solar flares have in fact been reported by Reiter (1969, 1971), but were accompanied by increases in current density (see also Cobb, 1967). A global model of atmospheric electricity has been developed by Hays and Roble (1979) and Roble and Hays (1979), who also modeled the changes produced by a weak PCA event and a Forbush decrease. The results suggest that the observed solar flare effects are not likely to be understandable as a simple

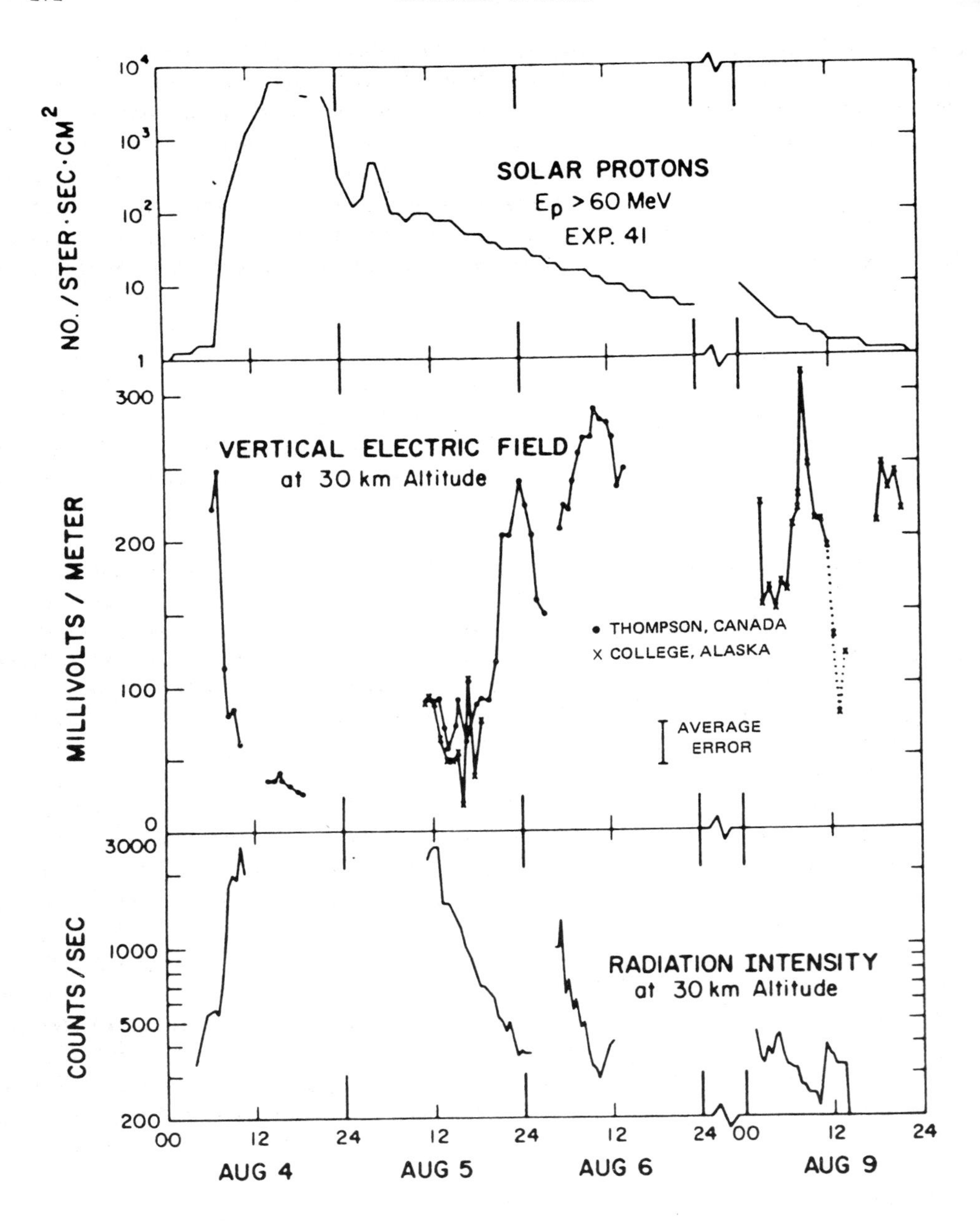

Fig. 7. Vertical electric field measured at 30 km altitude during the solar particle event of August 1972 (center). Solar proton flux at energies above 60 MeV measured by Explorer 41 (top), and radiation intensity measured by balloon-borne X-ray detectors (bottom) (Holzworth and Mozer, 1979).

response to conductivity changes, but do not necessarily require a change in global thunderstorm activity, as suggested by Reiter (1969, 1971).

it should be emphasized that our understanding of atmospheric electrical phenomena in general is fairly rudimentary, and there is no real assurance that the 'classical' picture of the fair weather field is correct. The perturbations produced by solar particle events may be extremely useful in validating the theory, since they are accompanied by calibrated changes in atmospheric conductivity.

5.3. Potential Impact on Climate

There has been a recent revival of interest in the old and much debated question of a link between solar activity and terrestrial weather and climate. The literature contains a large number of alleged statistical correlations between solar activity and meteorological parameters, but many of these are of doubtful validity (Pittock, 1978), and have helped to give the subject a somewhat unsavory reputation. Some of the correlations, on the other hand, are striking and difficult to dismiss, thus helping to sustain widespread interest. Until a plausible mechanism connecting some aspect of solar activity with the lower atmosphere appears, however, the field will continue to suffer from a lack of credibility.

Solar particle events have frequently been considered as likely candidates for providing the desired link. The direct effects of solar particles extend deeper into the atmosphere than any other direct manifestation of solar activity, with the possible exception of electric fields. The total energy flux they represent, however, is negligible compared to the input of solar electromagnetic radiation. To give an example, the total input of energy to the polar caps in the form of protons capable of reaching the stratosphere (i.e. with energies above 30 MeV) during the intense event of August 1972 was about 6×10^5 erg cm^{-2} (Reid, 1977b). This is equivalent to the average radiative energy supplied by the Sun in less than 2 s.

Comparisons of this kind are frequently made, but need some qualification. Among the factors that tend to enhance the importance of solar particles are (1) the fact that the Earth on the average radiates away exactly as much energy as it receives from the Sun, so that the comparison should be with the small radiative imbalances that might exist locally rather than with the total radiative budget, (2) the confinement of the solar particle effects to the polar regions, where the Sun's radiative input is weak or nonexistent, and (3) the fact that the bulk of the particle energy is deposited in the upper stratosphere, where the direct solar radiative input is small. Even with these caveats, however, the direct energy input is fairly negligible, and proponents of a climatic link have tended to search for some kind of trigger mechanism that can use solar particles to redistribute or release stored energy that would not be available otherwise. The mechanisms that have been suggested fall into three broad categories: electrical, radiative, and dynamical.

In the last section we discussed the changes in atmospheric electrical parameters that accompany solar particle events. There have been claims that the development of thunderstorms may be affected by the magnitude of the background electric field, which would allow solar particles to influence global convective storm activity. Changes in galactic cosmic-ray flux, such as Forbush decreases, would have a similar effect. The mechanism has been discussed in some detail by Markson and Muir (1980), but remains in the realm of speculation at present.

The realization that solar particle events lead to depletion of stratospheric ozone suggests that there might be an effect on the global radiation budget, since radiation in the 9.6 μm band of ozone is an important feature of the infrared spectrum of the Earth. Recent awareness of the potential destruction of ozone by man's activities has led to a number of studies of the radiative and dynamical effects of such reductions (Ramanathan *et al.*, 1976; Reck, 1976; Ramanathan and Dickinson, 1979; Fels *et al.*, 1980). Because of the complexity of the radiative processes involved, it is not yet possible to say whether ozone depletions of the magnitude produced by solar particle events can have any significant effect on tropospheric temperatures. Substantial effects, however, are most unlikely.

Hines (1974) suggested that changes in middle atmosphere composition resulting from solar activity might lead to changes in the reflection and transmission properties of the middle atmosphere for planetary waves, thus leading to small changes in the planetary wave amplitude in the troposphere. Schoeberl and Strobel (1978) have used a zonally averaged quasi-geostrophic numerical model to study the effects of the August 1972 solar particle event on the general circulation. They concluded that the global effect would be negligible, largely because of the relatively small fraction of the area of the earth affected by the solar particles.

Mechanisms other than those mentioned above have been proposed from time to time, including for example the 'seeding' of high cirrus clouds by the ions formed by solar particles. An extensive discussion of the entire subject can be found in Herman and Goldberg (1978). Until both a convincing mechanism and some convincing evidence for an effect appear, however, the subject will have difficulty in progressing beyond the stage of speculation.

5.4. Solar Energetic Particles and the Evolution of the Atmosphere and Biosphere

The study of energetic solar particle events has a short history. The earliest documented event, seen only on ground-based cosmic-ray detectors, was that of February 1942 (Forbush, 1946), while the first event to be studied extensively was the high-energy event of February 1956. In a total time span of about 40 years, it seems unlikely that we have sampled the full range of events that the Sun might have produced in its lifetime, longer than our sample period by a factor of more than 10^8. Many events have now been seen, having a wide variety of fluxes and energy spectra. Two stand out from the rest, however, as illustrating extremes – the event of August 1972, whose total energy flux was much larger than that of any other, and the event of February 1956, which has exceptionally large fluxes at relativistic energies, although the total energy flux was not outstanding. Both of these events apparently produced terrestrial effects that were unique: the August 1972 events caused substantial and long-lasting depletions of stratospheric ozone, amounting to about 15% in the high-latitude upper stratosphere (Heath *et al.*, 1977), while the February 1956 event briefly increased ground-level neutron fluxes by a factor of about 90 at some locations (Pomerantz and Duggal, 1974) and added about 10% of an entire solar cycle's supply of C^{14} to the atmosphere (Mendell *et al.*, 1973). Presumably there is some limit to what the Sun can produce in a single flare, but we have little idea where that limit lies.

Another factor that has certainly varied in the past is the strength and orientation of the geomagnetic field, which presently shields most of the global atmosphere from direct solar particle effects, confining these effects to the polar regions. The intensity of the Earth's dipole moment has varied in an apparently cyclical way during the past 10 000 years, having gone through a maximum some 50% higher than its present value about 2000 years ago, and a minimum of about half its present value about 6000 years ago (Cox, 1968). Changes of this magnitude, however, would have a relatively small effect on solar particle shielding, since the minimum latitude to which a particle can penetrate depends only on the one-sixth power of the dipole moment (Siscoe and Chen, 1975). In addition, however, the dipole component of the field has undergone many reversals of polarity. The transition between polarities is short, geologically speaking, but probably occupies a significant length of time in terms of the lifetime of terrestrial organisms. Since the geological time span is brief, it is difficult to measure, but current estimates place it at a few thousand years (Cox *et al.*, 1975). During this period the dipole field strength probably decreases drastically, and may disappear entirely, removing most of the Earth's magnetic shield against solar energetic particles and low-energy cosmic rays. It is not unlikely that the Sun could produce several 'giant' flares during such a period of vulnerability, causing ozone depletions much larger than that of August 1972 and high-energy particle fluxes considerably greater than those of February 1956. Such events could have severe environmental consequences, as was pointed out by Reid *et al.* (1976), who suggested these effects as a possible explanation for the observed coincidence between polarity reversals and the extinction of certain marine micro-organisms (Hays, 1971).

An apparent correlation between polarity reversals and climate changes has been pointed out (e.g. Wollin *et al.*, 1971), but no satisfactory mechanism for producing such a correlation has yet been proposed. If some link between energetic particle fluxes and climate were established, the link with polarity reversals might follow from the reduction in geomagnetic shielding at these times. The association of polarity reversals with climate change might also play an important part in the marine extinctions. Once again, however, the lack of a plausible physical mechanism has led to a certain amount of skepticism.

6. Conclusion

We have seen that energetic particles of solar origin can have a significant impact on the terrestrial environment. The wide range of energy spectra and particle fluxes seen in these events leads to a correspondingly wide range in the nature and magnitude of the effects, and leads naturally to speculation about the possible impact of large individual events in the distant past (or future).

The more modest events that have been observed during the past four decades have had their major effects in the middle and upper atmosphere, with practical consequences largely confined to the relatively sophisticated fields of communications and radar operations. If the future growth of technology takes us in the direction of increased human activity in space, however, the direct radiation exposure aspects of these events may become a serious consideration. Even omitting such exotic possibilities, the interaction between solar energetic particles and the terrestrial environment is sufficiently wide-ranging and complex that it is likely to remain a fruitful area of study for years to come.

References

Alcaydé, D.: 1968, *Ann. Geophys.* **24**, 1031.
Argemi, L. H.: 1964, *Ann. Geophys.* **20**, 273.
Armstrong, A. H., Harrison, F. B., Heckman, H. H. and Rosen, L.: 1961, *J. Geophys. Res.* **66**, 351.
Bailey, D. K.: 1957, *J. Geophys. Res.* **62**, 431.
Bailey, D. K.: 1959, *Proc. IRE* **47**, 255.
Blake, J. B., Martina E. F. and Paulikas, G. A.: 1974, *J. Geophys. Res.* **79**, 1345.
Brasseur, G. and Nicolet, M.: 1973, *Planet. Space Sci.* **21**, 939.
Brown, R. R.: 1964, *J. Atmos. Terr. Phys.* **26**, 805.
Brown, R. R.: 1967. *J. Atmos. Terr. Phys.* **29**, 317.
Chupp, E. L., Hess, W. N. and Curry, C.: 1967, *J. Geophys. Res.* **72**, 3809.
Cobb, W. E.: 1967, *Mon. Weather Rev.* **95**, 905.
Cox, A.: 1968, *J. Geophys. Res.* **73**, 3247.
Cox, A., Hillhouse J., and Fuller, M.: 1975, *Rev. Geophys. Space Phys.* **13**, No. 3, 185.
Crutzen, P. J.: 1970, *Quart. J. Roy. Meteorol. Soc.* **96**, 320.
Crutzen, P. J., Isaksen I. S. A., and Reid, G. C.: 1975, *Science* **189**, 457.
Crutzen, P. J. and Solomon, S.: 1980, *Planet. Space Sci.* **28**, 1147.
Crysdale, J. H.: 1960, *IRE Trans.* **CS-8**, 33.
Dalgarno, A.: 1967, *Space Research* **VII**, 849.
Damon, P. E., Lerman J. C., and Long, A.: 1978, *Ann. Rev. Earth Planet. Sci.* **6**, 457.
Davis, J. R.: 1974, *IEEE Trans.* **COM-22**, 484.
Domingo, V. and Page, D. E.: 1971, *J. Geophys. Res.* **76**, 8159.
Dragt, A. J., Austin, M. M., and White, R. S.: 1966, *J. Geophys. Res.* **71**, 1293.
Dungey, J. W.: 1961, *Phys. Rev. Lett.* **6**, 47.
Edgar, B. C., Miles, W. T., and Green, A. E. S.: 1973, *J. Geophys. Res.* **78**, 6595.
Edgar, B. C., Porter, H. S., and Green, A. E. S.: 1975, *Planet. Space Sci.* **23**, 787.
Eriksen, K. W. and Landmark, B.: 1961, *Arkiv. Geofys.* **3**, 489.
Evans, L. C. and Stone, E. C.: 1969, *J. Geophys. Res.* **74**, 5127.
Fanselow, J. L. and Stone, E. C.: 1972, *J. Geophys. Res.* **77**, 3999.
Fehsenfeld, F. C. and Ferguson, E. E.: 1969, *J. Geophys. Res.* **74**, 2217.
Fehsenfeld, F. C., Schmeltekopf, A. L., Schiff H. I., and Ferguson, E. F.: 1967, *Planet. Space Sci.* **15**, 373.
Fels, S. B., Mahlman, J. D., Schwarzkopf, M. D., and Sinclair, R. W.: 1980, *J. Atmos. Sci.* **37**, 2265.
Ferguson, E. E.: 1974, *Rev. Geophys. Space Phys.* **12**, 703.
Ferguson, E. E., Dunkin, D. B., and Fehsenfeld, F. C.: 1972, *J. Chem. Phys.* **57**, 1459.
Field, E. C.: 1969, *J. Geophys. Res.* **74**, 3639.
Forbush, S. E.: 1946, *Phys. Rev.* **70**, 771.
Frederick, J. E.: 1976, *J. Geophys. Res.* **81**, 3179.
Freier, P. S., Ney, E. P., and Winckler, J. R.: 1959, *J. Geophys. Res.* **64**, 685.
Gall, R., Jimenez J., and Camacho, L.: 1968, *J. Geophys. Res.* **73**, 1593.
Good, A., Durden, D. A., and Kebarle, P.: 1970, *J. Chem. Phys.* **52**, 222.
Greenhill, J. G., Fenton, K. B., Fenton, A. G., and White, K. S.: 1970, *J. Geophys. Res.* **75**, 4595.
Hays, J. D.: 1971, *Bull. Geol. Soc. Amer.* **82**, 2433.
Hays, P. B. and Roble, R. G.: 1979, *J. Geophys. Res.* **84**, 3291.
Heath, D. F., Krueger A. J., and Crutzen, P. J.: 1977, *Science* **197**, 886.
Herman, J. R. and Goldberg, R. A.: 1978, *Sun, Weather, and Climate*, NASA SP-426, Washington, D.C.
Herzberg, L.: 1960, *J. Geophys. Res.* **65**, 3505.
Hines, C. O.: 1974, *J. Atmos. Sci.* **31**, 589.
Holzworth, R. H. and Mozer, F. S.: 1979, *J. Geophys. Res.* **84**, 363.
Hultqvist, B. and Ortner, J.: 1959, *Planet. Space Sci.* **1**, 193.
Hynds, R. J. and Morfill, G.: 1976, *J. Geophys. Res.* **81**, 2445.
Imhof, W. L., Reagan, J. B., and Gaines, E. E.: 1971, *J. Geophys. Res.* **76**, 4276.
Johnston, H. S.: 1975, *Rev. Geophys. Space Phys.* **13**, 637.

Lanzerotti, L. J.: 1968, *Phys. Rev. Lett.* **21**, 929.
Lauzerotti, L. J.: 1972, *Rev. Geophys. Space Phys.* **10**, 379.
Larsen, T. R., Jespersen, M., Murdin, J., Bowling, T. S., Van Beek, H. F., and Stevens, G. A.: 1972 *J. Atmos. Terr. Phys.* **34**, 787.
Leinbach, H.: 1967, *J. Geophys. Res.* **72**, 5473.
Leinbach, H. and Reid, G. C.: 1959, *Phys. Rev. Lett.* **2**, 61.
Leinbach, H., Venkatesan, D., and Parthasarathy, R.: 1965, *Planet. Space Sci.* **13**, 1075.
Libby, W. F.: 1955, *Radiocarbon Dating*, (2nd edn), Univ. of Chicago Press, Chicago.
Lingenfelter, R. E. and Flamm, E. J.: 1964a, *J. Atmos. Sci.* **21**, 134.
Lingenfelter, R. E. and Flamm, E. J.: 1964b, *J. Geophys. Res.* **69**, 2199, [correction in *J. Geophys. Res.* 69, (1964) 4201].
Lingenfelter, R. E. and Ramaty, R.: 1970, in I. U. Olsson (ed.), *Radiocarbon Variations and Absolute Chronology*, Wiley Interscience, New York, p. 513.
Lockwood, J. A. and Friling, L. A.: 1968, *J. Geophys. Res.* **73**, 6649.
Maeda, K. and Aikin, A. C.: 1968, *Planet. Space Sci.* **16**, 371.
Markson, R. and Muir, M.: 1980, *Science* **208**, 979.
Megill, L. R., Adams, G. W., Haslett J. C., and Whipple, E. C.: 1971, *J. Geophys. Res.* **76**, 4587.
Mendell, R. B., Verschell, H. J., Merker, M., Light, E. S., and Korff, S. A.: 1973, *J. Geophys. Res.* **78**, 2763.
Meyer, P., Parker, E. N., and Simpson, J. A.: 1956, *Phys. Rev.* **104**, 768.
Michel, F. C.: 1965, *Planet. Space Sci.* **13**, 753.
Michel, F. C. and Dessler, A. J.: 1965, *J. Geophys. Res.* **70**, 4305.
Michel, F. C. and Dessler, A. J.: 1975, *J. Geophys. Res.* **80**, 2309.
Narcisi, R. S. and Bailey, A. D.: 1965, *J. Geophys. Res.* **70**, 3678.
Narcisi, R. S., Philbrick, C. R., Thomas, D. M., Bailey, A. D., Wlodyka, L. E., Baker, D., Federico, G., Wlodyka, R., and Gardner, M. E.: 1972a, in J. C. Ulwick (ed.), *Proc. COSPAR Symp. on the Solar Particle Event of November 1969*, AFCRL Special Rept No. 144, p. 421.
Narcisi, R. S., Sherman, C., Philbrick, C. R., Thomas, D. M., Bailey, A. D., Wlodyka, L. E., Wlodyka, R. A., Baker, D., and Federico, G.: 1972b, in J. C. Ulwick (ed.), *Proc. COSPAR Symp. on the Solar Particle Event of November 1969*, AFCRL Special Rept No. 114, p. 411.
Ness, N. F.: 1965, *J. Geophys. Res.* **70**, 2989.
Nicolet, M.: 1975, *Planet. Space Sci.* **23**, 637.
Oelbermann, E. J.: 1970, *J. Franklin Inst.* **290**, 281.
Ortner, J., Egeland, A., and Hultqvist, B.: 1960, *IRE Trans.* **AP-8**, 621.
Paulikas, G. A.: 1974, *Rev. Geophys. Space Phys.* **12**, 117.
Paulikas, G. A. and Blake, J. B.: 1969, *J. Geophys. Res.* **74**, 2161.
Peterson, J. R.: 1976, *J. Geophys. Res.* **81**, 1433.
Pieper, G. F., Zmuda, A. J., Bostrom, C. O., and O'Brien, B. J.: 1962, *J. Geophys. Res.* **67**, 4959.
Pittock, A. B.: 1978, *Rev. Geophys. Space Phys.* **16**, 400.
Pomerantz, M. A. and Duggal, S. P.: 1974, *Rev. Geophys. Space Phys.* **12**, 343.
Porter, H. S., Jackman, C. H., and Green, A. E. S.: 1976, *J. Chem. Phys.* **65**, 154.
Ramanathan, V., Callis, L. B., and Boughner, R. E.: 1976, *J. Atmos. Sci.* **33**, 1092.
Ramanathan, V. and Dickinson, R. E.: 1979, *J. Atmos. Sci.* **36**, 1084.
Rapp, D., Englander-Golden, P., and Briglia, D. D.: 1965, *J. Chem. Phys.* **42**, 4081.
Reagan, J. B., Meyerott, R. E., Nightingale, R. W., Gunton, R. C., Johnson, R. G., Evans, J. E., Imhof, W. L., Heath, D. F., and Krueger, A. J.: 1981, *J. Geophys. Res.* **86**, 1473.
Reagan, J. B. and Watt, T. M.: 1976, *J. Geophys. Res.* **81**, 4579.
Reck, R. A.: 1976, *Science* **192**, 557.
Reid, G. C.: 1976, in D. R. Bates and B. Bederson (eds), *Advances in Atomic and Molecular Physics*, Vol 12, Academic Press, New York, p. 375.
Reid, G. C.: 1977a, *Planet. Space Sci.* **25**, 275.
Reid, G. C.: 1977b, in B. Grandal and J. A. Holtet (eds), *Dynamical and Chemical Coupling Between the Neutral and Ionized Atmosphere*, D. Reidel, Dordrecht, p. 191.
Reid, G. C. and Collins, C.: 1959, *J. Atmos. Terr. Phys.* **14**, 63.
Reid, G. C., Holzer, T. E., Isaksen, I. S. A., and Crutzen, P. J.: 1976, *Nature* **259**, 177.

Reid, G. C. and Leinbach, H.: 1959, *J. Geophys. Res.* **64**, 1801.
Reid, G. C. and Sauer, H. H.: 1967a, *J. Geophys. Res.* **72**, 197.
Reid, G. C. and Sauer, H. H.: 1967b, *J. Geophys. Res.* **72**, 4383.
Reiter, R.: 1969, *Pure Appl. Geophys.* **72**, 259.
Reiter, R.: 1971, *Pure Appl. Geophys.* **86**, 142.
Roble, R. G. and Hays, P. B.: 1979, *J. Geophys. Res.* **84**, 7247.
Ruderman, M. A. and Chamberlain, J. W.: 1975, *Planet. Space Sci.* **23**, 247.
Rusch, D. W., Gerard, J.-C., Solomon, S., Crutzen, P. J. and Reid, G. C.: 1981, *Planet. Space Sci.* **29**, 767.
Sandford, B. P.: 1961, *Nature* **190**, 245.
Sandford, B. P.: 1963, *Planet. Space Sci.* **10**, 195.
Schoeberl, M. R. and Strobel, D. F.: 1978, *J. Atmos. Sci.* **35**, 1751.
Sellers, B. and Stroscio, M. A.: 1975, *J. Geophys. Res.* **80**, 2241.
Simpson, J. A.: 1960, *J. Geophys. Res.* **65**, 1615.
Singh, V. and Singhal, R. P.: 1978, *J. Geophys. Res.* **83**, 1653.
Siscoe, G. L. and Chen, C.-K.: 1975, *J. Geophys. Res.* **80**, 4675.
Smart, D. F., Shea, M. A., and Gall, R.: 1969, *J. Geophys. Res.* **74**, 4731.
Solomon, S., Rusch, D. W., Gerard, J.-C., Reid, G. C., and Crutzen, P. J.: 1981, *Planet. Space Sci.* **29**, 885.
Sternheimer, R. M.: 1959, *Phys. Rev.* **115**, 137.
Stone, E. C.: 1964, *J. Geophys. Res.* **69**, 3577.
Swanson, E. R.: 1974, in J. A. Holtet (ed.), *ELF-VLF Radio Wave Propagation*, D. Reidel, Dordrecht, p. 371.
Swider, W. and Dean, W. A.: 1975, *J. Geophys. Res.* **80**, 1815.
Swider, W. and Keneshea, T. J.: 1973, *Planet. Space Sci.* **21**, 1969.
Thorne, R. M.: 1977, *Science* **195**, 287.
Thorne, R. M.: 1980, *Pure Appl. Geophys.* **118**, 128.
Ulwick, J. C.: 1973, *Space Res.* **13**, 581.
Van Allen, J. A., Fennell, J. F., and Ness, N. F.: 1971, *J. Geophys. Res.* **76**, 4262.
Warneck, P.: 1972, *J. Geophys. Res.* **77**, 6589.
Weber, E. J., Mende, S. B., and Eather, R. H.: 1976, *J. Geophys. Res.* **81**, 5479.
Weeks, L. H., Cuikay, R. S., and Corbin, J. R.: 1972, *J. Atmos. Sci.* **29**, 1138.
Westerlund, S., Reder, F. H., and Abom, C.: 1969, *Planet. Space Sci.* **17**, 1329.
Willim, D. K.: 1974, in J. A. Holtet (ed.), *ELF-VLF Radio Wave Propagation*, D. Reidel, Dordrecht, p. 251.
Winters, H. F.: 1966, *J. Chem. Phys.* **44**, 1472.
Wollin, G., Ericson, D. B., and Ryan, W. B. F.: 1971, *Nature* **232**, 549.
Zipf, E. C. and McLaughlin, R. W.: 1978, *Planet. Space Sci.* **26**, 449.

Aeronomy Laboratory,
National Oceanic and
Atmospheric Administration,
Boulder, Colorado,
U.S.A.

INDEX